The Projective Cast

The MIT Press   *Cambridge, Massachusetts*   *London, England*

The Projective Cast *Architecture and Its Three Geometries* *Robin Evans*

First MIT Press paperback edition, 2000
© 1995 Massachusetts Institute of Technology
All rights reserved. No part of this book may be reproduced in any form by any electronic or mechanical means (including photocopying, recording, or information storage and retrieval) without permission in writing from the publisher.

This book was set in Caslon and Futura by Graphic Composition, Inc. and was printed and bound in the United States of America.

Library of Congress Cataloging-in-Publication Data

Evans, Robin, 1944–
    The projective cast : architecture and its three geometries / Robin Evans.
        p. cm.
    Includes bibliographical references and index.
    ISBN 978-0-262-05049-4 (hc. : alk. paper)—978-0-262-55038-3 (pb. : alk. paper)
    1. Architecture—Composition, proportion, etc.   2. Geometry.
I. Title.
NA2760.E93   1995
720′.1—dc20
                                      94-42924
                                        CIP

For our children Daniel and Bonnie

Contents

List of Illustrations   ix

Foreword   xix

Acknowledgments   xxi

Introduction: Composition and Projection   xxv

Part One

Chapter One

Perturbed Circles   3

Chapter Two

Persistent Breakage   55

Part Two

Chapter Three

Seeing through Paper   107

Chapter Four

Piero's Heads   123

Chapter Five

Drawn Stone   179

Part Three

Chapter Six

The Trouble with Numbers   241

Chapter Seven

Comic Lines   273

Chapter Eight

Forms Lost and Found Again   323

Chapter Nine

Rumors at the Extremities   337

*Conclusion:
The Projective Cast   351*

Notes   372

Illustration Credits   402

Index   404

List of Illustrations

1. Sant'Eligio degli Orefici, Rome, Raphael Sanzio, designed 1509, interior. Page 7.

2. Sant'Eligio degli Orefici, plan. Page 8.

3. Sant'Eligio degli Orefici, section showing nine centers. Page 10.

4. Tempietto, San Pietro in Montorio, Rome, Donato Bramante, 1502, section. Page 11.

5. Leonardo da Vinci, *Place for Preaching*. Page 12.

6. Baptistery, Pisa, section and plan. Page 13.

7. Santa Maria delle Grazie, Milan, Donato Bramante, late 1480s, dome. Page 15.

8. San Giorgio Maggiore, Venice, Andrea Palladio, 1565–1580, dome. Page 15.

9. G. P. Lomazzo, *Choirs of Angels*, 1565–1571, Foppa Chapel, San Marco, Milan. Page 20.

10. Diagram of dual-centered organization implied by sixteenth-century dome frescos. Page 21.

11. Francesco Botticini, *Assumption of the Virgin*, altarpiece, c. 1475. Page 22.

12. Giovanni di Paolo, *The Creation of the World and the Expulsion from Paradise*, 1445. Page 24.

13. Sandro Botticelli, illustration for Dante's *Divine Comedy, Paradiso*, canto 23. Page 26.

14. San Lorenzo, Florence, Filippo Brunelleschi, 1421, sacristy. Page 28.

15. Chigi Chapel, Santa Maria del Popolo, Rome, Raphael Sanzio, c. 1516, dome. Page 31.

16. Raphael Sanzio, *Disputà*, Stanza della Segnatura, Vatican, 1511. Page 32.

17. Sallustio Peruzzi, drawing of Sant'Eligio, c. 1560–1565. Page 39.

18. Leonardo da Vinci, solid sphere. Page 40.

19. Renaissance modification of the Ptolemaic system of celestial orbits. Page 41.

20. Leonardo da Vinci, map of Imola, 1502–1503. Page 45.

21. Marco Fabio Calvo, map of ancient Rome, 1527. Page 46.

22. Richard Meier, sketch plan of the Atheneum, New Harmony, Indiana, 1975–1979. Page 56.

23. Sigfried Giedion, *Space, Time and Architecture*, 1941: Picasso's *L'Arlésienne* and Gropius's Bauhaus. Page 58.

24. Esprit Jouffret, projections of the pairs of octahedra constituting the hypersolid h_2. Page 61.

25. Caspar David Friedrich, *The Wreck of the "Hope,"* 1823–1824. Page 64.

26. Georges Braque, *Mandola*, 1909–1910. Page 64.

27. Pablo Picasso, *Guitar* maquette, 1912. Page 65.

28. Raymond Duchamp-Villon, Maison Cubiste, 1912. Page 65.

29. Vergilio Marchi, Model of Futurist House, 1916. Page 66.

30. Villa Savoye, Poissy, Le Corbusier, 1928–1931, view out of ramp window. Page 68.

31. Walter Gropius, Kapp-Putsch Monument (Memorial to the March Victims), 1920–1922, Weimar. Page 69.

32. Baker Dormitory, MIT, Cambridge, Massachusetts, Alvar Aalto, 1947–1948, north side. Page 72.

33. Baker Dormitory, plan. Page 72.

34. Watts Towers, Los Angeles, Sam Rodia, 1921–1945. Page 77.

35. The City of Composite Presence, David Griffin and Hans Kolhoff, from Rowe and Koetter, *Collage City*, 1978. Page 77.

36. Forum des Halles, Paris, C. Vasconi and G. Pencreach, 1972–1979. Page 81.

37. Humphry Repton, *A Design to Exemplify Irregularity of Outline in Castle Gothic*. Page 82.

38. Parc de La Villette, Paris, Bernard Tschumi, 1982–1989, *folie* staircase. Page 88.

39. Parc de La Villette, Paris, *folie* in front of J. L. Baltard's Slaughterhouse Hall (1863–1867). Page 88.

40. Gordon Matta-Clark, one of the photographs in his contribution to the *Idea as Model* exhibition, Institute of Architecture and Urban Studies, New York, 1976. Page 90.

41. Rooftop Lawyers' Office, Vienna, Coop Himmelblau, 1983–1988. Page 93.

42. Hans Scharoun, *Kirche als Fels*, 1910. Page 95.

43. Philharmonie, Berlin, Hans Scharoun, 1956–1963, plans. Page 96.

44. Philharmonie, exterior from west. Page 97.

45. Philharmonie, auditorium. Page 97.

46. Philharmonie, foyer. Page 98.

47. Philharmonie, foyer. Page 99.

48. Philharmonie, foyer. Page 99.

49. Pablo Picasso, *Weeping Woman*, 1937. Page 103.

50. *The Principles of Parallel Projection*, from Daniel Fournier, *A Treatise on the Theory of Perspective*, 1761. Page 109.

51. Alberti's perspective construction. Page 110.

52. Raphael Sanzio, *The School of Athens*, 1511, Stanza della Segnatura, Vatican. Page 111.

53. Watercolor of Villa Madama (anon.), eighteenth century. Page 112.

54. Villa Madama, Rome, Raphael Sanzio, 1517–1521, plan. Page 112.

55. *Christ Disputing with the Elders in the Temple*, fourteenth century, Dragonelli Chapel, San Domenico, Arezzo, detail of tiled floor. Page 113.

56. Sant'Andrea, Mantua, Leone Battista Alberti, designed 1470, detail of facade. Page 114.

57. The Arch of Augustus and the church of San Michele, Fano. Page 115.

58. Andrea Palladio, final facade study for San Petronio, Bologna, 1579. Page 116.

59. San Giorgio Maggiore, Venice, Andrea Palladio, 1564–1580, detail of facade. Page 117.

60. Philharmonie, Berlin, Hans Scharoun, 1956–1963, published sections. Pages 120–121.

61. Bernardino Pinturicchio, one of the series of scenes from the life of Aeneas Silvius Piccolomini. Page 124.

62. Snapshot (author). Page 124.

63. Albrecht Dürer, perspective machine. Page 128.

64. Albrecht Dürer, aids to draw the perspective of a reclining woman. Page 128.

65. Jacques Androuet du Cerceau, perspective construction. Page 132.

66. Raphael Sanzio, *Marriage of the Virgin*, 1504. Page 134.

67. Leonardo da Vinci, *Adoration of the Magi*, drawing, 1481. Page 136.

68. Leonardo da Vinci, *Adoration of the Magi*, c. 1481. Page 137.

69. Three early methods of perspective construction. Pages 138–139.

70. Solid dodecahedron, from Luca Pacioli, *De divina proportione*, 1496. Page 143.

71. Piero della Francesca, *The Flagellation of Christ*. Page 145.

72. Luca Cambiaso, pen and ink figure study. Page 146.

73. Piero della Francesca, *De prospectiva pingendi*, plan-and-vanishing-point construction of an unaligned octagon. Page 149.

74. Piero della Francesca, *De prospectiva pingendi*, a building projected with the aid of a vanishing point. Page 150.

75. Piero della Francesca, *De prospectiva pingendi*, rotated orthographic projection of a cube. Page 152.

76. Piero della Francesca, *De prospectiva pingendi*, perspective of a rotated cube. Page 152.

77. Piero della Francesca, *De prospectiva pingendi*, orthographic projections of a head. Page 153.

78. Piero della Francesca, *De prospectiva pingendi*, Page 153.

79. L. B. Alberti's finitorum, from *De statua*. Page 155.

80. Piero della Francesca, *De prospectiva pingendi*, orthographic projections of a tilted head (elevations). Page 157.

81. Piero della Francesca, *De prospectiva pingendi*, orthographic projections of a tilted head (plans). Page 157.

82. Piero della Francesca, five women from *The Proving of the True Cross*, San Francesco, Arezzo. Page 160.

83. Piero della Francesca, *The Baptism of Christ*, detail. Page 160.

84. Five women from *The Proving of the True Cross:* schematic rotations. Page 161.

85. Five women from *The Proving of the True Cross:* rotations in detail. Page 162.

86. Plan showing orientations of feet, bodies, and heads of the three foreground figures in Piero's *Flagellation*. Page 163.

87. Piero della Francesca, Brera altarpiece. Page 164.

88. Piero della Francesca, *Nativity*. Page 165.

89. Vanishing points of the stall roof in Piero's *Nativity*. Page 165.

90. Piero della Francesca, *Dream of Constantine*, detail, San Francesco, Arezzo. Page 166.

91. Leonardo da Vinci, *A Town Overwhelmed by a Deluge*, c. 1515. Page 168.

92. Peter Paul Rubens, copy of Leonardo's *Battle of Anghiari*, c. 1603. Page 169.

93. Paolo Uccello, *Subsiding of the Flood*, Chiostro Verde, Santa Maria Novella, Florence, 1446–1448. Page 170.

94. Paolo Uccello, *Battle of San Romano: Niccolò da Tolentino at the Head of the Florentines*, 1456. Page 171.

95. Paolo Uccello, *Battle of San Romano: The Counterattack of Micheletto da Cotignola*, 1456. Page 172.

96. Paolo Uccello, *Battle of San Romano: The Unhorsing of Bernardino della Ciarda*, 1456. Page 172.

97. Paolo Uccello, perspective study of a chalice. Page 173.

98. Paolo Uccello, perspective study of a *mazzocchio*. Page 173.

99. Two perspectives of a disk with equal segments. Page 174.

100. Hôtel Bullioud, Lyons, Philibert Delorme, 1536, elevation of courtyard. Page 182.

101. Philibert Delorme, perspective of the *trompe* at Anet, 1549–1551. Page 183.

102. Philibert Delorme, *trait* for the *trompe* at Anet. Page 185.

103. Philibert Delorme, the *trompe* at Anet, relation between the *trompe* and the *trait*. Page 187.

104. Philibert Delorme, the *trompe* at Anet, the basic rotation and *rabattement*. Page 187.

105. Philibert Delorme, the *trompe* at Anet, developed surface of the underside. Page 188.

106. Philibert Delorme, the *trompe* at Anet, perspective. Page 188.

107. *Trompe* under the organ loft, seminary chapel, La Flèche, Jacques Nadreau, 1636. Page 189.

108. Francesco Salviati, *Bathsheba Going to King David*, Palazzo Sacchetti, Rome, 1552–1554. Page 191.

109. A. F. Frézier, *escalier vis de Saint-Gilles suspendu*. Page 192.

110. Notre Dame de la Couture, Le Mans, suspended stone stair, architect unknown, 1720–1739. Page 193.

111. Pulpit, Saint-Sulpice, Paris (second project), Charles de Wailly, 1789. Page 194.

112. Hendrick Goltzius, *Job in Distress*, c. 1616. Page 197.

113. Gérard David, *Christ Nailed to the Cross*. Page 198.

114. Philibert Delorme, frontispiece to the *Premier tome*, 1567. Page 199.

115. Diagram of the first 7 of 11 successive operations in Girard Desargues's universal method for stonecutting. Page 202.

116. J.-B. de La Rue, frontispiece to *Traité de la coupe des pierres*, first published 1728. Page 204.

117. A block of stone and its *trait*, from Abraham Bosse, *La Pratique du trait*, 1643. Page 206.

118. François Derand, *trait* for an oblique arch descending through a cylindrical tower into a dome. Page 207.

119. Philibert Delorme, rusticated temple. Page 211.

120. Hubert Robert, *L'Arc de Triomphe d'Orange*, 1783. Page 212.

121. Saint-Paul-Saint-Louis, Paris, François Derand, 1627–1641, interior. Page 215.

122. Abbaye de Notre Dame de la Couture, Le Mans, architect unknown, 1720–1739, cloister vaults and pilaster. Page 216.

123. Sainte-Geneviève, Paris, J. G. Soufflot, crypt vaults, 1758–1764. Page 216.

124. Sainte-Geneviève, Paris, J. G. Soufflot, choir vaults, completed 1779. Page 217.

125. Halle au Blé, Paris, Nicolas Le Camus de Mézières, 1762–1766, section showing suspended double spiral stairs. Page 218.

126. Halle au Blé, cross section. Page 219.

127. Halle au Blé, section showing grain store vault. Page 219.

128. Philibert Delorme, *trait* for a modern star vault. Page 222.

129. Gloucester Cathedral, choir vault, William Ramsey (?), 1337–1367. Page 227.

130. Gloucester Cathedral, detail of severy in choir. Page 227.

131. Gloucester Cathedral cloister, fan vaulting, begun 1378. Page 230.

132. Peterborough Cathedral, retrochoir, John Wastell (?), early sixteenth century, intrados of twenty intersecting fans. Page 232.

133. Peterborough Cathedral, retrochoir. Page 233.

134. Henry VII Chapel, Westminster Abbey, Robert and William Vertue, 1500–1520, pendant fan vaults. Page 234.

135. Extrados of Henry VII Chapel vaults, drawn by Robert Willis, 1842. Page 235.

136. François Blondel, harmonically proportioned Attic base with its corresponding chord. Page 242.

137. Proportional analyses of Santa Maria Novella, Florence by Wölfflin, 1889, and Wittkower, 1949. Page 248.

138. Proportion in perspective, showing the generation of harmonic ratios when the viewing point O is one bay width distant from the picture plane. Page 252.

139. Santo Spirito, Florence, Filippo Brunelleschi, started 1436, the nave. Page 253.

140. Albrecht Dürer, figure from *Vier Bücher von menschlicher Proportion*, 1538. Page 255.

141. Vittore Carpaccio, *Birth of the Virgin*, 1504–1508. Page 257.

142. Girard Desargues's theorem of four-point involution. Page 258.

143. Albrecht Dürer, perspective adaptation of equal measures. Page 260.

144. Albrecht Dürer, figure proportioned with the perspective adaptation shown in the previous figure. Page 260.

145. Albrecht Dürer, diminishing spiral stair with risers proportioned by perspective adaptation, Page 261.

146. Comparison of musical scales visualized as the subdivisions of a monochord. Page 264.

147. Notre-Dame du Haut, Ronchamp, Le Corbusier, 1950–1955, view from south. Page 274.

148. Le Corbusier, *The Modulor*, red and blue series. Page 275.

149. Le Corbusier, charcoal sketch plan of Ronchamp, June 1950. Page 279.

150. Hanning's diagram of 1943 and its errors. Page 280.

151. Michelangelo's Capitol with *traces régulateurs* superimposed. Page 280.

152. Spanish pot, from *L'Art décoratif d'aujourd'hui*. Page 282.

153. Hans Erni, *Die neuen Ikarier*, 1940. Page 283.

154. Ronchamp, east end of roof shell. Page 283.

155. Ronchamp, view into cowl of red tower. Page 285.

156. Le Corbusier's revision of Maillart's diagram. Page 286.

157. Le Corbusier, *Apollo and Medusa* cameo, first published 1948. Page 288.

158. Le Corbusier, plate from *Poème de l'angle droit*. Page 288.

159. Ronchamp, plan. Page 289.

160. Church at Lourtier, Switzerland, Alberto Sartoris, 1932. Page 293.

161. Ronchamp, interior. Page 294.

162. Iannis Xenakis's *pans de verre ondulatoires*, monastery of La Tourette, 1955–1959. Page 297.

163. Philips Pavilion, Brussels International Exhibition, Le Corbusier and Iannis Xenakis, 1958. Page 299.

164. Iannis Xenakis, ruled surfaces of sound in *Metastasis*, 1954. Page 299.

165. Ronchamp, twisted interior surface of south wall. Page 300.

166. Ronchamp, the south tower, drawing by Olek, December 1953. Page 301.

167. Ronchamp, wire and paper model. Page 303.

168. Ronchamp, roof shell, drawing by André Maisonnier, September 1952. Page 304.

169. Four projections of Guarini's conoid. Page 306.

170. The U.S. aircraft carrier *Lexington*. Page 308.

171. Le Corbusier, diagram from sketchbook D17, 1950. Page 309.

172. Antoine Pevsner, *Construction dynamique*. Page 311.

173. André Maisonnier, elevation of the Ronchamp roof as a ruled surface, March 1951. Page 313.

174. Interior of Ronchamp showing underside of roof shell. Page 313.

175. Woman at prayer, from Le Corbusier, *Ronchamp*. Page 318.

176. Plate from Le Corbusier, *Aircraft*. Page 319.

177. Developing the surface of a cone with any base that has been truncated by a sphere, from Gaspard Monge, *Géométrie descriptive*, 1838. Page 325.

178. Courtyards, from J. N. L Durand, *Précis des leçons d'architecture*, 1802–1805. Page 326.

179. A. F. Frézier, conoidal *arrière-voussoirs*, from *La Théorie et la pratique de la coupe des pierres*, 1739. Page 329.

180. Ruled surfaces, from Thomas Bradley, *Elements of Geometrical Drawing*, 1862. Page 330.

181. Stern of the *Great Eastern*, photograph by Robert Howlett, 1857. Page 331.

182. Casa Batlló, Barcelona, Antoni Gaudí, 1904–1906, detail of masonry. Page 332.

183. Sagrada Familia, Barcelona, Antoni Gaudí, 1884–1926, finial. Page 332.

184. Le Corbusier, 1928 sketch of Gaudí's 1909–1910 roof for the Parochial School of Sagrada Familia. Page 333.

185. Theo van Doesburg, *Counter-Compositions V*, 1924. Page 338.

186. Theo van Doesburg, *Color Construction in the Fourth Dimension of Space-Time*, 1924. Page 340.

187. Theo van Doesburg, version of the hypercube. Page 342.

188. Alternative axonometric representation of the hypercube. Page 343.

189. El Lissitzky, axonometric of the Proun Room, 1923. Page 343.

190. Einstein's analogy of spherical space, using stereographic projection. Page 345.

191. Einstein Tower, Potsdam, Erich Mendelsohn, 1919–1921. Page 347.

192. Masolino, The Foundation of Santa Maria Maggiore by Pope Liberius. *Page 356.*

193. Chapel of the Santissima Sindone, Turin, Guarino Guarini, 1667–1690, section. Page 361.

194. Chapel of the Santissima Sindone, view into dome. Page 362.

195. Luigi Moretti, plaster cast of interior space of Guarino Guarini's Santa Maria della Divina Providenza, Lisbon. Page 365.

196. Santa Maria della Divina Providenza, Lisbon, Guarino Guarini, designed c. 1681, begun 1698, section. Page 365.

197. Projection and its analogues: The Arrested Image. *Page 367.*

Foreword

When my husband, Robin Evans, died in our house on 19 February 1993, he was on the point of finishing this, his second book. I had gone out, leaving him happily engaged editing the notes for the final chapter. I returned to find him dead. He had died totally unexpectedly at the age of 48. He had had a cerebral hemorrhage.

The day started off auspiciously. Bob (as he was known to us) had been offered the post of Professor of Architectural History at the Bartlett School, University College, London. We met for lunch and discussed how good it would be for him to have such a position, and, maybe, stay in England. He had been spending part of each year at Harvard, which he enjoyed very much, but I know Bob was pleased to feel that his work was being properly recognized. And the book was about to be completed. I think, and hope, he died contented.

Our two children, Daniel, 18, and Bonnie, 15, have lost the most wonderful father. I have lost the best of husbands and the best of friends. I know that Bob will live on, though, in his books, essays, and in the memories of all his family, his friends, and the people he taught. I was overwhelmed by the number of students who wrote to tell me what an important influence Bob had been in their lives. Bob was such a gentle, modest, unassuming person that I think many of his family and friends were unaware of the real significance of his work.

Fortunately many people were not, and were determined that Bob's work would get published. My most sincere thanks go to Kate Heron, who with her partner Julian Feary had been our friend for years, and helped to galvanize all the other people into action who have been instrumental in getting this book into print. John Bold, a friend, a neighbor, and an architectural historian, has worked tirelessly with her. These people, and many others, have helped me so much in the months since Bob's death. I will never be able to repay them for their kindness.

In the acknowledgments to Bob's first book, *The Fabrication of Virtue* (Cambridge University Press, 1982), Bob thanks me for "generosity and patience that bordered on the indulgent." I thank him here for indulging me for some thirty years with his kindness, goodness, and love.

Janet Evans

Finsbury Park

1994

Acknowledgments

Robin Evans had made his revisions to the main text of this book and was still working on the footnotes when he died. Final revisions would no doubt have been made before the book went to press, since Bob's ideas and his expression of them continued to evolve as he sought ways of making still clearer the communication of the highly complex issues and arguments to which his working life as author and teacher had been devoted. In his final notes (on his computer) he reminded himself that "I am lacking a proper ending, that is, something that sounds like a parting word, thought or parting shot."

He continued to try to pin down the elusive relationship that exists between buildings and their representations:

> PROPOSITION: *That a work of architecture is more than the sum of its representations. Is this to glorify the architectural object, putting it in a position unattainable by photographs, drawings and writings? No, because pictures and words are* always less *than what they refer to. Referential art is, by its very nature, reduced from its referents. Thus to say that a building is more than its pictures is not to say that it is more* art-like *than its pictures. It does suggest that it is more difficult to make a building art-like than a picture because the perceptions of the building are more* in themselves but less *manageable, less capable of full orchestration.*

After Bob died, Julian Feary, Wilfried Wang, and ourselves agreed with Janet Evans that we should gather up the text and illustrations and ensure their safe passage through the necessary stages to reach publication, enlisting the invaluable help of Robin Middleton and Mary Wall. It has been a privilege to do so.

Outstanding work on the endnotes and illustrations, as well as editing of the text, has been a combined effort involving many of Bob's friends and colleagues in England and in the United States over many months. We would especially like to thank those already mentioned for their considerable efforts as well as Peter Carl, Mary McLeod, and Dorothea Dietrich for help with particular notes, and Jeffrey Kipnis, Rodney Place, Ed Robbins, Fred Scott, Grahame Shane, and Paul Shepheard for relevant anecdotal advice. We also wish to express our appreciation of the work of the staff at the MIT Press who have been consistently supportive of this enterprise, and particularly that of the editor Matthew Abbate who has been sensitive and discreet in all his interventions, and of the designer Mimi Ahmed for an elegant design and patience in discussing the quality of printing. We are grateful to Richard Difford, one of Bob's former students at the University of Westminster, for his compilation of the index and his alertness in the final scrutiny of the text. We have all been encouraged throughout by the warmth and positive support of Janet Evans.

Bob worked for some ten years on this book and during that time published articles and gave lectures that were essential to the subject. Each lecture or lecture series and each essay retains its autonomy, yet has contributed to the development of this book. Bob has acknowledged students and colleagues in the endnotes here and refers to his own previously published work. Bob lectured widely in the U.S. and the U.K., and special mention must be made of his students at the Harvard Graduate School of Design, who were participants in his courses and audiences at lectures in the GSD Harvard Visitor series; at the University of Westminster in London, where two successive years of architecture students took part in twelve-part lecture series on projection; and at Cornell University, where he delivered the Preston H. Thomas Memorial Lectures in 1988. In addition, significant individual lectures were given at Columbia University, the University of the South Bank in London, the Architectural Association, the Bartlett School at the University of London, and the Royal Institute of British Architects.

Two of the many scholarly institutions with which Bob was involved during his peripatetic teaching career have made further essential contributions to this publication. Harvard University Graduate School of Design has provided generous financial assistance, and the University of Westminster has given welcome practical support.

In an enterprise of this kind, it is inevitable that we will have failed to acknowledge many of those whom Bob himself would have thanked and failed to find many appropriate references and illustrative credits which he intended. We apologize for any of these sins of omission and would be grateful if readers of this book would point them out to us via the publishers so that they may be corrected in any future editions.

John Bold

Katharine Heron

London 1994

Introduction:
Composition and Projection

Geometry has an ambiguous reputation, associated as much with idiocy as with cleverness.[1] At best there is something desperately uncommunicative about it, something more than a little removed from the rest of experience to set against its giant claim of truth. Flaubert, in *The Dictionary of Accepted Ideas,* defines a geometrician as "travelling on strange seas of thought—alone."[2] And when Joseph Conrad wished to characterize the futile effort of concentration made by the earnest but mentally retarded youth Stevie in *The Secret Agent,* he would describe him as "seated very good and quiet at a deal table, drawing circles, circles, circles; innumerable circles, concentric, eccentric, a coruscating whirl of circles that by their tangled multitude of repeated curves, uniformity of form, and confusion of intersecting lines suggested a rendering of cosmic chaos, the symbolism of a mad art attempting the inconceivable."[3]

There have been, and there still are, architects with seemingly unlimited faith in the power of geometry. They search for shapes and measures which they hope will divulge the mystery of their calling and at the same time lock the mystery into place as a professional secret, or even a personal secret. We may arm ourselves against such naivety and yet concede that all architects will from time to time adopt the posture of Stevie, looking much the same as he when embroiled in the reveries of design work. In this posture they may become susceptible to the same delusions of which we can so readily imagine Stevie to have been victim. There are good reasons why they might. Without the architect's faith that geometrically defined lines will engender something else more substantial yet discernible through the drawing, without faith in the genetic message inscribed on paper, there is no architecture. It has often been said that architecture is more than mere building. In this sense it is considerably less.

Geometry is one subject, architecture another, but there is geometry in architecture. Its presence is assumed much as the presence of mathematics is assumed in physics, or letters in words. Geometry is understood to be a constitutive part of architecture, indispensable to it, but not dependent on it in any way. The elements of geometry are thus conceived as comparable to the bricks that make a house, which are reliably manufactured elsewhere and delivered to site ready for use. Architects do not produce geometry, they consume it. Such at least would be the inevitable conclusion of anyone reviewing the history of architectural theory. Several key Renaissance treatises commence with a brief résumé of geometric figures and definitions borrowed from Euclid: point, line, plane, triangle, rectangle, and circle. Sebastiano Serlio, for example, began his *First Book of Architecture* (1545, English translation 1611) by affirming "how needfull and necessary the most secret Art of Geometrie is." Without it the architect is no more than a stone despoiler, he said, and then went on to explain how what he called the flowers picked from Euclid's garden would endow building with reason.[4] His peculiar metaphor, whereby what we understand to be at the root of architecture is described as its ornament, gives the impression that the foundation is in some sense an accessory or afterthought; an afterthought because buildings could and did exist without it, a foundation in that geometry offers certainty in situations beset by doubt.

The job of a foundation is to be as firm as a rock. It is supposed to be inert. Dead things are easier to handle than live ones; they may not be so interesting but they are less troublesome. From the point of view of the architect seeking firmness and stability, the best geometry is surely a dead geometry, and perhaps that, by and large, is what architecture is made with. What I mean by a dead geometry is an aspect of geometry no longer under development from within. Triangles, rectangles, and circles as defined in Euclid have been pretty well exhausted as subjects of geometrical enquiry. As these elements lose their mystery, interest in them subsides, but in this state of devaluation they become more valuable elsewhere because their behavior is completely predictable. Consequences can be foreseen. Dead geometry is an innoculation against uncertainty.

Yet the architect's attitude to this stabilizing geometry has always been two-faced. Toward the lay world its presence is traditionally advertised with pride, while within the profession architects tend to be suspicious of its power over what they do. Its value may be in its deadness, but if it is not kept under control it may revive, like a monster, or the morbidity may spread, like a disease.

The ideal is of a vital and creative art supported on the dead certain truth of geometry. The very statement is enough to make us think twice. Is the geometry in architecture really so reliable? It is, as we shall see, difficult enough to say where the geometry in architecture is exactly. Reports come from several locations. Either it is mobile, which is a sign of life, or it is multiplied and harder to categorize.

But the entrenched idea of the firm foundation has itself been underpinned by other definitions that may be no less insupportable. For instance, it fits neatly with the perception that geometry is a rational science, while architecture—the art of architecture—is a matter for intuitive judgment. According to this credible-sounding distinction, geometry gives architecture a reasonable ground but does not confine it to rationality. The creative, intuitive, or rhetorical aspects of architecture can therefore ride on the back of its geometric rationality. That is what

Guarino Guarini, the seventeenth-century mathematician and architect, conveyed with his concise definition: "Architecture, though dependent on mathematics, is nevertheless an art of adulation."[5] Whilst this division between base and superstructure has been constructed into a demonstrable truth in a large quantity of historical buildings, it is neither universal nor necessary. Serlio's flowers suggest as much, and Guarini's own architecture threatened the dependence he had announced by bringing a new and far less predictable geometry into play. Either science was interfering with art or it was hard to tell the difference between science and art.

Geometry used to be called the science of space. For various reasons this definition was discarded, so geometry no longer has an obvious subject matter. The question arises, how then is it a science? What is it a science of? Some mathematicians have even proposed that geometry, together with the rest of mathematics, should be reclassified either as a humanity or as an art, since it is said to be guided by an aesthetic sense. "A mathematician, like a painter or a poet is a maker of patterns," wrote G. H. Hardy, typically.[6] The role of intuition in mathematics has also been extensively discussed over the past century.[7] As a result, many professional mathematicians are not only possessed of the idea that the ultimate justification of their work is not mere truth but beauty; they also regard intuition as essential to the performance or appreciation of mathematics of any sort. There is no need to justify these ideas. I only want to present them as running counter to the ordinary understanding of what geometry is, and running parallel to the ordinary understanding of what art is.

The most fleeting acquaintance with recent writings on the nature of mathematics will convince anyone that the definition of architecture as an art born of science because founded on geometry would make little sense viewed from the mathematicians' side of the fence. Viewed from that side there does not appear to be much of a fence. From the mathematician's point of view the definition might be rewritten thus: architecture is an art born of another art because it is based on geometry, which is a visual art. This rewritten definition should not pass unchallenged, because we cannot be certain that architecture is an art, or that geometry is basic to it, or that the beauty in geometry has anything to do with the beauty in architecture, but at least it allows us to disabuse ourselves of a prejudice that still directs the understanding of geometry from within architecture.

The following chapters show that geometry does not always stabilize architecture; that the geometry in architecture was not always dead at the time of its employment, although it may have died later; and that in architecture expired geometry sometimes gained a life after death. They show also that the perception of geometry's role has been vastly affected by a collective oversight. The first place anyone looks to find the geometry in architecture is in the shape of buildings, then perhaps in the shape of the drawings of buildings. These are the locations where geometry has been, on the whole, stolid and dormant. But geometry has been active in the space between and the space at either end. What connects thinking to imagination, imagination to drawing, drawing to building, and buildings to our eyes is projection in one guise or another, or processes that we have chosen to model on projection. All are zones of instability. I would now claim that the engaging questions of architecture's relation to geometry occur in these zones. Composition, which is where the geometry in architecture is usually sought, may still for convenience be considered the crux of the matter, but it has no significance in and of itself. It obtains all its value via the several types of projective, quasi-projective, or pseudo-projective space that surround it, for it is only through these that it can be made available to perception. That is the thesis of this book.

The distinction between composition and projection in architecture has its counterpart in mathematical geometry. First came a geometry whose idealities were well adapted to the measuring of things. This was organized into a consistent body of propositions by the Greeks and obtained its classic exposition in Euclid's *Elements*. Euclidean geometry was concerned with the ratios and equalities of lines, areas, and angles. However abstract, however contemplative in spirit, however remote from practical application, it must surely have arisen from, and easily translates back into, the tasks of shaping artifacts, laying out buildings, and surveying land. Later came a geometry no longer concerned with measuring the intrinsic properties of objects: projective geometry.

Attention shifted, at first slowly and cautiously, from the object per se to its images: shadows, maps, or pictures. It is easy to appreciate intuitively that any rigid object will propagate a variety of possible images of itself in space, that these images will alter by continuous deformation, not by fits and starts, and that while there can be no fundamental image, we would nevertheless expect to recognize some kind of permanent identity from several such images. It is equally easy to appreciate intuitively that the images of this rigid object are elastic. Though consistent in their deformations, they do not conserve measured lengths or angles. In Euclidean geometry it is always as if the figures in the books could be applied like templates directly to a material, whereas the figures of projective geometry belong to some absconded, mercurial item that remains out of reach. The key realization in the development of projective geometry was that while figures deform according to the point of view, lines of sight do not deform. So rigidity is transferred from objects to the medium of their transmission, which is most easily imagined as light. That is why Henri Poincaré put the contrast in terms of physical subject matter: "One would be tempted to say that metrical geometry is the study of solids, and projective geometry that of light."[8] One would be tempted to add, following William Ivins and others, a sensual discrimination: metrical geometry is a geometry of touch (haptic) because congruity of figures is assessed by whether they feel the same when put together, while projective geometry is a geometry of vision (optic) because congruity is assessed by whether they look the same from a given standpoint.[9] Neither characterization is completely true, as Poincaré himself went on to show,[10] but they give a first rough indication of the difference, and they enable us to see why architectural composition is such a peculiar enterprise: a metric organization judged optically, it mixes one kind of geometry with the other kind of assessment. Perhaps this is reason enough for the confusion surrounding it.

For several centuries (from the fifteenth to the eighteenth) the development of projective geometry derived some of its impetus from architectural procedures and even from architects. However, my main concern in this book is not with the once fertile relation between architectural projection and mathematical geometry, but with the relation between projection and architecture, which is less well understood. I never intended to write a summary history of geometry and architecture through the ages. It could be argued that the most intense interaction between the two subjects occurred during the seventeenth century, which is touched on but not dwelt on in what follows. Instead of a synoptic survey I chose to concentrate on several quite specific kinds of interaction, often focusing on individual buildings to do so. The scope is largely confined to Europe from the fifteenth to the twentieth century. Coverage is limited and incidental, but it is not intended as accidental or arbitrary. An episodic treatment of this sort has no advantage unless the episodes intimate something other than the fact of their own unique occurrence. I have sometimes tried to indicate aspects of this extra intelligence, but my hope would be that the reader might more easily gain in the reading what I have been unable to state as conclusions in the writing, and I say this not to exonerate myself from the task of generalization, but merely to express the hope that it will be a book like so many others I have read.

The history of architectural projection is just beginning to be investigated. It has played a very small part in the development of architectural theory. Only two well-known architects gave it a significant place in their writings—Philibert Delorme and Guarino Guarini—and modern commentaries on their work have consistently ignored or marginalized this aspect of what they did. General discussions on the subject have nevertheless developed to the point where a consensus can be identified: insofar as projection alters architecture it is to be regarded with suspicion. This consensus has been reached because projection is thought to be an agency proper to the science of engineering and alien to the art of architecture. Either projection is acceptable because it is transparent, or it passes between the creative imagination and the item created like a dark cloud, reinforcing the already enormous prejudice against anything technical. This view is challenged by a historical narrative that extends further back than the nineteenth century. It is wise for architects to remain cautious of projection, but it would be foolish of them to disregard it.

Is this not reminiscent of things heard elsewhere? The way that architecture is divided between geometric drawing and building may be compared to the division between writing and speech. And has it not been demonstrated that there is a tremendous philosophical prejudice against writing that encourages us to think of speech as authentic, with writing a questionable copy of speech, secondary, second-hand, second-rate despite its universal currency? Has not this prejudice been challenged? And are we in architecture not just as prejudiced against geometric drawing? Yes, on all counts. We would nevertheless be well advised at this stage to resist the temptation that presents itself and which has already proved irresistable to some. We must not assume that a certain resemblance gives us leave to treat the two situations as identical, taking terminology, arguments, and conclusions lock, stock, and barrel from literary theory, plastering them onto architecture, and calling the result a theory of our subject. Likeness is not identity; orthographic projection is not orthography; drawing is not writing and architecture does not speak.

A lot can be learned from literary theory, not least circumspection, also a sufficient confidence that the subject for which a theory is being sought is itself worthy of some modest consultation in the matter. In architecture the trouble has been that a superior paradigm derived either from mathematics, the natural sciences, the human sciences, painting, or literature has always been ready at hand. They have supplied us with our needs at some cost. We beg our theories from these more highly developed regions only to find architecture annexed to them as a satellite subject. Why is it not possible to derive a theory of architecture from a consideration of architecture? Not architecture alone but architecture amongst other things. If we take the trouble to discriminate between things, it is not just to keep them apart but to see more easily how they relate to one another. Architecture can be made distinct but it is not autonomous. It touches so much else, and across its borders there is continuous activity. A crucial source of intelligence for such a theory would therefore be the numerous transactions between architecture and other topics, for instance geometry.

The Projective Cast

Part One *Chapter One* **Perturbed Circles**

Wölfflin and Wittkower

Between them, two art historians, Heinrich Wölfflin and Rudolf Wittkower, established the centralized church as a characteristic architectural expression of the Renaissance. In *Renaissance and Baroque* (1888) Wölfflin enlarged on Burckhardt's earlier observation that the centralized plan, "last in the realm of absolute architectural forms,"[1] had finally reached maturity then. It was portrayed for the first time as the epitome of its age:[2]

> *The ideal type of the Renaissance church was central in plan and surmounted by a dome. In this form the age found its most perfect expression. . . . Every line, inside and outside, seems to be conditioned by one central regularising and unifying force, and it is this that accounts for the static and restful quality characteristic of buildings of this kind. The four arms of the cross are in balance, undisturbed and unmoving; the light from the dome is distributed equally to all parts of the building. A state of fulfillment, of perfect being reigns throughout.*[3]

These limpid, domed structures embodied complete unity and self-sufficiency, and shared the calm and beauty of what Wölfflin called the "grand style" in which clear and pure forms were "born easily, free and complete."[4]

The sense that Wölfflin claimed to make of these buildings was derived directly from the architectural forms themselves. There was more than a hint of divination in his descriptions. The expression of the age was the expression in its buildings; to gain insight into the one was to gain insight into the other. It would appear from the way Wölfflin wrote in *Renaissance and Baroque* that inasmuch as history was a cultural issue it was a formal issue.[5] At this stage, Wölfflin's formalism, far from being abstract and desiccated, was almost mystical in the degree to which true expression was seen to shine through stones.

As products of Wölfflin's heightened consciousness his descriptions were brilliant indeed. But turning the inferences drawn from these formal observations into acts of communion with the spirit of a remote epoch was altogether another thing.

Rudolf Wittkower's *Architectural Principles in the Age of Humanism* (1949) also identified centralized churches as the climax of Renaissance architecture and treated them as expressions of Renaissance mentality. But Wittkower took issue with the view that their meaning was directly accessible to modern sensibilities. Surprisingly, Wittkower made no direct allusion to Wölfflin, although his treatment of the same subject matter provides the highest possible contrast.[6] For Wittkower the centralized churches issued from deeply religious motives. The quality of their expression could not be gleaned by direct inspection, however intense, because Renaissance architecture was not an architecture of pure form; it was an architecture of symbolic form. So its meaning was not cast in its superficial appearance, but lay embedded in underlying ideas of cosmic order and harmony. There was no way either, of synthesizing these two contending interpretations. It was not just a matter of shifting focus in search of meaning; the meanings derived from the one approach were necessarily irreconcilable with those derived from the other. "We maintain," wrote Wittkower, "that the forms of the Renaissance church have symbolical value or, at least, that they are charged with a particular meaning *which the pure forms as such do not contain.*"[7]

What followed was an exposition of this encoded meaning. Platonic-Christian cosmology, number mysticism, and theories of natural harmony were disinterred from the fabric. Wittkower's seminal publication, immensely influential amongst the succeeding generation of architects and architectural historians, proved its case beyond reasonable doubt insofar as it was a case against Burckhardt, Ruskin, and all those who, from various motives, had characterized the Renaissance as exclusively humanizing and aesthetic.

Wittkower later became less insistent on the primacy of symbolic meaning. Wölfflin also modified his stance. A few years after the publication of *Renaissance and Baroque,* he was treating the sense in art as largely independent of the spirit of its time.[8] Both made their most sweeping claims while writing of the centralized churches, which itself requires explanation. One might expect opposed schools of interpretation to feed on different material, but here they are, sharpening their teeth on the same bone.

Wittkower's interpretation of Renaissance centralized churches was, like Wölfflin's, based on an assumption about what kind of ulterior reality architecture expresses and where that expression was to be sought within architecture. For Wölfflin the forms of centralized buildings contained directly assayable sense, the cumulative import of which betrayed the mood of the age. For Wittkower, on the other hand, the supporting documentation was the repository of the legitimate meanings. Meanings lay behind forms. Forms were signs. The legibility of these signs was always at risk because it was based on cultural conventions, not universal psychology. The architectural historian stood over them as guardian and expositor, protecting and perpetuating their meanings. The symbolic in *Architectural Principles* is the authorized. Things are what they were said to be. By this definition meaning is constructed through stipulation, and on the authority of those who recorded and propagated it.[9] Substituted for the dubious divinations of formal criticism are the recommendations, opinions, and explanations supplied by architects, priests, theologians, and other interested parties involved in the equally dubious enterprise of attaching specific meanings to selected elements of composition such as number, ratio, and shape.

Wittkower exposed a vast and underexploited reserve of material that has enriched the study of architectural history no end, and yet a distinction might still be made between what he was writing about and how he wrote about it in this early but influential work. The trail in search of meaning led him back through buildings to texts. The story of Renaissance architecture was henceforth to be the story of the incorporation of prescribed symbols into monuments. A fascinating ecology of words and things was mapped out: an ecology that, viewed from within its own frame of reference, gave the impression of being entirely self-sufficient. The point is, it requires an act of complicity between the historian and his sources to maintain vision within the frame.

Most of the publications and documents cited by Wittkower reveal writers attempting to stabilize and fix architectural meaning, often prior to its manifestation in buildings. Why would such efforts be made? One plausible reason for restricting and dictating meaning in this way would be to facilitate recognition of things not otherwise evident; another would be to avoid some perceived difficulty or embarrassment that would arise if significations were not so categorically nailed in place. The rationale of fixed meanings may therefore be constructive or obstructive in its effects, and so we are faced not just with the problem of interpreting the signs, but of interpreting their very presence. Was there some informing hope or anxiety animating the symbolization of architecture, and if so, was it peculiar to the Italian Renaissance?

There is not the slightest hint of any such anxiety presented either by Wittkower or his successors, nor is it very much more obvious in the sources they cite. In what follows I will try to show that it did exist. Hard to discern within the frame because a great deal of attention had been paid to its tranquillization and diffusion, the source of disquiet is easier to perceive from outside; not far outside—a mere step. And even within, the composure is not as complete as appearances would suggest.

Mind you, composure was one of the few things about which Wölfflin and Wittkower were in agreement. There were several qualities identified in Wölfflin's description of the centralized churches that did not lend much weight to the prevailing idea that the Renaissance was an age of rampant, worldly humanism. These were qualities such as stasis, restfulness, serenity, the undisturbed and unmoving. Wittkower did not reject these; he amplified them by referring them to theories of universal harmony from which he thought they may more justly be derived. Similar descriptive terms are to be found scattered amongst the more lyrical passages in recent literature on Renaissance architecture, which might suggest there is a constant sense that can be got direct from the centralized churches that corresponds to their assigned meaning; a sense that can be gauged by modern sensibilities as well as dredged from historical documents. The uncontested qualities would include harmony, serenity, and repose. There is one other, easy to miss because it appears not to be a quality but the defining property—centrality.

Centers

Where is the center of a centralized church? At first sight the question may appear fatuous. When we look at a centralized plan the center is immediately evident: the point on the sheet of paper where the axes of symmetry cross. But a centralized church does not possess an unequivocal point that can be identified with the same alacrity.

Take Sant'Eligio degli Orefici for example. Sant'Eligio is a small church on a Greek cross plan built for the Corporation of Goldsmiths in Rome. It was designed in 1509, probably by Raphael, though some still argue for Bramante's involvement. After Raphael's death it was continued by Baldassare Peruzzi, who supervised construction of the dome and lantern from 1533 until his own death in 1536.[10] Sant'Eligio exhibits the ineffable harmony and serenity, agreed as typical by Wölfflin and Wittkower, to a remarkable degree. One can see exactly what they meant in a place like this. Yet while there is nothing ineffable about a center, a list of possible options will have to be compiled to broach the subject of its location.

1 Sant'Eligio degli Orefici, Rome, Raphael Sanzio, designed 1509, interior.

Is the center of Sant'Eligio the point on the floor beneath the dome, equivalent to the center of the plan? Is it some way up, in the eyes of an observer who might stand on this spot? Is it the monstrance that stands on the high altar facing the entrance? The first is the least taxing choice. The second makes sense if, like Wölfflin, we are impressed by the experience of centrality. The third is undoubtedly the principal liturgical center. All three are displaced from the middle of the enclosed space.

2 Sant'Eligio degli Orefici, plan (after Sanguinetti).

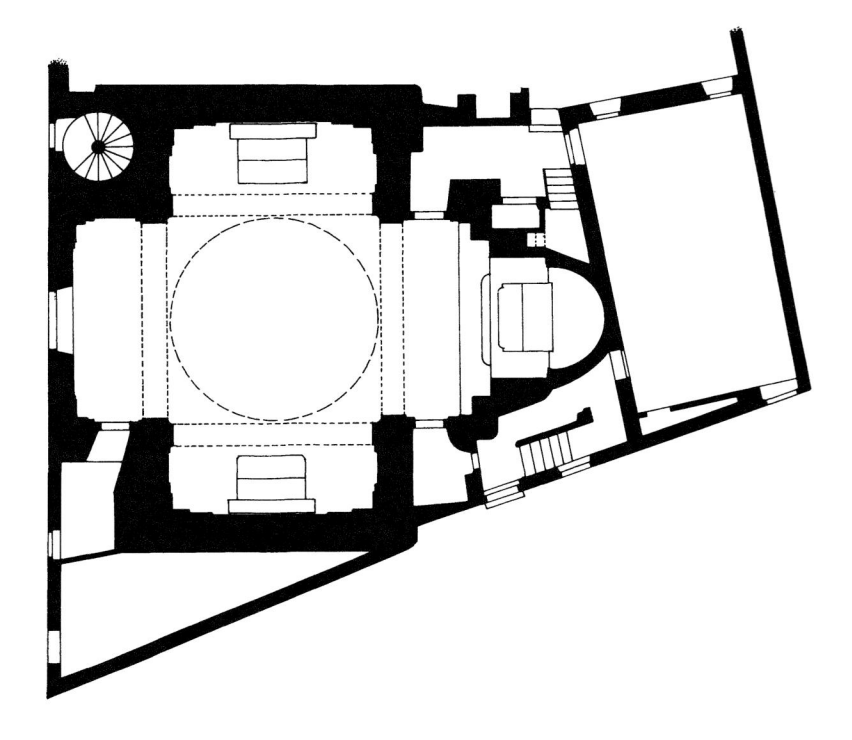

Where then is the middle of the enclosed space? The interior of Sant'Eligio is divided by a continuous cornice. Below are pilasters which, though abstract and compressed, convey the salient features of column and lintel structure. Above are barrel vaults held on flat, hooplike arcuations that extend unmodified from the pilasters below. The continuity of structure, suggesting that the walls and vaults are joined with long ribbons of structural stiffening, only serves to emphasize the apparent change in constitution as the strips cross the cornice line. The structure represented beneath the cornice is classical; the structure above is not. Below are flat planes and straight lines; above are surfaces derived from the circle and the sphere. Below the light is dim, above bright. So could the center be at the level of the cornice? This point is pretty close to the volumetric center, and is in the same plane as the centers of the four crossing arches. Its candidacy is strengthened by the treatment of the transition between the crossing arches and the base ring of the drum above. The ring of the drum does not rest on the four arches. It hovers some way above them. The area between, which would normally comprise four separate pendentive triangles, fuses into one continuous encircling surface of stucco. The surface is part of a sphere with its center on the point in question. From the abstracted standpoint of analysis, this may be a strong contender for centrality, but I doubt if many people would identify it as such in the building. It has no real presence there.

More perceptible is a location in the body of the drum where all the light converges from the four windows around the drum itself and the lantern over the crown of the dome. Could this be the center? Or is it identical with the geometrical center of the hemispherical dome? Is it in the ring of the oculus at the zenith of the dome? Could it be beyond this in the miniature cap over the lantern, or even in the orb mounted on the pinnacle outside?

These are the nine options. Who could say which should be accorded preeminence? The equivocation is thorough. I want to show that it is anything but accidental. Indeed, I want to show that we underestimate these works in our refusal to countenance anything as devious as equivocation to explain the way they are. We are familiar enough with ambiguity, and this multiplying of centers is, if we are willing to consider it, a kind of ambiguity. But it is not much like the disruptive ambiguity with which we in the twentieth century are familiar—an ambiguity that forces attention toward the paradoxes of multiplied meaning; quite the opposite. At Sant'Eligio a serene unity is conveyed that prevails over any sense of its own illusory character. Against all odds the impression given is very much like that identified in Wölfflin's eulogy: "every line inside and outside, . . . conditioned by one central, regularising and unified force."

Inspection of other centralized churches from the fifteenth and sixteenth centuries yields similar results. The number of contending centers is always in the same range. Even Bramante's small memorial temple at San Pietro in Montorio, more concentric in plan, less complex structurally, has eight distinct centers.[11] The Tempietto is in fact an extremely ingenious superimposition of two domed interiors. In the floor of the main chapel is a grating through which can be seen, on the ground some way below, an impost hole, allegedly the place of St. Peter's inverted crucifixion. The eyes are also directed to this same spot through a window in

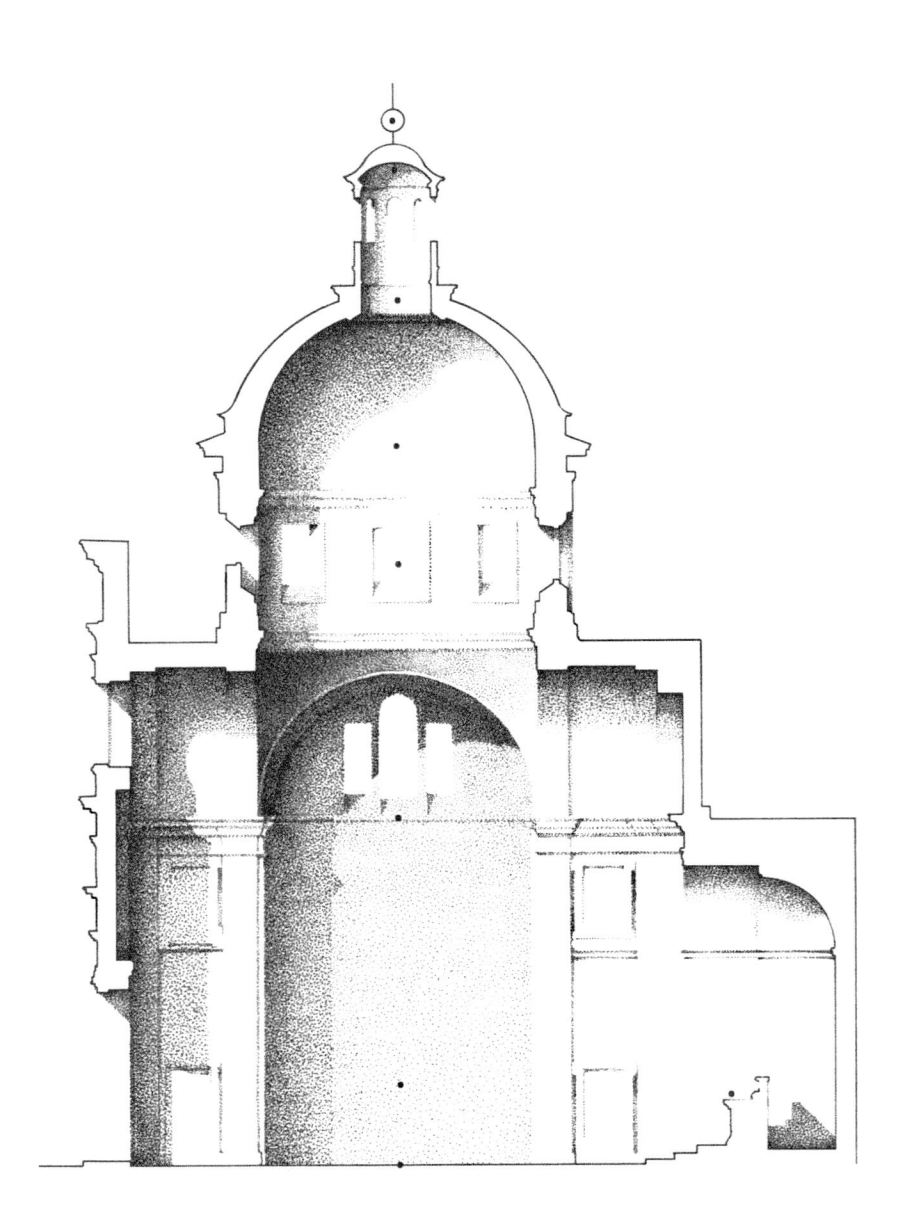

3 Sant'Eligio degli Orefici, section showing nine centers (drawn by the author).

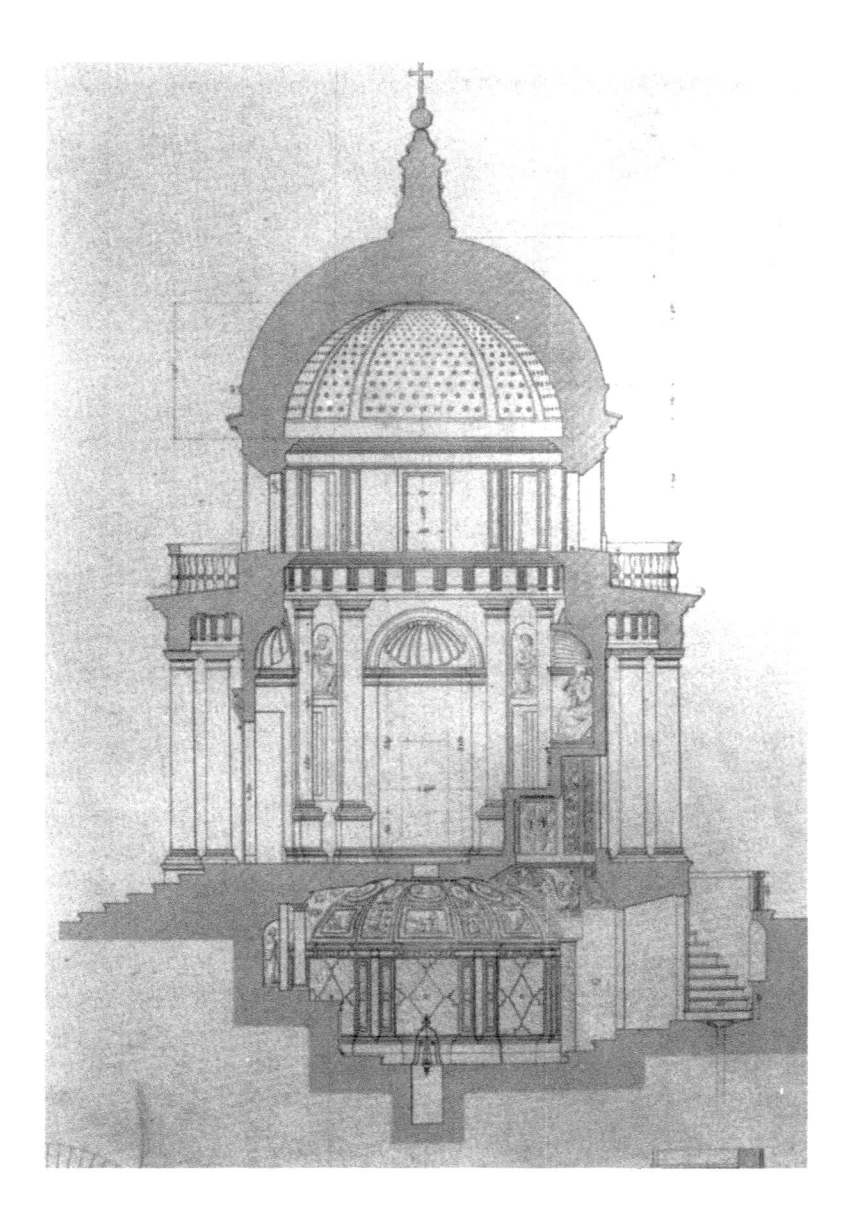

4 Tempietto, San Pietro in Montorio, Rome, Donato Bramante, 1502, section (from Letarouilly, Edifices de Rome moderne*).*

the back of the rotunda, which aims the view down through the cavity behind the hollow altar table, divulging the sacred site by making it visible from outside the building. The impost is in a crypt, discreetly entered via stairs behind the colonnade. If the crypt were included in the calculation the total number of centers would rise to eleven.[12]

Exceptions are few. Only one proposal from this period could be said to have attempted the gathering of centers into one unitary point, and it is hardly typical. This was the "place for preaching in" sketched by Leonardo around 1490. The preacher was meant to stand on top of a column, which would have brought him to the geometrical center of an encompassing spherical auditorium. But even this improbable arrangement for the propagation of the word was capped with a smaller, eccentric hemispherical dome.[13]

5 *Leonardo da Vinci,* Place for Preaching, *Bibliothèque de l'Institut de France, Paris.*

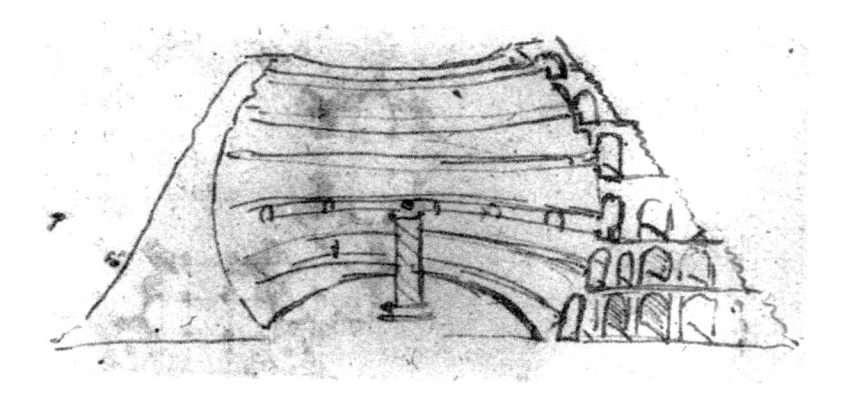

There has never been a completely centralized building. Some, though, are more focused than others. The surprising thing is that the centralized churches of the Renaissance are less focused than either medieval or antique prototypes. The most renowned ancient domed structure, source of inspiration and fascination for so many architects, including of course Raphael and Peruzzi, was the Pantheon. The Pantheon is a hemisphere on a drum that raises the center of the dome sphere way beyond reach. Claude Perrault, who wanted to show that this venerable monument was unworthy of its great reputation, claimed that the four layers of recessed coffering were not arranged concentrically, as he felt they should be, but were distorted so that they would appear in a perfectly concentric pattern when viewed from the middle of the floor. It is as if the center has been displaced to a point about five feet above ground at eye level. "This change," said Perrault, "makes these hollow pyramids appear from below, in the middle of the Temple, just as they would, were one raised to the centre of the Vault."[14] In fact the caissons do not focus on a point, as Perrault maintained. Certainly they are arranged eccentrically, but not in so simple and relentless a pattern. Perrault's misapprehension nevertheless shows that he was aware of the intrinsic ambiguity in the building. He thought that the architect had been intent on reducing the impact of the ambiguity. In this respect he was probably right. Only four putative centers can be identified in the Pantheon: floor, eye level, center of hemisphere, oculus.[15]

In Renaissance studies, domed, centralized buildings are sometimes treated as if they were peculiar to the Renaissance period. Even Wolfgang Lotz, usually wary of sweeping generalization, says that while there were centrally planned medieval buildings, they were not independent edifices like the freestanding monumental churches at Prato, Todi, and Montepulciano.[16] This is an odd statement from so attentive a historian

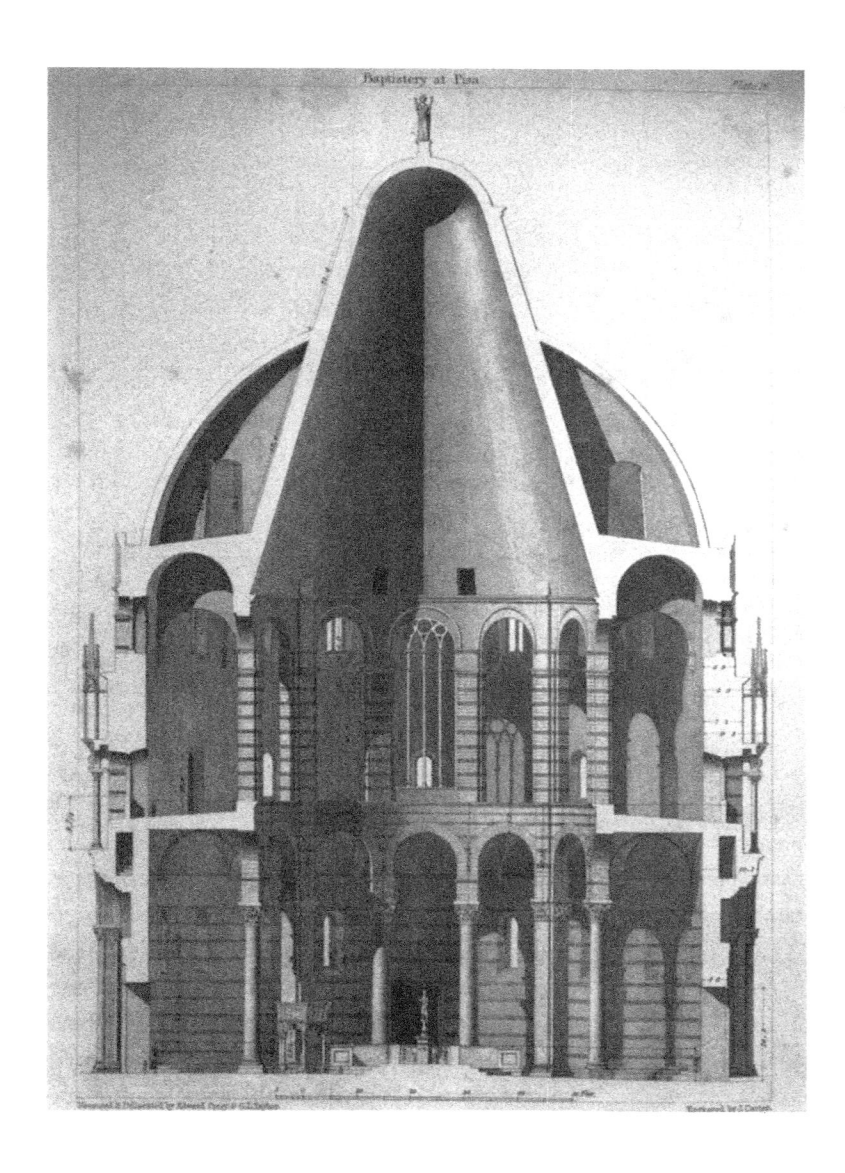

6 Baptistery, Pisa, section and plan (from Cresy and Taylor, Architecture of the Middle Ages in Italy, *1829).*

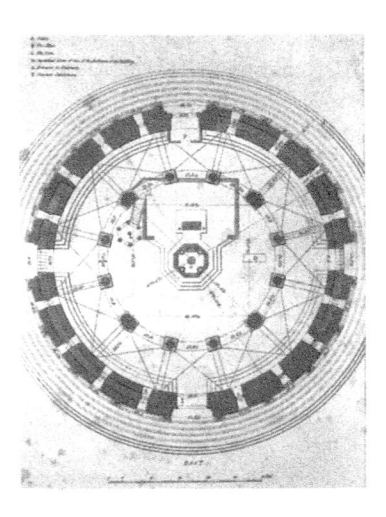

PERTURBED CIRCLES **13**

of Italian architecture, since in the precincts of many Italian cathedrals and churches are detached, centralized domed medieval buildings—baptisteries.

Unlike the centralized churches, where, in practice, the altar was positioned eccentrically, baptisteries were constructed around centralized gatherings. Relics of the days when baptism was a thrice yearly public event requiring immersion, they were raised around baths set in the middle of the floor. Most were on polygonal plans[17] with surmounting vaults or domes that bore directly down onto massive walls (as at Florence, Parma, and Pistoia) or, in more complex cases, onto an inner ring of columns and piers (as in the Orthodox baptistery at Ravenna, the Lateran baptistery at Rome, and at Pisa).[18] The weight of the crowning vaults was always brought straight down to the ground.

In contrast to the medieval baptisteries and to Roman precedents like the Pantheon, relatively few of the centralized buildings of the Renaissance brought the circle of the dome springing directly to the ground.[19] Eschewing the simple assurance of direct support, domes hung in apparent defiance of gravity. This is surely the distinguishing feature of Renaissance examples. It is especially notable in realized buildings. There were plenty of theoretical projects by Leonardo, Serlio, Cataneo, and others that opted for simple downward transmission of loads from dome to tambour to footing, whereas real buildings were often far more daring.[20] As a result of the intrepid removal of supporting masonry, there is an insistent upward displacement of attention. Brunelleschi's gigantic segmental cupola over Santa Maria del Fiore in Florence was raised over open exedrae. From then on one might say that the ambition was to evacuate as much structure as possible from under the dome while pushing the shapes of vaults into more and more consistently circular and spherical configurations.[21] For this reason we may be inclined to think that there is more sphericity and circularity in Renaissance architecture than even Wittkower (who was perplexed at "the contrast between the fervent eulogy of the circle and its restricted use in practice") seemed to recognize;[22] it is held aloft in the building, not inscribed in the plan.

The characteristic elements of Renaissance centrality—the spherical pendentive, the circular drum, and the circular lantern—conspire to multiply and extend the number of centers. All three elements had made earlier appearances, but their combination only became typical after Brunelleschi's Old Sacristy at San Lorenzo (1419–1428).[23]

7   Santa Maria delle Grazie, Milan, Donato Bramante, late 1480s, dome.

8   San Giorgio Maggiore, Venice, Andrea Palladio, 1565–1580, dome.

What then is to be made of the unexpected multiplicity of sites, each of which might plausibly be claimed as central, within a class of buildings whose very designation is thus brought into question? The designation itself is not the issue, because differences between apparent and actual constitution would remain, under whatever title. But then it could be said that the difference between apparent and actual centrality is itself only made available by analysis, and indicates no more than the satisfactory artistic resolution of a diverting formal puzzle, incapable of sustaining any greater significance that might be brought to bear on it. The multiplication of centers is a matter of no importance that becomes trivial when made much of. Perhaps it is obvious and I am laboring the point. That may explain why no one else mentions it. But then why is there no indirect acknowledgment of it either? If it were taken for granted, it would show up as an assumption, but it does not. Read closely, read between the lines and there are still few hints of recognition. It is always as if the single center were an architectural fact, compromised only by the clergy's insistence that the altar be placed at the far end of the church. The absence of acknowledgment suggests very strongly, after all, that the difference between apparent and actual constitution has not been so easily seen.

To the two highly ingenious existing descriptions of Renaissance centralized churches I have added a third rather prosaic one. There are now one symbolic description and two formal descriptions. Let us redress the balance by returning attention to the symbolic. We will find that the multiplication of centers in church architecture begins to make wider sense as soon as we look beyond the privileged assignments of meaning so clearly expounded by Wittkower. Those remain important, but they are not the whole story.

Something similar to the distinction between apparent and real operates in the domain of symbols and icons. Neither symbol nor icon makes the slightest sense without something to refer to. The referent need not be real, but it is bound to be distinct from the symbol or icon that does the referring. So there must always be something outside. Now what I want to do is to consider this "something outside" from another point of view. We have become used to seeing it through the medium of our chosen symbols as if this exclusive and partial view were complete in itself, when in fact it cannot be properly comprehended until the nature of its bias is understood. The portrait is not the person. A church is not the cosmos, but if it was held to represent the cosmos, what did it emphasize about it, and how else was the same thing being described in these same years?

We may talk loosely about "representing the architecture of the universe," yet a building such as Sant'Eligio solicits formal interpretation because of its virtually complete reliance on quite abstract architectural means, as if declaring its independence of other sources of intelligence. That is why Wittkower's choice of the centralized church as the battleground against formalism was so strategically bold, for this area lay deep in enemy territory. Iconographic investigations of meaning tend to be conducted on things lending themselves to that approach, things more or less encrusted with icons. Sant'Eligio, its upper reaches stripped bare of images, is typical of many earlier Renaissance churches. Wittkower pointed to this tendency.[24] For him the deliberate elimination of conventional images practiced by humanist architects like Brunelleschi, Alberti, Michelozzo, and Sangallo the Elder was of the greatest significance. These men were not iconoclasts. They were expelling one set of images from the temple to expose others more integral to architecture and, presumably, more effective in the conveyance of the intended message: number, proportion, and shape, none more clearly exposed than the perfect, reposeful spheres and circles overhead.

The question is not whether the centralized churches were symbols of universal order, but how they were so. And this is a particularly difficult question to answer because nothing is more slippery than a symbol. The thing to notice is that architectural symbolism may be described in two ways: it can be translated or its manner of operation can be described. Wittkower presents us with more translation than operational description. When he tells us that the dome represents the sky, he is translating. He tells us nothing of how the symbol operates. We are simply given the information "this means that" with words instead of architectural symbols. When, on the other hand, he says that Marian churches were round because of "the roundness of the universe over which she [Mary] presides," [25] he tells us a little about the kind of symbolization that took place. In this case the symbol is round because the thing it symbolizes is round. Symbols may be arbitrary assignations of meaning where no resemblance is called for between the symbol and what it symbolizes. They may also be quite closely tied to the thing symbolized, either by causal connection or by resemblance.[26] As we shall see, much hinges on the degree of arbitrariness and the degree and manner of resemblance.

When a symbol shares defined formal properties with its referent, we call it a model. Thus we could point to a sphere rotating on a rod and say it models the diurnal motion of the earth in its movement and orientation but not in its speed or size. A model can usually be tested for truth or falsehood. But most symbols are not constrained by this rigorous requirement of specified formal equivalence. When we say a six-pointed figure is star-shaped, we do not mean that it is the shape of a real star, or even of the twinkling we see. There is no question of truth or falsity; there is only a question of acceptability. There is some question of resemblance but it is vague and indefinite.

The symbolization of the universe in the churches is of this sort: vague and indefinite in respect of resemblance. If they were intended as precise models of Heaven's fabric, Renaissance churches would look like Renaissance armillary spheres, which they do not. Anyway, the pressing demands of structure pulled large masonry buildings into shape more decisively than any symbolic program or mimetic intention. Yet the resolution into circular and spherical forms is enough to indicate that centralized churches shared some formal properties with the universe they represented. The observation that they are neither precise models nor arbitrary signs, although it may be obvious, turns out to be of immense importance.

Let us now return to the deferred question of how the things represented by architecture were being described outside of architecture, and let us start with the simple issue of the center. If the centralized churches represent the universe, what is at the center of the universe they represent? To begin with, compare these two statements. Says Wittkower: "For the men of the Renaissance this architecture with its strict geometry, the equipoise of its harmonic order, its formal serenity and, above all, with the sphere of the dome, echoed and at the same time revealed the perfection, omnipotence, truth and goodness of God."[27] Says Pevsner: "The [centralized building] has its full effect only when it is looked at from the one focal point. There the spectator must stand and, by standing there, he becomes himself the measure of all things."[28] Wittkower was careful not to put God at the center in the way that Pevsner put man there, but even so, these statements are in danger of canceling each other out. Cosmic symbolism and anthropocentricity are not easily reconciled, as Wittkower surely knew, so he argued for the former, an option that would be harder to justify if it were shown that any effort to specify occupation of a definitive center would prove equally fruitless, as indeed it does.

The cosmology long accepted within the Catholic faith was adapted from Ptolemaic astronomy. The Earth was at the center and the moon circled around it. Beyond were the other planets, the sun amongst them, all nested within the vast sphere of stars fixed on a crystalline surface whose ethereal nature was demonstrated by the awesome celerity of daily rotation round the immobile Earth. Beyond this was the outermost sphere, the *primum mobile*. The greatest difference between the Ptolemaic and the Christian account was that Ptolemy saw the celestial bodies as powers in their own right, whereas the Christian orbs were pushed by a host of spirits.

The animation of astronomy with Christian souls began in earnest with a fifth-century work traditionally attributed to Dionysius the Areopagite, *On the Celestial Hierarchies*.[29] But more than anyone else it was Dante, still very much admired in the fifteenth and sixteenth centuries,[30] who turned the universe into a plenum of spirits. The teeming population of the *Divine Comedy* was presented in formation, giving shape to the universe it moved. Thomas Kuhn, the historian of science, has singled out Dante as the great synthesizer of astronomy and theology.[31] In the *Convivio* and the *Divine Comedy* Dante was also the great synthesizer of figures: figures as persons, figures as numbers, and figures as shapes. After the assimilation of his majestic poetic construction demonstrating the orderly correspondences between them, any one kind of figure could stand for the others. Thus latent within shape would be theology, and any construction purporting to represent the heavens would bring with it an invisible population of graded souls.

Mapping their relative positions within either real or fictional space was to prove no easy matter. The universe described by Dante in the *Convivio* was the astronomers' nest of spherical shells. This was not the prospect that was opened up to the reader of the *Divine Comedy*, which, in a fashion markedly similar to the centralized churches, disclosed successive rings along an axis of descent and ascent, this axis being the route of Dante's journey and also the line of his narration of it. Two kinds of model might therefore be imagined: the one a complete nest of spheres seen from outside, the other a partial view of the same from within. The partial view from within cuts from the lower, heavier layers of being through to the ethereal outer layers, which is the formation revealed in Dante's epic by virtue of his being raised through the rings. This latter—the less privileged vision—would be that permitted to us who, after all, live inside not outside the system. Even then, it would only become apparent if we could somehow see the plenum of souls and at the same time see through them.

The tendency to divest churches of icons and frescos, notable in the early Renaissance period, was reversed in the early sixteenth century. Images began to float back into the shell, providing us with valuable collateral evidence. Only by cutting through a universe packed with souls could its contents be made manifest. The dome frescos of the heavenly host developed during the sixteenth century by Correggio, de Ferrari, Vasari, Lomazzo, and numerous other painters[32] illustrate just such a privileged cut from an underprivileged position. To make the frescos comprehensible we have to relinquish the idea that the dome represents the shape of the heavenly sphere. For if it does that, it must do something else as well. The hemisphere of the dome, a model of the vault of heaven as seen from the earth, has to be thought of as pushing through a larger model of the same formation, like a great lens exposing superimposed strata of angels, archangels, thrones; dominions, virtues, principalities; powers, cherubim, and seraphim, leading finally to a glimpse of God's infinite effulgence at the apogee of the dome. The painters' solution to the problem of showing the depth of universal space as well as its outer shape by virtual duplication of the spherical figure was elegant, consistent, and effective. If we left the matter here we might convince ourselves that architecture, painting, poetry, and theology, in their concordance, were further demonstration of the essential harmony of the Renaissance world picture.

A large altarpiece illustrating this arrangement of the duplicated sphere, produced in the 1470s by the Florentine artist Botticini, has, rather disconcertingly, a coffin as centerpiece.[33] Its presence can be explained as necessary to the subject, the assumption of the Virgin into heaven after her death, but it presages a more general predicament. The Botticini altarpiece had been commissioned, and was probably designed, by Matteo Palmieri, who had composed a poem called the *Città di Vita* inspired by Dante's *Divine Comedy*. The painting is known to depict the world as conceived in the poem, where Palmieri describes a descent into the low center of hell (*Scende in basso centro inferna*).[34] Dante was more lurid. The center of his universe was located on Lucifer's frozen anus.[35] This was the spot on which the unregenerate weight of the whole cosmos pressed; the center of gravity. And it was much the same in the Aristotelian, as in the Ptolemaic, as in the Neoplatonic, as in the Christian version. Marsilio Ficino, it is true, did away with the devil, but his Platonized characterization of evil was a cold, heavy clay into which not the smallest

9 *G. P. Lomazzo, Choirs of Angels, 1565–1571, Foppa Chapel, San Marco, Milan.*

10 *Diagram of dual-centered organization implied by sixteenth-century dome frescos (drawn by the author).*

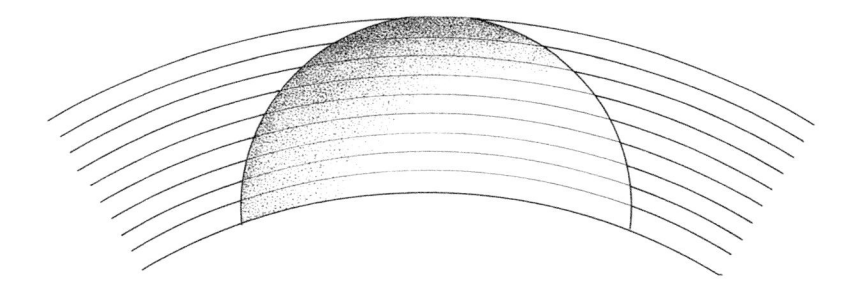

scintilla of spirit could penetrate.[36] Since all things sought their level, the core of the universe was a deadness. These visions are hardly anthropocentric. They are, to use Lovejoy's apposite term, diabolocentric.[37]

Man's allotted place was in any case never at the center but in a middling region, on a kind of spherical shoreline where matter and spirit met—the surface of the earth.[38] Man's own constitution, body and soul, was held to be a microcosm of this larger mixing. The center was corrupt, as in a maggoty apple, and man's eccentric, superficial position was uncongenial, estranged and full of yearning or distraction. Thus we might put paid to any simple anthropocentrism, without prejudicing the observation that, whatever their symbolism, anyone could occupy what seems

11 Francesco Botticini, Assumption of the Virgin, *altarpiece, c. 1475, National Gallery, London.*

like the center of Renaissance domed structures in the same way that, looking up on a clear night, we indulge ourselves in the illusion of our own centrality among the stars, conscious of our solipsism.

Moving back to the celestial hierarchy represented in the dome frescos, note the introversion that had taken place. The outer ring of the heavenly host was equivalent to the innermost shell of the nest of spheres. The whole set of hierarchical relations had been turned inside out. "Neither is it the orbed form that makes the heavens but their undeviating order," wrote Pico della Mirandola.[39] Order may be more fundamental than form but, within art, we have no trouble at all with the idea that this undeviating order could be cleverly reversed so as to preserve the orbed form. It was reversed in consequence of the painters' stratagem described above. The effulgence in Lomazzo's fresco of the angelic choir in the Foppa Chapel, San Marco, Milan, or the Florence Duomo after its painting by Vasari and Zuccari, show the most extended thing as the most localized, while the most local is the most extended. In domical pictures of heaven this remained an iconographic constant from the fifth century to the eighteenth, the spatial credibility of which was vastly enhanced by the introduction of perspective foreshortening, because perspective recession made it possible to combine a single, disklike centrality with an infinitely distant vanishing point that was right in the middle and yet inordinately far away.[40] An anomaly exposed through the application of painting to architecture was resolved by further exploitation of the same combination of media. Yet, however satisfying, it is obvious that such a resolution is localized within and specific to the illusion. Where, other than in perspective, could infinity resolve into a point?

Envelopment and Emanation

When attention is drawn to it, the reversal of the celestial hierarchy in dome frescos seems quite startling, although the matter was so deftly handled that we do not notice it otherwise. Again, we might imagine that it was the skillful resolution of a problem as local to the meeting of architecture and painting as was its resolution, but this would be quite wrong. In fact the very same problem existed outside of art in less tractable form. Within Christian theology two spatial characteristics were attributable to God: envelopment and emanation. With sufficiently vague and mystical ideas about the architecture and motion of the universe, conflict between these may be averted, but they will come into conflict as soon as any attempt is made to specify relative position unequivocally. How was this potential conflict handled in the fifteenth and sixteenth centuries?

Envelopment was a consequence of God's limitless extension. In medieval and Renaissance thought the physical universe was understood to be finite but God's presence was infinite. He was outside of it. Reduced to a naive anthropomorphism, this infolding potency of the Creator was shown as a figure holding the universe like a large plate, as in Piero di Puccio's fresco in the Campo Santo at Pisa,[41] or pushing the rim of the concentric wheels of stars and planets around the broken and ectopic hub of Earth, as in Giovanni di Paolo's *Expulsion from Paradise*.[42] Raphael also showed it thus in a ceiling panel of the Stanza della Segnatura, only with a female personification of Astronomy pushing the celestial sphere. In representing the attribute of indefinite extension a problem was necessarily raised as to how infinite potency might also be shown. On what principle could an engulfing force operate? This is not a "scientific" question corroding a blissfully prescientific scene; the difficulty was intrinsic and recognized to be so. For instance, it was this very difficulty that lay behind the ingenious cosmogony proposed by the thirteenth-century theologian Robert Grosseteste, according to which the universe began with a great expansion from a luminous source that expelled from itself all light and power to the furthest extent, leaving the Earth as the spent remainder of a once omnipotent locus. In this second state, the heavens then turn their light inward toward the original, dead center, as if in elastic rebound.[43]

The patterns of force with which we are most familiar, perhaps because they are easiest to picture and locate, are emanations from centers, as when, for example, a stone thrown in a pond makes expanding ripples. In Giovanni di Paolo's *Expulsion* this was acknowledged by capping one symbolic expedient with another. An aura of golden radiance extends from the enveloping Deity's head. Because radiation is centrifugal, and because effective power cannot easily be imagined otherwise, the universe so construed has two centers: one a vital radiance, the other a dead core. In the *Divine Comedy*, which conveyed ideas of spatial relation with unprecedented clarity, a comparable enigma of dual centeredness is encountered in canto 28 of the *Paradiso*. Dante, having ascended through the planets and stars, was rising into the broadening empyrean toward the *primum mobile* when God was finally made visible to him as an infin-

12 Giovanni di Paolo, The Creation of the World and the Expulsion from Paradise, *1445, Robert Lehman Collection, The Metropolitan Museum of Art, New York.*

itesimal point of staggering brilliance surrounded by nine circles of flame which were the hierarchy of angels. "From this point hang the heavens and all nature" (*Da quel punto depende il cielo e tutta la natura*), he was told. Up to this level the schema of envelopment had prevailed, where

Each sphere with God's own love is more instilled
The further from its center it appears.[44]

But now, presented with a complete reversal of everything he had so far witnessed, Dante confessed himself thoroughly perplexed. His confusion was dispelled only after his guide, Beatrice, had instructed him on the difference between the greatness of God's love and the power of God's love. Greatness was signified by peripheral distribution of enveloping influence; power was signified by central radiance. The image, though far more sophisticated in expression, relied as did Giovanni di Paolo's on the paradoxical identity of envelopment and emanation; an identity that could not be accomplished directly in human experience, but only indirectly, in representation, and even then only with some difficulty. Botticelli's illustrations to this part of *Paradiso* dealt with the paradox by multiplying it. In the drawing for canto 23, for example, the artist resorted to four separate, simultaneous concentric arrangements on the same sheet. The signs of the zodiac indicate the expanding spheres through which the poet rises (geocentric). Dante and Beatrice are themselves ringed in a kind of tondo within the picture (anthropocentric). To the right, the Virgin is surrounded by a circle of twelve stars (Mariocentric). Above this the sunlike globe with the face of Christ is surrounded by flamboyant souls (emanation).[45]

If the sphericity, centrality, and gradation of the cosmos were principal evidences of God's ordered creation, if God's power was to be conceived as a radiance (which was, as we shall see, under the influence of Neoplatonism, more and more the case) and if his presence was to be detected in light, what part would the sun play? A look at Renaissance paintings of religious subjects will show intact the entire range of luminous devices of Byzantine descent, only to some extent naturalized. To be spiritual was to glow; every soul a miniature of the great Aureole. Nimbuses, mandorlas, halos, and coronas abound. The analogy with light was there in the Bible and was repeated and expounded by countless Christian writers through the centuries. Of all things perceptible, light was God's nearest approximation. Jacques Derrida claims that the whole history of Western philosophy might be rewritten as a photology,[46] a captivating thought that would certainly apply to this branch of theology, but with one restriction: it must not be acknowledged as the sun's light.

13 Sandro Botticelli, illustration for Dante's Divine Comedy, Paradiso, *canto 23, Uffizi, Florence.*

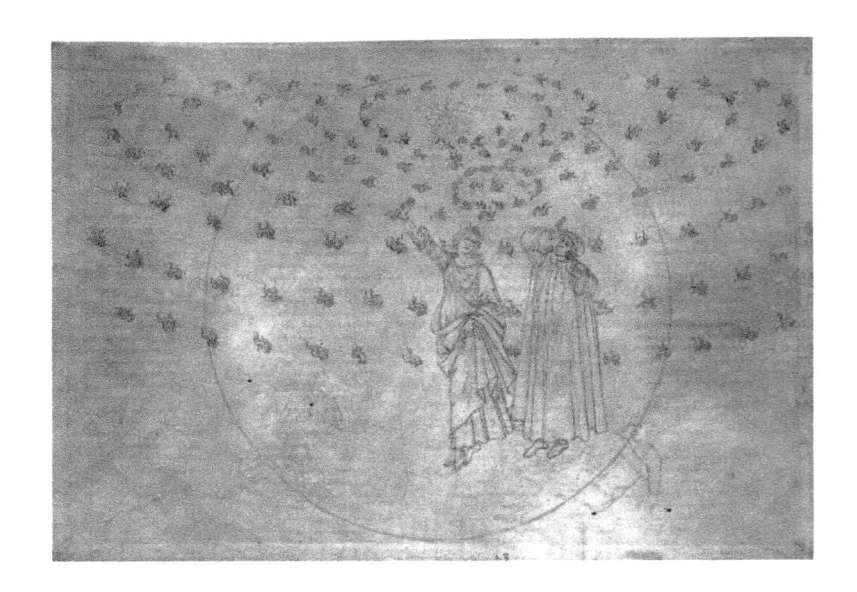

Despite the unabated propagation of the light metaphor in literature and painting, explicit reference to the solar image was restrained during the Middle Ages.[47] Uncomfortably close to heresy, it was first forced into close contact with Christianity by the convert Emperor Constantine, *Sol Invictus Imperator*.[48] It was Constantine who raised centralized domed buildings over the holy sites in Jerusalem, and it was he who, in the words of Walter Lowrie, seemed to confuse Christianity with sun worship.[49] Afterward, his nephew, Julian the Apostate, distinguishing between them, reinstated solar paganism in opposition to Christianity. This was an inauspicious beginning. To a lesser or greater degree all celestial identifications were henceforth tainted, but especially so the sun. Never quite extinguished, its presence was clouded out by the generalized prose of illumination. Similarly in painting. And yet from what else could those radiating gold disks behind God's emissaries—Mary, Christ, the Lamb, and the Dove—have been derived? What else could they be like but the sun?

One of the most extravagant Renaissance images of this sort was the *Death and Assumption of the Virgin,* with its giant solar badge, painted by Gerolamo da Vicenza in 1488.[50] Less extravagant but comparable was Brunelleschi's treatment of the oculus rings in the Old Sacristy of San Lorenzo and the Pazzi Chapel. Under the scallops round the edge of the domes were twelve lunettes supplementing the light from the central lantern. Their pattern resembles a common pictorial convention that showed the twelve Apostles transfigured into stars around a luminous

Virgin.⁵¹ Considered as symbols, the lunettes oscillate between resemblance and convention. The stars from the customary representations had been abstracted by Brunelleschi, but their light (which in the pictures of stars was a representation of a metaphor) was decidedly real through the windows. So one might say that the likeness was not formal but substantial, insofar as it resembled not what the symbol referred to (an Apostle) but what was portrayed in another kind of symbol for the same thing (a picture of a star). The final turn in this alternation between resemblance and convention, form and substance, is to be found in the dark ribs of *pietra serena* that radiate from the lantern oculus, boosting the expansion of real light with a graphic parody of the same phenomenon. Just before reaching the oculus ring, each rib divides and turns round on itself to join the companion molding from the ribs to right and left. This produces a dark, radiant waving around the circle, like a conventional image of the sun, in negative.⁵²

I should emphasize that all the evidence from resemblance is circumstantial insofar as it indicates intent. Resemblance need not involve deliberate intent, but can call things to mind covertly in a culture that insists on deducible signification. Intelligence obtained by means of perceived resemblance can be put beyond the reach of legislation because it depends on recognition, not definition. Definition has to catch up with recognition. There is always a lapse of time, and in the interval, resemblance has its day.

In geocentric cosmology the sun plays a relatively insignificant role, circulating in the fourth sphere amongst the planets. Its demotion—a considerable intellectual triumph over sensible appearance—doubtless made the assimilation of Ptolemaic astronomy into Christianity that much easier. In between the planets, in the middle regions of spirit, the sun was a transducer of ineffable light. But while its dazzling fire was described by Plotinus, St. Augustine, and St. Thomas as an already degenerating visibility on its way down toward corruptible substance,⁵³ Renaissance writers, once again placing far more emphasis on the sun itself, were more inclined to prevaricate over its status and position within the celestial hierarchy. Marsilio Ficino, the key figure at the Platonic Academy in Florence who plays so large a part in our conception of Renaissance art, instigated the revision.

Ficino, his love of allegory notwithstanding, was somewhat less willing to consider the sun's visible splendor as a mere sign than were his Aristotelian adversaries. Ficino wrote four works on light and the sun: *De sole* (1454); *Quid sit lumen* (circa 1476); an expanded version of this, *De lu-*

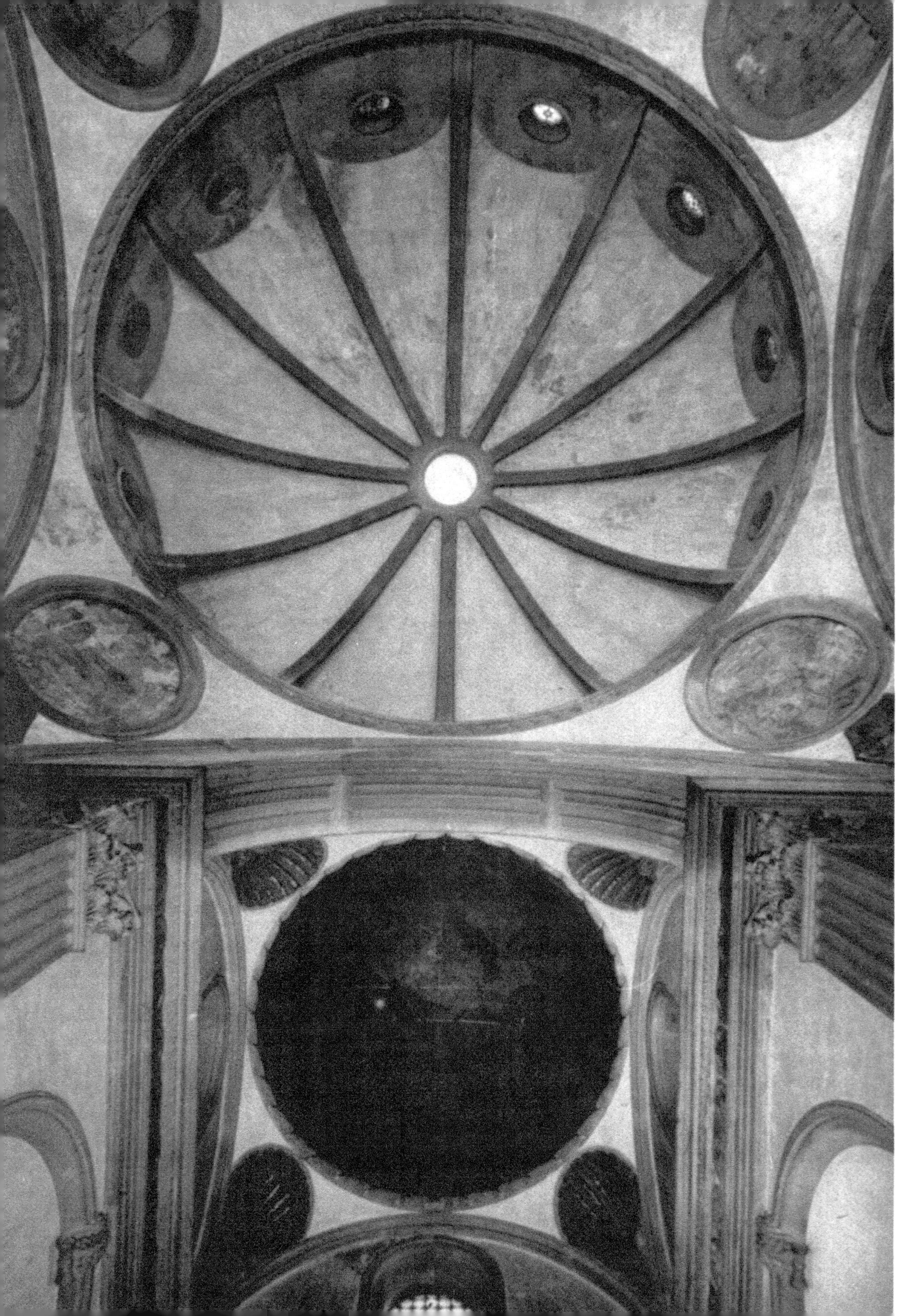

14 San Lorenzo, Florence, Filippo Brunelleschi, 1421, sacristy.

mine (1492); and *Orphica comparatio solis ad Deum* (1479).[54] Ficino fixed the sun at the focus of his light metaphysics, saying, for example, that although the splendor of the sun was its light, the sun preceded its light, as the father was before the son.[55] Allegorical as ever, the sun was now the origin, not just an emissary. Likewise when Ficino, concluding the *Comparatio*, told the reader he may look up and imagine "the supercelestial one who has pitched his tent in the sun," or when he speculated on the possibility that "perhaps light is itself the celestial spirit's sense of sight or its act of seeing," adding, "the sun can signify God himself to you, and who would dare to say the sun is false,"[56] or when in *De sole* he says the sun is an image or a statue of God and that "the sun regulates and guides all things celestial like a veritable lord of the sky,"[57] we can sense the import of these works.

There was not much distance between Ficino and Copernicus in terms of the praise they bestowed on the sun, except that Copernicus augmented the increasingly focused symbolism by trying to put the sun at the geometric center of the universe, which resolved the dilemma of choice between envelopment and emanation in favor of the latter. It also turned Ficino's symbol into a model, with the most important thing in the most important place. As Copernicus himself wrote in a renowned passage: "At rest, however, in the middle of everything is the sun. For in this most beautiful temple, who would place this lamp in another or better position than that from which it can light up the whole thing at the same time?"[58] The metaphor of the round temple was used to persuade us that the sun belongs in the center.

I am not suggesting that the centralized churches were examples of incipient Copernicanism. I am suggesting that the architect of Sant'Eligio and the architect of the new universe were dealing with the same problem at the same time. The first documentary evidence that Sant'Eligio was under construction comes in 1514, the year that Copernicus circulated his first description of the heliocentric theory.[59] Of course, they dealt with the problem in very different ways. The solutions offered by Copernicus in what is called technical astronomy were disruptive because they upset the delicate balance maintained between allegory and fact in the theological account of the heavens. In short: he converted the uncertain, slippery characteristics of a symbol into the more decisive, unequivocal characteristics of a model. Ficino, in contrast, insisted that he must not be taken too literally. His first words to the reader of *De sole* were, "this book is allegorical and anagogic, rather than dogmatic." His last were to warn that, despite what he had said, the celestial sun must not be adored as author of everything, because it only points toward a Supercelestial One.[60]

How stark the contrast between the historians of science recounting the Copernican revolution and the architectural historians presenting the serene domed architecture of the same period. What Copernicus wished to achieve through a hazardous reconstruction of the cosmic hierarchy was achieved in Sant'Eligio (as in Ficino's writing and as in the dome frescos) by equivocation. Unity they all aspired to, but it was not the same kind of unity. Different things were conserved; different things sacrificed.

Equivocation in Painting and Architecture

In the bright lantern oculus of Brunelleschi's domes with their umbrellas of rays; in the oculus of Sant'Eligio on which is inscribed "*Astra Deus, nos templa damus, tu sidera pande*" (God [gives] the stars, we give the temples, spread thou the heavens); in the illusionistic mosaic oculus on the cusp of Raphael's Chigi Chapel dome, through which peers a skyborn personification of God (his posture as ambiguous as his location);[61] and in Lomazzo's effulgent sunlike deity at Milan we see how one point of equivocation, a center displaced onto a peripheral surface, was treated.

The location of God the Father as a figure or as a light was often at the meridian of the dome. Does this solve the problem by deciding on one of our nine optional centers? Have we found what we are looking for? What then of the other eight locations identified in Sant'Eligio? Are they supernumerary? Are they part of an autonomous play of forms contained within the demands of masonry structure, indicative only of some general mood, as Wölfflin would have it? Perhaps this is how they had evolved—the question is hard to decide. Yet in one of his paintings Raphael resorted to the same multiplication of centers. The correspondence between the painting and the building cannot be used to corroborate Raphael's authorship of Sant'Eligio, since the properties in question were not unique to this church. It nevertheless shows that, in Raphael's hands, the deliberate equivocation about centers was tied to the specific issue of Eucharistic doctrine as well as the general issue of the universe's form.

15 Chigi Chapel, Santa Maria del Popolo, Rome, Raphael Sanzio, c. 1516, dome.

The *Disputà* was part of a cycle of frescos for the Stanza della Segnatura at the Vatican undertaken by Raphael in 1508/9, about the same time as Sant'Eligio was designed.[62] The *Disputà* illustrates the drawing down of the Holy Spirit into the human body by celebration of the Eucharist, the

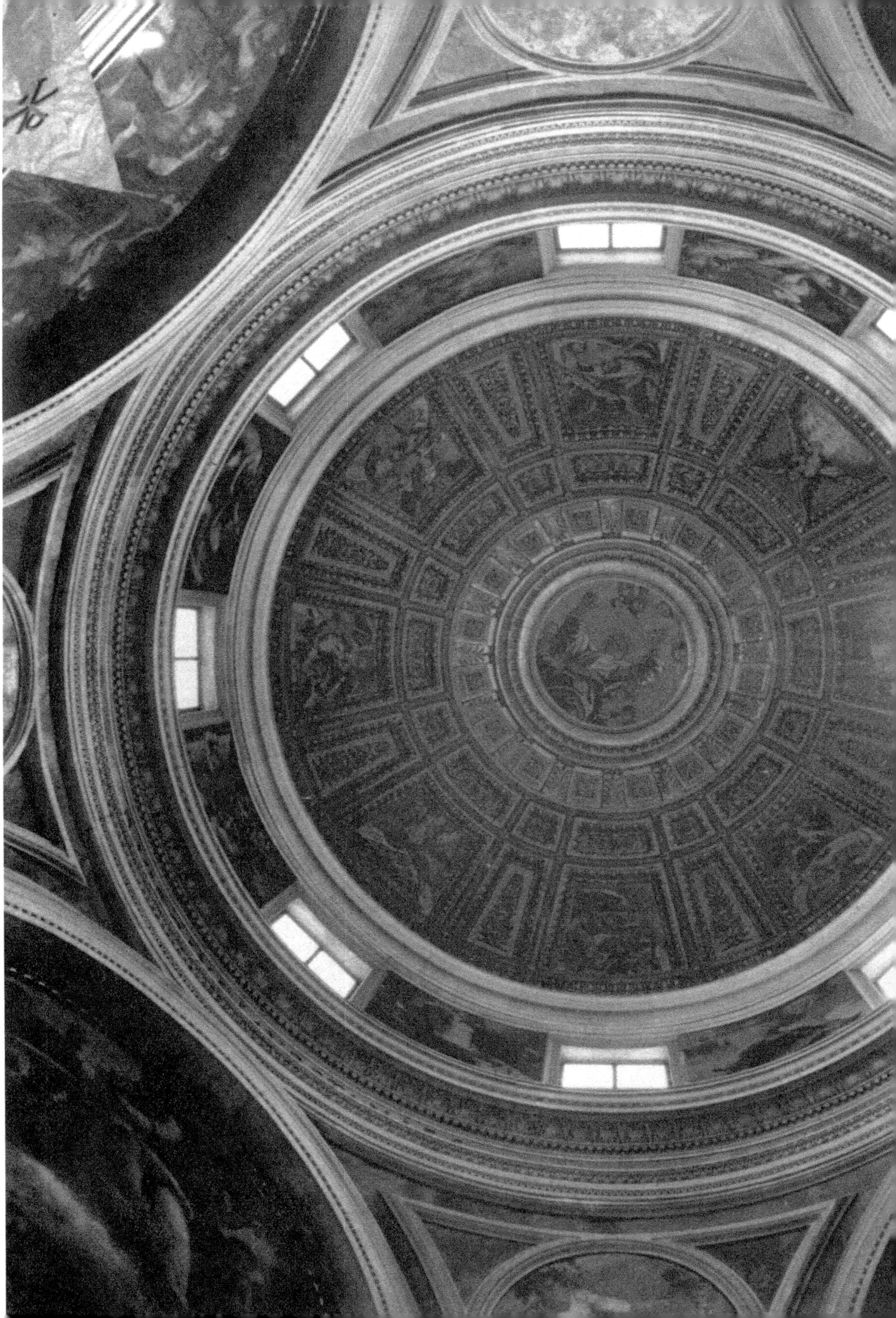

THE PROJECTIVE CAST   **32**

16 Raphael Sanzio, Disputà, *Stanza della Segnatura, Vatican, 1511.*

mystery of transubstantiation performed during Catholic mass in the very churches we have been investigating. "The Dispute" seems a strange title. It is said to be the result of Vasari having employed the ambiguous verb *disputare* in his description. An accident of language, we are told, should not be allowed to color our perception of its actual quality. The fresco illustrates "unity not discordance," as required by the theme, says Holmes.[63] "There is no strife or disputation," says Pastor.[64] "It is no 'dispute,' it is rather an exchange of ideas in the Italian sense," says Fischel. If there are signs of struggle, the struggle is stylistic not theological, according to Crowe and Cavalcaselle.[65] Wölfflin also thought the title inappropriate to this "calm gathering," observing that "there is no dispute in this assemblage, hardly even any speech."[66] Wölfflin went on to say that it would in any case be a mistake to identify the story of the painting with what it expressed. The sense of calm was conveyed directly by the painting, so he concluded that "historical learning" is not essential to our understanding of it.[67]

Everyone seems to agree that the title is a mistake that can be traced back either to Vasari's misreading of the picture or later misreadings of Vasari. As to the second possibility, there can be little doubt about what Vasari meant; he saw disputation. It is not just one word but a whole passage about the figures on either side of the altar that conveys this:

> *But he [Raphael] demonstrates far more art and skill in the Christian saints and doctors who, in groups of six, three, and two, arguing the issue, their expressions showing a certain curiosity and an anxiety in wanting to find the truth about what is in doubt, signal their discussion with their hands, by the action of their bodies, by inclining their heads to hear, knitting their brows, and showing astonishment in many different ways, all varied and appropriate.*[68]

Two reasons may be given for Vasari's "misreading." First, it may not be a misreading; it may be a fair account of what can reasonably be inferred by looking at the dramatic gestures of the mortals portrayed in the fresco. This I think is true. If it is, then it puts Wölfflin's complaint that too much time is spent on what the figures in the fresco mean, too little on what they are, in a most peculiar light. Vasari, doing just what Wölfflin would recommend, looking straight at the painting, saw a scene of animated dispute, where Wölfflin would see quiet unanimity. The reader will recognize that more recent interpretations of the *Disputà*, whether formalistic or iconographic, insist that it is untroubled by strife, and will recall that similar agreement was reached on the centralized churches.

The second reason is that Vasari was writing about the *Disputà* some forty years after completion of the fresco. During that period the Eucharist had been thrust into political and theological prominence by the Reformation leaders. Luther's first attack on the Eucharistic doctrine of transubstantiation came in 1518.[69] So it could be that Vasari was projecting more anxiety back into these figures than was either intended or expressed by Raphael. Bear in mind though, that if Vasari was capable of retrospective projection, so are we.

True, it is hardly credible that a major work commissioned by Julius II for the papal chambers would portray dogma as disputable. Eucharistic doctrine had nevertheless been the subject of rancorous controversy already in the ninth and the thirteenth centuries.[70] The arguments always flared up around the same difficulty: were the body and blood of Christ actually or only symbolically present in the bread and wine of the Eucharist; if actually present, then how so, and if only symbolically present, then what kind of symbolic presence was it?

Raphael's fresco manages to show the Eucharist as an accomplished fact and, at the same time, as a historical accomplishment forged within the Church by popes and saints; it manages to show the architecture of the Church not as a building but as a constellation of figures; it manages to synthesize the depiction of mundane argument (the figures on the floor described by Vasari) with a sense of profound serenity. It seems to be overwhelmingly centripetal, until you try to discover the location of the focus.

The architectonic character of the *Disputà* has often been remarked. Geymüller compared the shape made by the throng of souls to a chapel in Bramante's proposal for St. Peter's; Sommer projected a cross section of the same scheme by Bramante onto the fresco to demonstrate, unconvincingly, the concordance between them,[71] and Hersey detects the more general lineaments of a domed apsidal church.[72] While attempts to prove exact congruences of shape and identical proportions may be improvident, there is a resemblance. There is also a more abstract formal correspondence between centralized churches and the *Disputà*. Given Raphael's astonishing ability to synthesize in the realm of appearance things that normally resist integration, this may not be surprising, but it is revealing.

The ambiguities of position made use of by Raphael in the *Disputà* were pictorial, not architectural. Raphael uses a combination of frontality, axiality, perspective recession, and superimposition to define relative position with sufficient uncertainty when the fresco is scrutinized, while giving the impression of unity and clarity at a glance. These are quite different means than those employed at Sant'Eligio. Even so, the correspondence between centers, almost point for point identical with the internal structure of the church, is striking.

Set in a broad landscape, the mystery of the *Disputà* is given architectonic definition by two rings of cloudy cherubs. The implied sphere defined by the rings is drawn up into a gold canopy. If we assume that the lower, denser circle of clouded souls dividing the celestial from the mundane world corresponds to the division made by the cornice line between the two species of architecture in Sant'Eligio, then the plummeting Holy Dove held in its frontal disk of radiation (a line of connection arrested in a focal point) would occupy the geometric center of this dividing plane. The position of the dove would therefore be equivalent to the center of the pendentive sphere. Four child angels holding texts of the four evangelists, who were customarily depicted on the four pendentives, for which there is no analogue in the fresco, hover around the dove. The ring of saints would then correspond to the ring of lights in the drum, and the higher ring of cloud would then be equivalent to the circle of the dome springing, into which light spills from a point external to the representation, as from a lantern above a dome. God the Father, seemingly central and holding a crystal orb in his left hand (a model within a model that would stand for the pinnacle orb mounted outside the church) is shown under this influx but on top of and behind the sunlike aureole around Christ. The precise position of Christ, though also seemingly central, is just as hard to determine, and yet, like the center of the fenestrated drum in Sant'Eligio, his body marks the most decisive focus of light. Below, on the altar table, the monstrance centers with the precision of a gun sight on the image of Christ's crucified body impressed on a wafer of bread.

Noticeably empty is the marble pavement, above which floats this spectral geometry of souls. The only evidence of the observer's displaced presence is the perspective vanishing point on the horizon at the base of the monstrance, which provides another center. The vanishing point, which is also the center of the semicircular lunette that frames the fresco, could easily have been made identical to the targetlike monstrance top, but Raphael chose to keep it separate.

Here we see Raphael using a formation that was applied to architecture and painting. We see that it relates to a quite specific function, since the *Disputà* fresco could be described as no more or less than an illustration of what, according to Catholic dogma, takes place in any church when a mass is performed. We are left in doubt only about which properties of the things portrayed are portrayed as real and which are symbols of spiritual things. This question, at the heart of the figurative arts, was at the heart also of the dogmatic theology of the Eucharist. As Calvin's successor, Theodore Beza, put it, there are only two possibilities: "either transubstantiation, or a trope."[73] Beza's exasperated polarization is inadequate to describe the subtleties of Eucharistic interpretation in this period, just as it is inadequate to describe the subtleties of the *Disputà* or Sant'Eligio, but it does indicate that all flourished in the same area of uncertainty.

So the *Disputà*, with its three luminous and five nonluminous centers, its apparently spherical space, and its graded, concentric hierarchy constructed with such resourceful polyvalence, might either be regarded as a painting modeled on architecture, or might help body out a Eucharistic-cum-trinitarian iconography for the abstract structure of domed centralized churches. The reader may well be asking why, having cast doubt on the validity of iconographic analyses that identify prescribed signification with meaning, I have spent time deciphering Sant'Eligio according to that very principle. The reason is this: if such a correspondence is demonstrated, then the clearly announced content of the painting can be projected into the vacant format of the architecture. Names can now be given to the multiplied centers of Sant'Eligio, and the original observation of multiplicity is perhaps made more credible. But what does this prove? It certainly does not prove that everything fits into place. It only tells us that things fall into place within a small region in the privileged domain of art. As with the dome frescos described above, the sublime coherence established within the *Disputà* betrays evidence of just what was problematic outside it.

Wherever the power of illusion is at its most intense; wherever the inversions, deceptions, and conflations occur, that is where the difficulties are likely to be most desperate beyond the borders of the illusion.

In the church, as in the fresco, Raphael treated cosmic and theological themes that were the source of considerable disquiet at the time. He constructed real objects with apparent properties—the floating lightness of the vaults in Sant'Eligio that weigh about 800 tons,[74] the space inside the surface of the *Disputà*—and, playing with the difference between apparent and real, he transformed the disquiet into serenity. We witness a transformation of mood (I use Wölfflin's word) effected by formal, not iconographic means. The transformation only makes sense, however, if we introduce something from outside. Something always has to be introduced from outside. Wölfflin introduced universal psychological responses. We have brought in the symbolism of the Eucharist, and Christian/Platonic cosmology, as did Wittkower, but have seen them in a somewhat different light, because we are no longer looking at them through the screen of art as if art disclosed the unvarnished truth. Art does not transmit meaning without modifying it. The significance of art does not lie in the meaning it is said to convey, or the meaning it is meant to convey, but in its *alteration of meanings that have been constituted elsewhere*. One way of approaching this third sort of sense is to compare the sense derived from forms with the sense conveyed by symbols. Wittkower was right, these two senses are not the same, but neither are they unrelated. We may therefore accept both Wölfflin and Wittkower so as to discern an interference pattern that is dependent on both but different from either.

Neither the centralized churches nor the *Disputà* were inspired by the constructive seeking out of equivalences and similarities that one might expect in a search for true representations, but by a constant creative dissembling, conjuring up local unities and correspondences in an effort to depict a larger system whose unity was incapable of spatial demonstration. If rejoined with the meaning that Wittkower wanted to restore—the stipulated meaning—these works of art become works of artifice and artfulness, which, in giving coherence to figments, stand as evidence of their existence and proof of their unification. An architecture meaningful in this sense must stand under the cloud that has hung over painting and poetry for millennia: the closer it aligns itself with truth, the more mendacious it becomes.

Circle and Sphere

How can anything as innocent-looking as a sphere or a circle be implicated in this impressive but suspect enterprise? It is widely acknowledged that within the physical corporeality of Renaissance art were unprecedented abstractions. Together with the realism of flesh and muscle, the new insistence on composition;[75] together with the fuller utilization of polyphonic sound, a further exploration of mensural proportion.[76] In architecture too a simultaneous solidification and abstraction can be observed. Strangely enough, though more corporeal in itself—dominated by its own weight—architecture is more abstract in the making than are the other arts. And it was through the abstracted means of design that Renaissance architects were able to work, as no other artists had, directly with the most mercurial of all media: geometric figures. In geometry, geometrical figures are not media, but in architectural design they are, since their task is to convey shape from one state to another. In this sense they are just as surely media as the inks with which they are drawn. Considered as media, they have the peculiar property of being changeless in themselves and volatile in relation to everything else.

In a universe construed after the fashion of Western metaphysics, with matter and spirit opposed, geometric forms move easily across the border between the visible and the invisible, the corporeal and the incorporeal, the absolute and the contingent, the ideal and the real.[77] As already pointed out, the anthropocentricity of Renaissance culture had less to do with man's geometric centrality than with his being in the middle of a nexus of communications between extreme states.[78] So man was dependent (as victim, beneficiary, recipient, or manipulator) on the flow of traffic through the web. Geometric figures, passing with such ease across the divide, were regarded as especially good couriers within the network. Ficino, in one of his numerous demonstrations of the incorporeal nature of beauty (involving color, light, and number as well as shape and order), asked his reader to envisage a building, and then to "try for a moment to abstract the matter, if you can. Abstract it in your thought. Then abstract the matter from the building and leave the order suspended [as it is]. Nothing will remain of the material body."[79] Alberti gave the same advice.[80] Are we then the spiders or the flies in the web? While we may follow Ficino's instructions, and dematerialize forms, is it not just as miraculous that they can be made by us with a material precision that accommodates the disinterring imagination?

Consider then the circle and the sphere and the part they play in Sant' Eligio, a building that lends itself well to imaginative dematerialization. Above the cornice, the rationality of the sphere and its sectioning appears to triumph over the logic of gravity. According to the surviving drawings of what is thought to be Raphael's original project, the four windows in the drum, which were built rectangular, were designed circular.[81] In that event, all the forms in the upper reaches of the interior would have been circular, cylindrical, and spherical. As built, the tectonic framework, emphasized in darker bands of stone, is identical to the lines of intersection between the spherical and cylindrical surfaces. The two things are precisely congruent, which is quite unusual in architecture. And all the spherical and cylindrical surfaces are integral to the structure of the building, although they seem to disavow their weight. Contrast Raphael's sophisticated use of the continuous spherical pendentive with late medieval efforts to bridge the same gap with squinches that clumsily exhibit the effort of bearing loads. It is a considerable achievement to get two independently defined properties, one geometrical, one structural,

17 Sallustio Peruzzi, drawing of Sant'Eligio, c. 1560–1565, Uffizi, Florence.

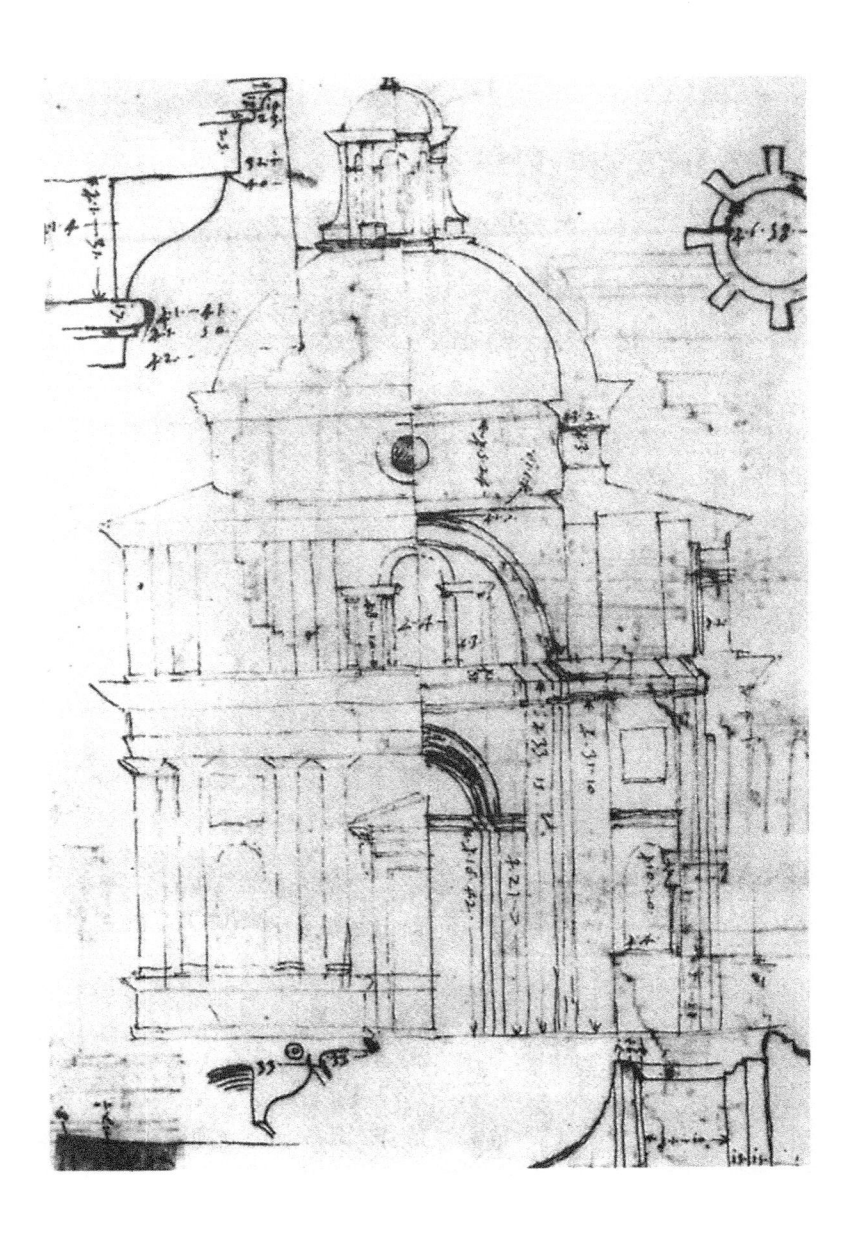

to coincide exactly, and for the resulting arrangement to express a lightness that turns against its own constitution in effortless self-refutation.

We should look more closely at the renowned perfection of circle and sphere. In what did it reside? Both the circle and the sphere comprise two elements, a center and a periphery. This we would learn from any treatise on either geometry or architecture. Their perfection was demonstrable in their unity, their symmetry, and the economy of their defini-

18 Leonardo da Vinci, solid sphere, from Luca Pacioli, De divina proportione, *1496.*

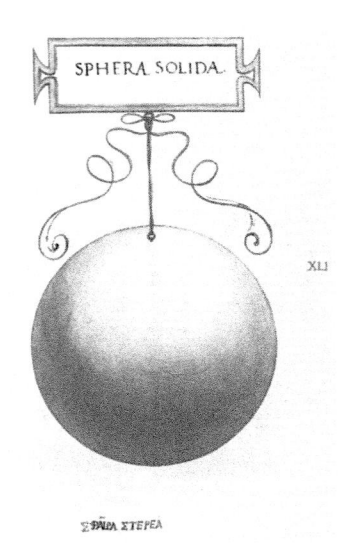

tion. A circle lies within all the points in a plane equidistant from a single point also within the plane.[82] A sphere, simpler still, is an enclosed surface of all the points equidistant from a given point.[83] The two component parts, focus and periphery, seemed indissolubly bonded within the definition and construction of these figures—especially of the circle, for compasses were the definition turned into an instrument; the piece of paper an abstract Euclidean plane given body. The unique conciseness of definition and simplicity of construction were, however, dependent on the introduction of a foreign element. The center is not integral to these figures, nor is it ultimately necessary for either their definition or their construction.[84] Euclidean geometry, the geometry by which Western thought has been permeated, is full of elements that, like the center of the circle, are ways of defining figures to which they themselves do not belong: axis, asymptote, *latus rectum*, tangent, chord, focus, and perpendicular. Only the center of the circle amongst all these stayed stubbornly tied to its figure—a historical circumstance, not a logical necessity.

The geometer's dilemma was similar to the theologian's. The same kind of difficulty kept turning up. It could not be resolved by aligning ideas with experience more closely, or tidying up an abstract definition. It was truly pervasive. So when Cusanus, exploring the properties of infinitely large geometric figures in his *Of Learned Ignorance* (1440), said that "the poles of the [heavenly] spheres coincide with the centre and there is no other centre than the pole, that is, blessed God himself," and when he concluded that the universe was a sphere, the center of which was everywhere and the circumference nowhere, he was not stating paradoxes that had been engendered by mathematical reasoning, as is so often supposed. The paradoxes were already established. He lent them credibility by obliterating the difference between center and edge.[85] Thus his deployment of mathematical rationality to alleviate theological difficulties is comparable to Raphael's deployment of composition in painting and architecture in the same cause.

From a practical point of view, doing away with the center of the circle would have seemed petulant. Nevertheless, in the centralized churches the centers are vacant and invisible. An iconography of centrality needs something on which to deposit its evidence. In an architecture of spherical and cylindrical surfaces this deposition would inevitably be eccentric and peripheral. In discussing the ocular God, attention has already been drawn to the contortion by which the spherical surface includes pictures of its own displaced center. This fusion of two elements into one was aided not only by perspectival illusions, but also by the pleonastic distribution of centers within these buildings.

THE PROJECTIVE CAST **40**

Again we may compare Raphael and Copernicus. I have already suggested that Copernicus broke through the ambiguities that sustained and protected art and theology with his bold revision of the cosmic scheme. It should be made clear, however, that this revision was necessary because Copernicus regarded something else as beyond question: circularity and sphericity.[86] It had always been understood that the basic Ptolemaic system of uniform motion within concentric circular orbits was inadequate to the needs of descriptive and predictive astronomy. Ancient authors—Apollonius, Hipparchus, and Ptolemy himself—had devised compound motions of smaller circles on larger circles (epicycles on deferrents), still other circles on these (epicycles on epicycles), and, most peculiar of all, eccentric points from which cyclic motion would appear uniform (equants).[87] At the end of the fifteenth century the celestial model was an overloaded complex of more or less eccentric, gyrating loci, threatening increased complication to save appearances and to save the circle. The circle had to be saved, it is easy for us now to see, because of the intellectual and aesthetic prejudice in favor of this figure. The pressure to maintain circularity may be judged from the untiring efforts of all astronomers up to and including Kepler to keep celestial motion within this pattern.

19 Renaissance modification of the Ptolemaic system of celestial orbits (after A. Koyré, The Astronomical Revolution*).*

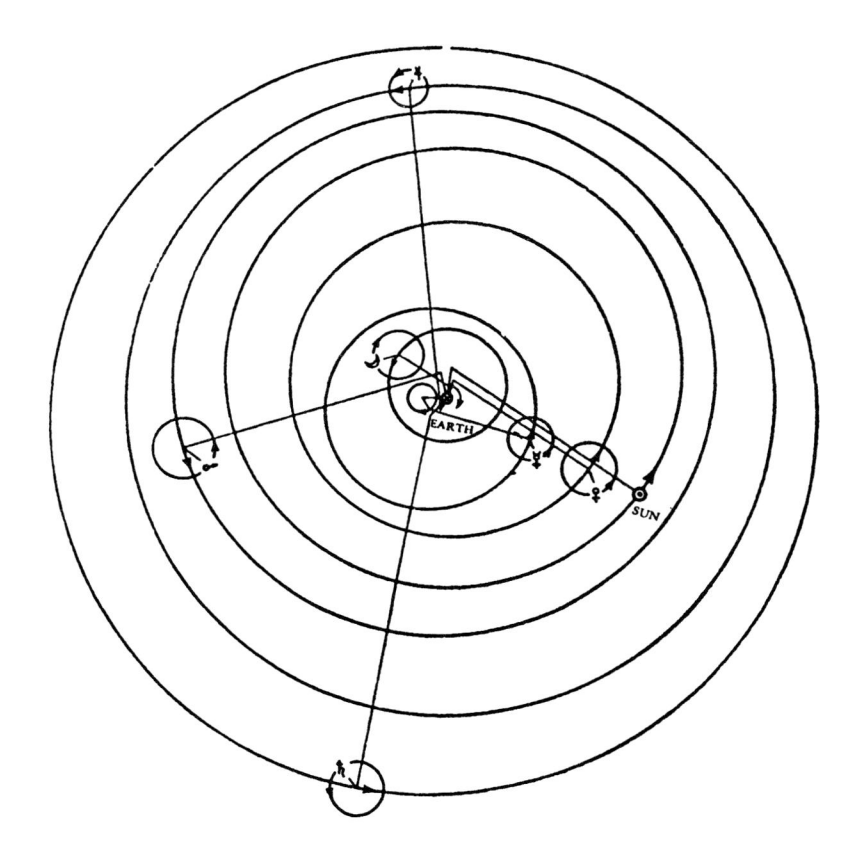

Copernicus, introducing his *Commentariolus* with a list of assumptions, felt it unnecessary to mention that the universe was spherical or that orbits were circular, so much were both taken for granted. It required the supreme intelligence of Kepler, capable of extraordinary self-critical introspection, to see through the circle. D. P. Walker and H. F. Cohen cite the same devastating passage from a letter of 1608: "For I too play with symbols . . . but I play in such a way that I do not forget that I am playing. For nothing is proved by symbols: nothing hidden is discovered in Natural Philosophy through geometric symbols; things already known are merely fitted [to them]."[88]

Through Kepler's insight we may understand why trees, birds' nests, human heads, and even human bodies were said to resemble spheres.[89] All of nature was presented as a devolution from the sphere and circle, through the regular solids and polygons, down to irregularity; an idea that could only be held onto by extremely aggressive idealization. One observes also throughout this period a symptomatic removal of epicycles and eccentricities from all but the most specialized astronomical illustrations, so preserving a picture of a simple rationality in a universe whose observed motions could only be made commensurate with it through an unseen, outlandish apparatus of modifications.[90]

The same might be said of the centralized churches, for were they not also ways of representing complicated arrangements fraught with difficulties as if they were simple, serene, and unified? Reject this and the situation is no happier, because if we accept at face value the unity and coherence that Wittkower tells us they are meant to represent, then what would stand in the way of our seeing the churches, as they have been seen before, as monuments to a fraudulent metaphysics; concocted legitimations that inflated a prosaic metaphor of social power to fill the universe?[91] Nothing. It was after all Alberti, the leading figure in Wittkower's account of centrality in ecclesiastical architecture, who compared the patriarch of his own noble family to a spider who spun his web so that every strand "finds its beginning, its root, its point of origin, in the center," where he waits, "vigilant and ready to hear and see everything."[92] It was after all Wittkower who observed with characteristic acumen that, in the centralized churches, the image of Christ Pantocrator had taken over from Christ the Man of Sorrows symbolized in Latin cross plans.[93]

Or we may gaze through a fog of sentiment, as is now the custom, and regard the stunning beauty of churches like Sant'Eligio as the product of an untroubled and complete consciousness irretrievably lost to modern man, who, knowing something of Pascal, inhabits a universe full of uncertainty. From what has been said so far it would not seem reasonable to regard the Renaissance world picture as either certain or coherent; nor would it seem any more reasonable to regard the centralized churches as attempts to copy such a picture in all good faith. Is it not more likely that the delectable fancy we indulge ourselves in, looking back at these buildings as if what they appear to represent had been there, is itself an aftereffect of their original, sublime mendacity? In this, historians, who have wanted to see the epoch reflected in its buildings, have probably been more gullible than the architects, painters, and patrons of the Renaissance, who may well have been more keenly aware of the suspension of their own disbelief.

It is often assumed that a symbol, while not necessarily representing a truth, represents a belief, and so speaks directly for a historical frame of mind. This questionable assumption has not been borne out here. What has been brought into view is an immense effort within the limbo of art to bring two worlds together with no apparent loss to either. This undertaking was accomplished in the hazy zone between the ideal and the real, the area where classical art is so effective, the area where the ambiguous symbol, neither model nor arbitrary sign, flourishes. Here architecture achieved a remarkable triumph.

From any other standpoint, the universal edifice appears anything but stable, anything but unified, anything but complete. Neither geometry nor cosmology nor theology could, in the event, turn ideal forms and relations into plausible models of reality without embarrassing contradiction. Although unity was presupposed, and although all were convinced that perfect geometrical constructions lay behind the diverse forms of the world, the excesses and defects, paradoxes and anomalies obtruded, so that if the principle held in one respect it failed in another. The attempt to define with greater precision a geometrical armature of unity and gradation within nature, and to find its presence in the shape, order, and motion of things, made the absence of these very properties all the more conspicuous.

Out of this strained and failing effort to make the world match its representation, against all probability, in the face of knowledge, these centralized churches arose calm and untroubled, as if the logic held, as if, in the purity of their forms, they did provide a receptacle for ideas, forces, and things that would not hold together elsewhere, as if there were no distance between the shape of the world and the way it was described, as if the ruler were the same figure as the subject. All these effects were won through the extreme precision with which architectural ambiguity[94] was used to conflate multiplicity into unity, not, I would emphasize again, the other way round.

It is a moot question how, in this blissful apotheosis of jeopardized qualities, the spirit or mood of the age was reflected. Certainly not as in a mirror. The architecture of Sant'Eligio is not an architecture that reflects a culture in all its fullness, but an architecture that supplements culture's incompleteness with a compensating image. It is not a representation so much as a complementary misrepresentation. It does not exemplify the spirit of the times but a spirit standing next to the times as then perceived; an impostor with a life of its own, feigning identity. Architectural historians, in an effort to let architecture speak, have tended to suppress the distinction that exists between a representation and what it represents. This suppression is no crime; it may even be useful in certain circumstances. In this case, though, it prevents us from recognizing the most astonishing aspect of these buildings: their successful merging of the forms of thought and the forms of things to which thought was applied, as if they were one and the same when they could already be seen to be different.[95]

If, in this account of centralized buildings, geometry is not operating quite as expected—neither entirely sterile, nor definitive, nor stable, nor clear—it still plays the role generally assigned to it within the history of architecture: that of governing a building's basic shape through restricted combinations of prismatic, cylindrical, and spherical volumes, their scale, ratio, and frequency regulated through rules of number. In this respect, too, a church like Sant'Eligio is a haven of reassurance. Elsewhere this role, so often construed as fundamental, was threatened not by some liberating or obliterating external agency but by geometry itself.

This began to happen as geometrical figures prized themselves apart from the things they were, in their perfection, said to define. Something of the sort was happening to the circle, here and there. Cartography offers the best examples. While architecture provided the opportunity to force substance into geometric shapes, the allied business of surveying provided an opportunity to divest geometry of this same task as soon as it was turned from the designing of perfect architectural forms to the recording of imperfect ones.

Significantly, those who had so artfully contrived the identity in architecture began to exploit the divorce in maps. The circular maps made by Alberti and Leonardo can be usefully compared with their circular

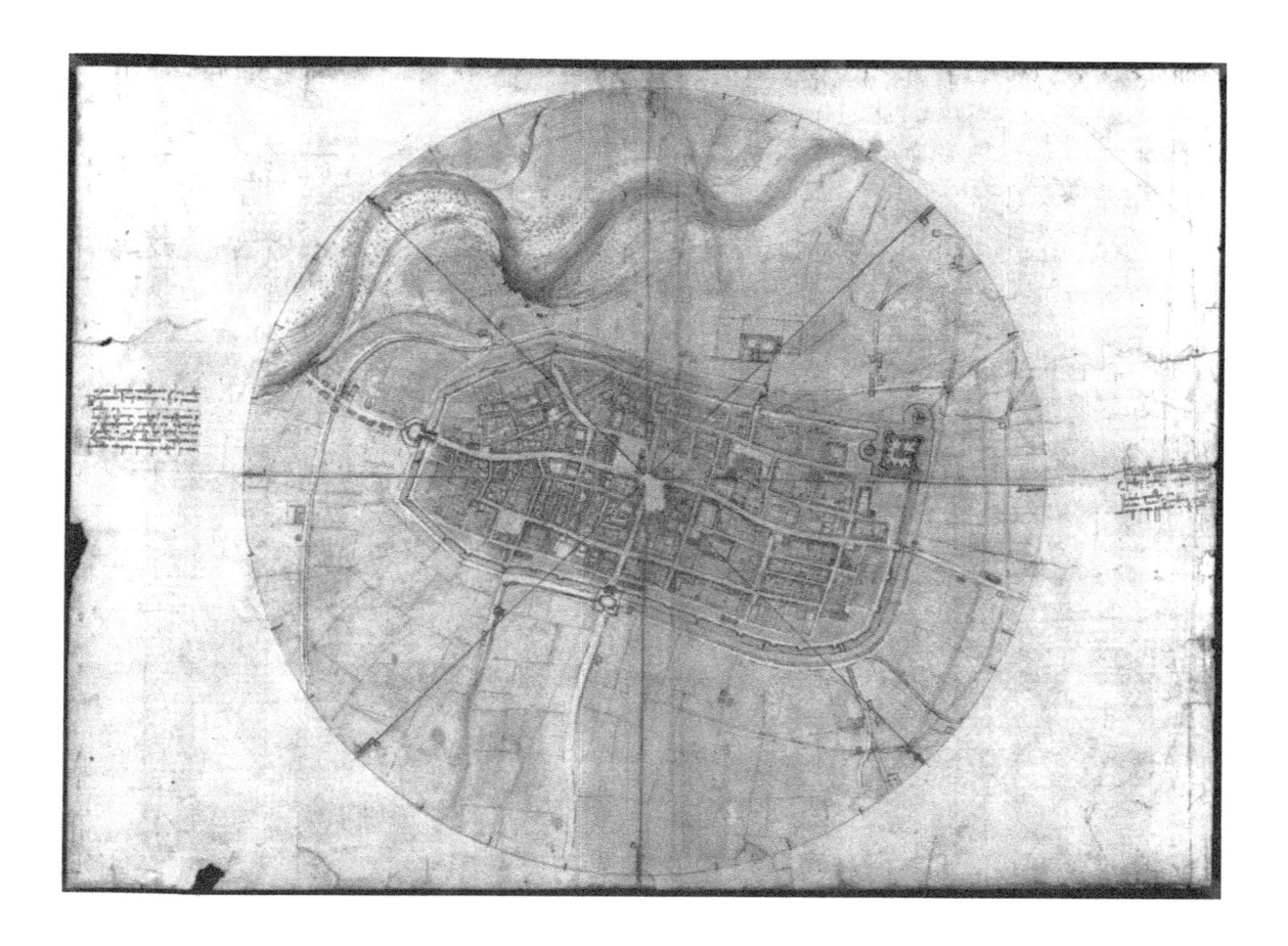

20   Leonardo da Vinci, map of Imola,

1502–1503, Royal Collection, Windsor.

domed buildings. Their maps started with an arbitrarily defined circle from the center of which equally spaced radii extended.[96] The center, which in earlier maps was fixed on a symbolically pivotal site—Jerusalem, Rome, St. Peter's—was still so in these, for Alberti centered his map of Rome on the Capitoline and Leonardo centered his map of Imola on its major crossroads. This, however, was the one and only point of correspondence between the geometrical figure and the shape of the settlement it surveyed.[97] Measurements were taken along the radial lines wherever they intersected with boundaries or buildings. This procedure allowed the plotting of any configuration, no matter how irregular. In Leonardo's map of Imola, drawn in 1502 for Cesare Borgia,[98] the tracery of measuring lines is still visible. The strict regularity of the measuring system now enabled the thing measured to free itself from the strictures of unwarranted geometrical interference; a city did not have to be round. Geometry, considered thus as an overlay rather than an underlay to reality, was in the odd position of gaining descriptive power as it relinquished its direct hold over the form of what it described.

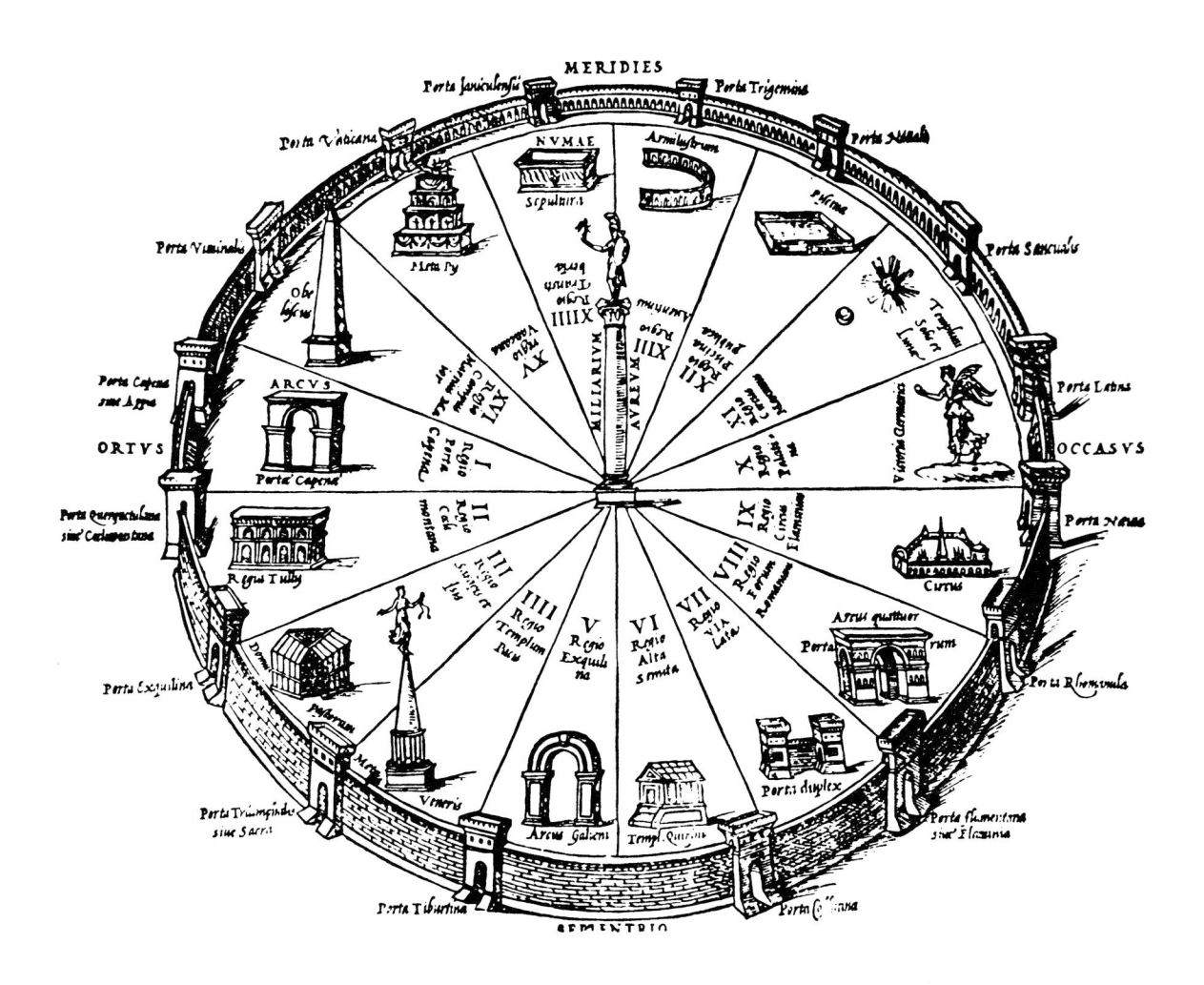

21 Marco Fabio Calvo, map of ancient Rome, 1527 (from Le piante di Roma, ed. A. Pietro Frutaz).

Raphael too was involved in surveying. His letter to Leo X describes the instrument that was of such inestimable value in his recording of ancient Roman ruins. It was flat and round, something like an astrolabe with a magnetic needle at the center. This circular instrument made it possible to describe any shape accurately. With it, he says, "we measure every sort of building of whatever form, round or square, or with strange angles and projections of whatever kind."[99] However, in contrast to what Raphael wrote, we have the two emblematic plans of Rome in perfect circular form made in 1527 by Marco Fabio Calvo, who worked with Raphael on the Roman survey.[100]

Writers on architecture would insistently reaffirm the dubious identity that art was to render plausible: the perfect universe was the perfect shape. There were good reasons why the distinction between underlying and overlaid geometries should not have been raised as pertinent to theory. The consequences were stupendous, since the question alone would entail acknowledgment of geometry's potential arbitrariness in the face of nature. We feel at home with this idea, but it was not easily admissible then. One effect of hanging great symbolic weight on geometry was that of inducing, by artificial means, reassurance as to its embeddedness in things.

On Centers and Circles

SOCRATES *Tell me then, is the universal fire nourished and generated and ruled by the fire in us? Or is the opposite the case, that the fire in me and you—in all creatures in fact—owes all this to that other fire?*

PROTARCHUS *Your question does not even deserve an answer.*

Plato, Philebus, *29c, trans. R. A. H. Waterfield (Harmondsworth, 1982).*

And not without point, the ancient theologians located goodness in the center and beauty in the circle. Or rather goodness in a single center, but beauty in four circles. The single center of all is God. The four circles around God are Mind, the Soul, Nature and Matter. . . . And the nature of the center is such that, although it is single, indivisible and motionless, it is nevertheless found in many, or rather in all, of the divisible and moveable lines everywhere. And of those invisible circles, that is, the Mind, the Soul and Nature, the circle of the visible world is an image.

Marsilio Ficino, Commentary on Plato's Symposium on Love *(1544), trans. Sears Jayne (Dallas, 1985), speech 2, caps. 2, 3.*

Let us now imagine that this castle, as I have said, contains many mansions, some above, others below, others at each side; and in the center and midst of them all is the chiefest mansion where the most secret things pass between God and the soul.... You need know only one thing about it [when the soul falls into a mortal sin]— that, although the Sun Himself, who has given it all its splendor and beauty, is still there in the center of the soul, it is as if he were not there for any participation which the soul has in Him, though it is capable of enjoying Him as is the crystal of reflecting the sun.

St. *Theresa of Avila,* The Interior Castle *(1588), First Mansion, trans. E. Allison Peers (New York, 1961).*

[7] The first step toward the investigation of the physical causes of the motions [of the planets] was that I should demonstrate that the focus of their eccentrics lies in no other place than the very center of the body of the sun, contrary to the belief of both Copernicus and Tycho. . . .

[27] It is incredible how much labor the moving powers constituted by this way of argument caused me in the fourth part [of the book], giving false distances of the planets from the sun, irreconcilable with the observations, when I tried to work out the equations of the eccentric. This was not because these moving powers were wrongly invoked, but because I had forced them to tramp around in circles like donkeys in a mill, being bewitched by common opinion. Restrained by these fetters they could not do their work.

Johannes Kepler, Astronomia Nova, Introduction (Heidelberg, 1609), trans. A. R. Hall and T. Salusbury, in M. B. Hall, ed., Nature and Nature's Laws (London, 1970).

Ever since Copernicus man has been rolling down an incline, faster and faster away from the center—whither? Into the void? Into the "piercing sense of his emptiness"?

Friedrich Nietzsche, The Genealogy of Morals, *trans. F. Golffing (New York, 1956), essay 3, sec. 25.*

We thus free ourselves from the distasteful conception that the material universe ought to possess something of the nature of a center.

Albert Einstein, Relativity: The Special and the General Theory, *trans. R. W. Lawson (London, 1920), 106–107.*

That which permits us to center our existence is also that which prevents us from centering it absolutely.

Maurice Merleau-Ponty, The Phenomenology of Perception *(1962), cited in Peter Dews,* Logics of Disintegration *(London, 1987), 34.*

Thus it has always been thought that the center, which is by definition unique, constituted that very thing within a structure which while governing the structure, escapes structurality. This is why classical thought concerning structure could say that the center is, paradoxically, within the structure and outside it. The center is the center of the totality, and yet, since the center does not belong to the totality (is not part of the totality), the totality has its center elsewhere. The center is not the center. The concept of centered structure—although it represents coherence itself, the condition of the episteme as philosophy or science—is contradictorily coherent. And as always, coherence in contradiction expresses the force of a desire.

Jacques Derrida, "Structure, Sign and Play" (1966), in Writing and Difference, *trans. Alan Bass (London, 1978), 279.*

Chapter Two **Persistent Breakage**

According to the orthodox opinion challenged in the previous chapter, centralized churches epitomize a universal stability and calmness of spirit typical of classical art. According to a corresponding orthodoxy, this order was shattered, not by modern architecture, but by the explosive reformulation of pictorial space effected by painters in the early twentieth century. Looking back, we have learned to see, in the abstract, fractured format of their canvases, the assassins of an outmoded vision and harbingers of the new.

A similar compositional fragmentation has since been episodically practiced in architecture. I shall try to define and then review its three major phases: firstly, the period in the 1910s and 1920s when faceting and fragmentation came to architecture as a revolutionary impulse borrowed from cubist painting; secondly, the postwar years when it appeared as a way of humanizing modern architecture; thirdly, the present day when it is said to signal the end of humanism and the end of modernity. It will already be clear that the motives for breakage have been diverse. I want to show that its persistence is more significant than the diversity of its interpretations, and that the revised relation between Renaissance architecture and Renaissance thought proposed in the first chapter may tell us something about the drive to break things up in our own times.

A fragment is part of something larger. Usually it has been broken off. However, in order to make any sense of the material about to be considered, a modified definition is needed: for our purposes a fragment is something that *looks as if* it has been broken. In architecture anything more than that is, with very few exceptions, either an accident or a disaster. For the most part, the only thing broken is a certain type of expectation that we have. This modified definition may steer toward formalism, with its exclusive concern for appearance, but a purely formal analysis of fragmentation would be impossible because a fragment, real or imaginary, must have a past. As soon as we identify something as broken, we become detectives of its history. What larger entity was it detached from? How did it end up in bits—or, with our modified definition, how are we to imagine it got that way?

22 Richard Meier, sketch plan of the Atheneum, New Harmony, Indiana, 1975–1979, Museum of Modern Art, New York.

Cubism, especially the painting of Picasso and Braque between 1907 and 1912, was the source of fragmentation in modern art. Modern architecture, on the other hand, is said to be total architecture, monadic to a fault, totalitarian even. Such is the consensus, and yet modern architecture has also been traced to cubism and characterized by fragmentation. The art historian Douglas Cooper reckoned that "Cubist methods influenced the art of camouflage during the First World War, and the design of modern architecture afterwards";[1] the philosopher Jean-François Lyotard chose Richard Meier's Atheneum building in New Harmony, Indiana, to typify what is fragmented and, for that reason, modern.[2] The Atheneum has a birth certificate to prove it: a little sketch plan (which I am told was made after the design had been finalized) that looks like a Braque drawing of a guitar, circa 1912. How could the mainstream of twentieth-century architecture be both fractional and total?

Cubism, Architecture, and the Sublimation of Fragments

One of the earliest historical treatments of the relation between cubism and architecture was in Sigfried Giedion's *Space, Time and Architecture* (1941). Giedion thought there was an irresistible force in history that gave identity to the products of an era. Presupposing that there was a correspondence between contemporary painting and architecture, it was then just a matter of telling what it was. This he did by juxtaposing two full-page illustrations, one of the workshop wing of the Bauhaus, designed by Walter Gropius (1926), and the other of *L'Arlésienne*, a painting by Picasso of 1911 to 1912.[3] Giedion says that they have three things in common: planarity, transparency, and simultaneity. Both are constructed from planar surfaces (the facets in the painting are comparable to the platelike slabs of the floors in the building); and both look transparent (the crystalline constitution of the painted image is comparable to the glazed curtain walls around the building).

The third characteristic, simultaneity, is dependent on the other two, but more crucial to an understanding of the comparison. It prompted Giedion to claim a spurious connection with the theory of relativity on the grounds that Einstein had begun one of his books with a discussion of this very term. What Giedion did not tell his readers was that Einstein's discussion undermined the normal understanding of simultaneity, showing how its scope was reduced in relativistic physics.[4] Yet Giedion's overenthusiasm for correspondence should not overshadow his more plausible observation as to how simultaneity might apply to architecture as well as painting. In the early documents of cubist criticism simultaneity generally meant the presentation of multiple impressions of a subject in one image.[5] *L'Arlésienne* is seen both *en face* and *en profil*. Several

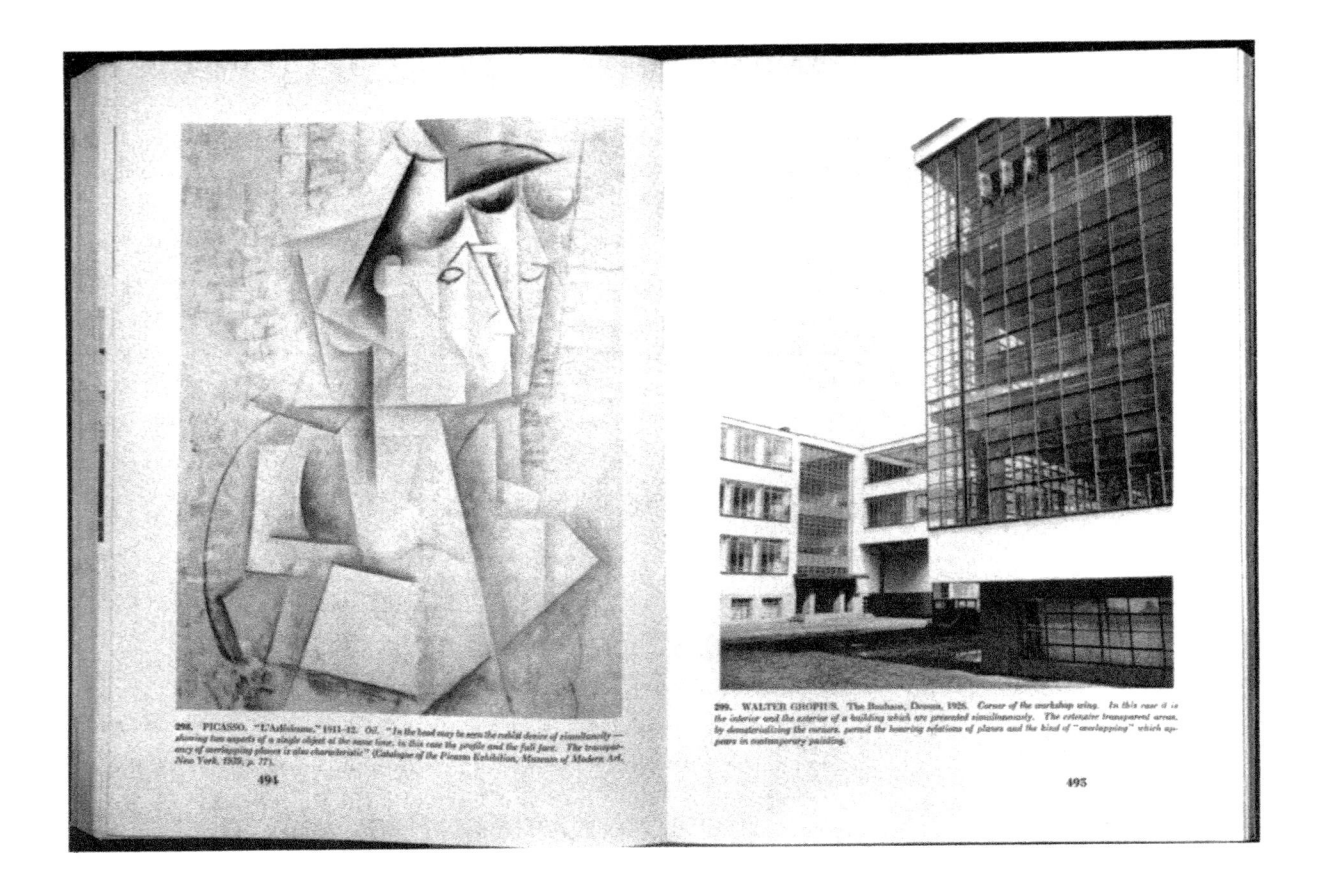

23   *Sigfried Giedion*, Space, Time and Architecture, *1941: Picasso's* L'Arlésienne *and Gropius's Bauhaus.*

views of several sides are shown at once. Giedion points out that this is exactly how we see the Bauhaus workshop wing. Because the walls are transparent, several sides of the building, as well as its inside and outside, are revealed at the same time.

But later in the book, Giedion illustrated the same idea with a building that was neither transparent nor planar. Reworking an image that had been published more lavishly by Thurman Rotan nine years earlier,[6] he collaged 11 photographs of the RCA Building in New York taken from various angles.[7] The operation performed on the opaque shaft of the skyscraper is supposed to be similar to that performed on the sitter for a cubist portrait. There is no fixed viewpoint, but instead a more complete collection of views. The result is equivocal, since each of the photographs, unified in itself, appears fragmented *as a result of the effort to show their totality.* It is interesting to compare this way of making knowledge more complete with the parable of the elephant in the cave, where several sages, feeling different parts of the animal, falsely construe the whole from their limited knowledge of a part. The story implies that in the full light of day, with vision restored, the unity of the complete beast would be clearly apprehended. By contrast, the cubist image, regarded as a conspectus, is made up of a plethora of aspects, their individual spatial unity disrupted in order to show everything at once.

The fractured simultaneity of cubist painting is quite different from the simultaneity that Giedion saw in the workshop wing of the Bauhaus, where there is not the slightest hint of fracture. On the contrary, the transparent Bauhaus building makes the space in and around it appear more isotropic and continuous than it would if the architecture were traditional, cellular, and compartmented.

We may suspect Giedion's motive for suppressing the fragmentary character of cubism in translation to architecture. He was certainly capable of highly selective and deceiving comparisons, but in this instance he was not acting peremptorily, nor was he alone. He was already perpetuating a tradition. It is significant that supportive writings about cubist painting from the 1910s and 1920s hardly ever suggest that anything was being broken. Even in Gertrude Stein's biography of Picasso published 30 years after the advent of cubism, while there is much to the effect that his work was a logical, systematic quest to reveal the quiddity of things, there is nothing to suggest that it was about brokenness, except for a brief passage toward the very end of the book where not the paintings but the twentieth century they are attuned to is described as "a time when everything cracks, where everything is destroyed, everything isolates itself."[8] Eunice Lipton has shown that, as far as Picasso criticism is concerned, the fractured, dismembered character of his work began to be recorded only in the thirties, when the busted images were generally understood to represent the Spanish tragedy.[9]

So why was this now universally acknowledged feature of cubist painting not in the foreground of its early explanations? It is a question of emphasis. If the stress is placed on the totality of the simultaneously presented image, cubism aims for wholeness; if the stress is placed on the dislocation of parts that this inevitably entails, then cubism ends up as fragmentation. Insistence on the logic of the simultaneously presented image reduces the fracturing and dislocation to a visual by-product of a synthesizing procedure, which may be why partisans of the new art sometimes compared the way they worked with the way engineers and architects drew.

When Gelett Burgess published his largely satirical account of new developments in French painting in the *Architectural Record* (1910), he recalled a conversation with Georges Braque in which the painter had said that to paint the complete woman, he needed to paint three figures, not one, just as an architect needs to make three drawings of a house; the plan, the side elevation and the front elevation. Then Braque gave the incredulous Burgess a drawing to demonstrate what he meant.[10] Picasso's statement that all the facets of one of his nude portraits could be cut from the canvas and reassembled to make a full-bodied model requires that we think of the painting as a more or less intuitive version of the geometric development practiced in technical drawing, as if it were an anthropomorphic equivalent of Mercator's projection.[11] The gallery owner and critic Daniel-Henry Kahnweiler, himself the subject of one of Picasso's best-known cubist portraits, was even more explicit. He believed that the cubist pictorial method made it possible to get rid of the distortion and illusion endemic in Western art:

> *With the representation of solid objects this could be effected by a process of representation that has a certain resemblance to geometrical drawing. This is a matter of course since the aim of both is to render the three-dimensional object on a two-dimensional plane. In addition, the painter no longer has to limit himself to depicting the object as it would appear from one given viewpoint, but wherever necessary for fuller comprehension, can show it from several sides, and from above and below.*[12]

Linda Dalrymple Henderson has unearthed some complicated technical diagrams of four-dimensional polyhedroids published by Esprit Jouffret in 1903, which use descriptive geometry and superimposed axonometric projections (both are varieties of geometric drawing) to show all the salient features of figures that could not be constructed in three-dimensional space. They look like cubism. Henderson thinks these were known to the Puteaux painters, perhaps to Picasso, but even if they were not, the likeness between the mathematician's diagrams and cubist compositions would remain significant.[13] Any complete representation of a complex solid (or hypersolid) in a single drawing involves disruption: witness the stereotomic drawings that were done to obtain the lengths and angles of the several faces of each block in a masonry vault (see chapter 6). A comprehensive graphic description necessarily destroys coherence. In pictures, totality and incoherence are synonymous.

That is why the analogy with technical drawing is revealing. Modern painting began as a violent campaign against perspective. Perspective was inadequate because it was partial vision; because it recorded appearance as distinct from reality; because it was not even the way we see but a mere convention. Perspective was distortion. All these arguments are

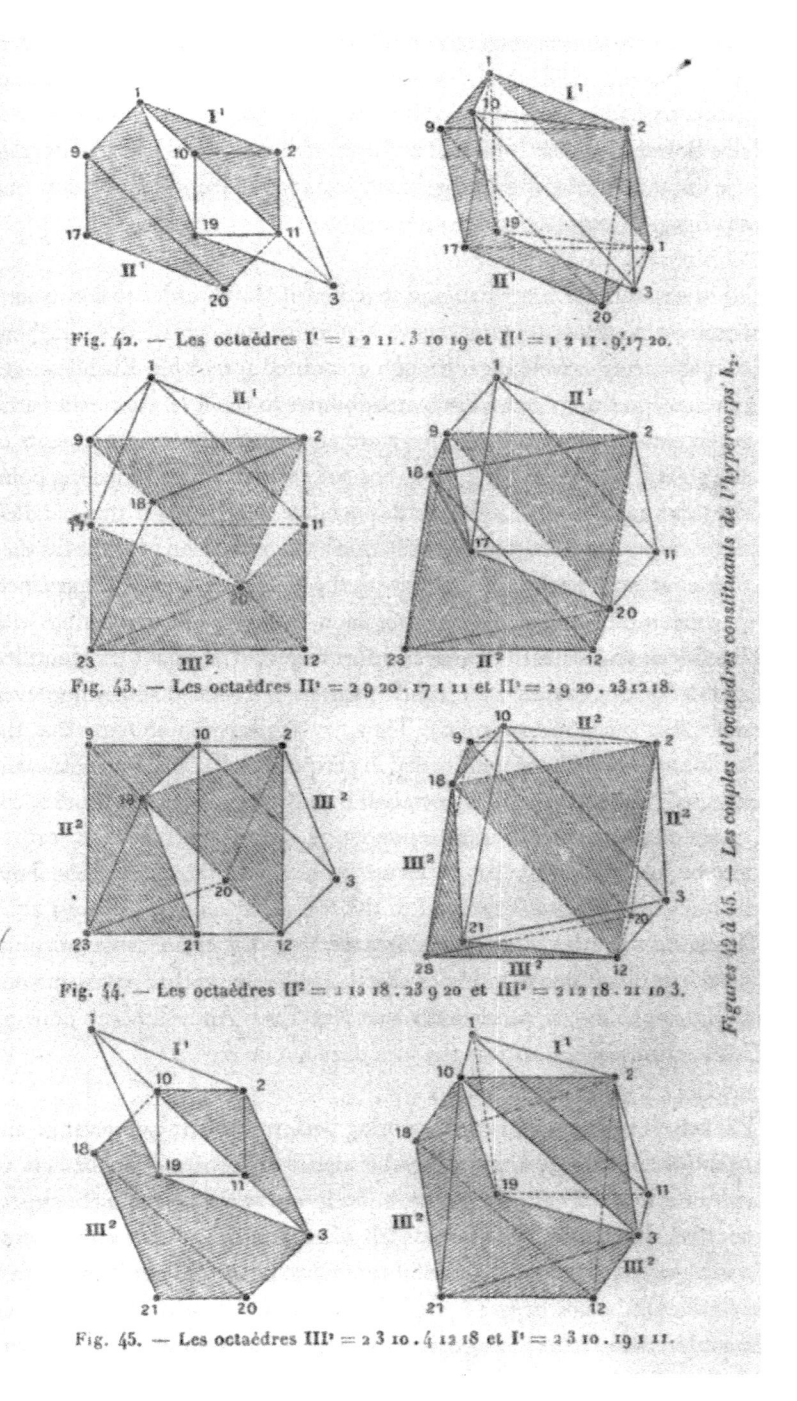

24 Esprit Jouffret, projections of the pairs of octahedra constituting the hypersolid h_2, from Traité élémentaire de géométrie à quatre dimensions, *1903.*

old, some as old as perspective itself. But a new one was slipped in: perspective was Euclidean. To be Euclidean was a fault because Euclidean geometry had been superseded by new geometries—those of Lobachevsky, Bolyai, and Riemann. Just as these mathematicians had transcended the limitations of traditional geometry, so too the painter must transcend the limits of perspective.

This was another hasty banging together of bits in order to construct a common platform for progressive history to run across. It so happens that the perspective image, though accounted for within Euclidean geometry, also brings certain incompatibilities to the fore. Anomalies arise when measurements of the image are compared with measurements of the objects depicted. The convergence of parallels into a vanishing point does not conform to Euclid's fifth postulate, which has it that parallels never meet, and the continuous dilation and contraction of bodies as they move back and forth plays havoc with the Euclidean idea of congruence. Architectural drawing, on the other hand, is more easily compatible with Euclidean postulates. Thus, in an effort to get rid of what they identified as Euclidean geometry, the cubist painters introduced something even more like Euclidean geometry. They nevertheless felt strongly that the artist's vision had been imprisoned in perspective. Escape was most easily effected by grasping, half in jest, half in earnest, onto the only other coherent, regulated method of representing space on a flat sheet: orthographic projection. Recent critics are prone to dismiss all this talk about technical drawing as incidental to the real achievements of cubist art.[14] They seem to have a good case because, from the entire range of cubist production, not one painting or sketch can be found that represents one thing completely or consistently that way. Even André Lhote's drawing made expressly to illustrate the idea does not do so.[15]

All acts of violence are illegible during performance. It can never be entirely clear whether they are preliminaries to construction or acts of spleen intent only on disfiguration. So it was with the war against perspective. Even now, how can we tell whether a cubist canvas represents a new way of seeing or is a gleeful exhibition of the old enemy's tortured and dismembered corpse? A vital clue to the cubists' own attitude can be found in their resort to analogies from technical drawing. Since it is perfectly clear that they were not using multiple orthographic projections, the analogy only makes sense if understood as a way of both acknowledging and justifying the fractured format of the picture plane. We have seen that the portrayal of an object in its totality requires the destruction of its picture's unity; conversely, the destruction of a picture's unity may, by inference, magic, or shared intuition, create the sense of a probable totality beyond the picture, and this is what the cubists believed they could accomplish with their kaleidoscopic compositions.

Technical drawing was an analogy, not an explanation, and as such it proved its worth. It helped reassert the importance of the object within a practice that was on the verge of annihilating it; it provided a precedent showing that overlaid multiplicity adds up to a unified picture, a precedent also for the collapse of pictorial depth into a shallow stratum; and, in its more complex demonstrations, it rendered objects transparent and gave suggestive instances of rotation and discontinuity. But although originating from engineering and construction, technical drawing did not provide a handy route back to architecture. Its influence was not reversible because the analogy was to pictures, and the visual properties of pictures of buildings are not those of buildings. The cubists focused on representation, giving us pictures of pictures rather than pictures of things, because they thought that pictures were more like perceptions, in which respect they too were perpetuating a long tradition. Lucretius thought that objects propagated thin films that peeled off like impalpable pictures, flying through the air so that we could perceive them that way. Two thousand years later, a few years prior to cubism, C. S. Pierce described the several views of an inkstand that had coalesced in his mind as "a generalised percept, a quasi-inference from percepts, perhaps I might say a composite photograph of percepts."[16]

The paradox of fractured totality is represented by analogy with technical drawing, but in cubism the analogy was doubly relocated, suggesting discontinuities of space (rather than of objects) and discontinuities in consciousness (rather than in pictures). The forces of consciousness were now portrayed as the forces of nature had once been portrayed. There are numerous romantic paintings of ectopic scenes: battles, wildernesses, natural disasters, and ruins of ancient grandeur. The difference between these and a still life by Braque is that the former are pictures of large subjects upon which enormous physical forces have been unleashed, whereas the latter is of an inert and untroubled subject, as inert and untroubled as anything could be. The ordinary, domestic nature of its content makes this phase of cubism all the more perturbing since we can attribute nothing of its disconcertingness to the familiar things that are being painted. Confronted with these small, dowdy, framed canvases of commonplace items, we have no alternative but to regard consciousness as a form of aggression perpetrated by the artist against an altogether defenseless and innocent reality. There is much to suggest that the early twentieth-century painter wanted to manufacture reality as the fifteenth-century painter manufactured pigment—by pummeling it into exis-

25 Caspar David Friedrich, The Wreck of the "Hope," 1823–1824, Kunsthalle, Hamburg.

26 Georges Braque, Mandola, 1909–1910, Tate Gallery, London.

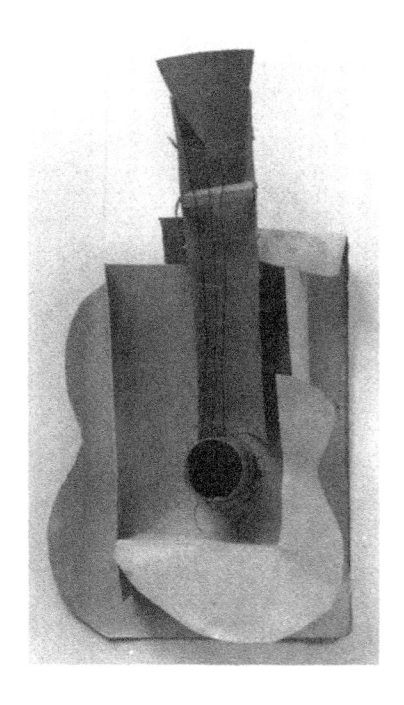

27 Pablo Picasso, Guitar *maquette*, 1912, Museum of Modern Art, New York.

28 Raymond Duchamp-Villon, Maison Cubiste, Salon d'Automne, Paris, 1912.

tence. So, in *Les Peintres cubistes*, Apollinaire does not hesitate to say of Picasso: "The great revolution of the arts, which he achieved almost unaided, was to make the world his new representation of it."[17]

Such Promethean efforts, affecting ideas and perceptions, need not greatly affect the physical environment so long as representation extends only to pictures and words. Architecture is the exceptional case because, substantial yet representational, it is more equivocally of the world and, at the same time, *about* the world than any other art form. This enables us to see why the transition from cubist painting to architecture was at first so easy and afterward so difficult. Insofar as a superficial compositional identity could be grafted from a canvas to a facade it was easy; insofar as architecture was being asked either to relinquish its tectonic organization or to model a new perception, it proved immensely difficult.

If cubist painting is about cognition, then a jump back into the third dimension should strip the object of any indication of the play of perception and memory inscribed in the process of picturing. Something else then has to be pulled out of the picture to indicate it. Picasso began experiments of this kind in 1912 with his ingenious cardboard maquettes of guitars, which really do look like drawings of his that have unaccountably spilled out into space.[18] Certainly, no one would mistake them for musical instruments. The *Maison cubiste* of Raymond Duchamp-Villon (1912),[19] the first effort at architecture amongst the cubist fraternity, is, by contrast, generally derided as a facile attempt to simulate the appearance of cubist shapes in architectural details.[20]

Although we know it only from photographs of mock-ups, no one would mistake it for anything but a project for a real house. Hilarious in its bourgeois irony, its bulky facets weighed down those parts normally reserved for decoration—doorcases, pediments, balconies, and gables—rendering weightless percepts into tonnage. The cubist enterprise was, in one sense, stifled under its own monument, in another, forged into a universally reproducible token of cubist style. Parts of otherwise conventional buildings could thus be made to resemble the modulated, folded, crystalline surfaces implied by the painters.[21]

The projects of Pavel Janák and other architects of the Prague Creative Artists' Collaborative in the 1910s,[22] and Kurt Schwitters's *Merzbau* in his house at Hanover (1923–1930),[23] typify the imitative response. These constructions represent paintings. The direction of mimesis is decisively reversed: real objects are made in imitation of their painted effi-

29 Vergilio Marchi, Model of Futurist House, 1916 (from Giovanni Lista, Futurism, *1986).*

gies. We should note that, reembodied, they were horribly confined by the requirements of structural stability. When Vergilio Marchi, working on the same lines, attempted a more thoroughgoing disruption in his Futurist House project (1916), he ended up, even in his small model, with something at once naively exuberant and excessively ponderous.[24] However great the effort to force the identity of architecture and painting, size and weight would draw architecture apart again. At this stage, the fate of projects more radical in their enthusiasm for fractured form was that they remained on paper, as did the great profusion of fantasies produced in Russia from 1912 onward, now published in Khan-Magomedov's indispensable study.[25]

So the fractured form, which had received scant written acknowledgment within the critical community, traveled well and traveled quickly, reaching into Gertrude Stein's interrupted prose, Marinetti's attempts to destroy syntax,[26] futurism, rayonism, constructivism, expressionism, stage and film sets, as well as architectural *projects,* but not into buildings. Those who did manage to develop buildings from cubist painting did so with more circumspection and with other means. It took a long time, even in those accelerated years. Not until the twenties did Le Corbusier, van Doesburg, and Mies van der Rohe find more fruitful and indirect derivations from modern painting. In their work the appearance of fragmentation was either expurgated or sublimated.

Le Corbusier thoroughly laundered cubism in his paintings before successfully creating a comparable, but not identical, sense of interleaving spaces and planes in his architecture. A notable quality of his purist canvases painted after the end of the First World War was the complete and consistent substitution of overlapping for fragmentation. In *Après le cubisme* (1918) he and Ozenfant condemned the confused dissociation found in contemporary art. A certain degree of deformation was justifiable in the interests of superior harmony, but that was all.[27] By contrast, Timothy Hilton, in his monograph on Picasso, uses one of Le Corbusier's paintings to illustrate what could happen when, as he put it, the juice was squeezed out of Picasso's ideas.[28] Le Corbusier would always maintain that cubism had been the basis of his work. And yet the justly celebrated layerings and ambiguities of the villa at Garches (1926–1928), which Rowe and Slutzky found to combine transparency with obscurity, like the best cubist art,[29] were not obtained by a simple transmission of forms from painting to architecture. Likewise, the equally momentous counter-compositions and architectural projects by van Doesburg from 1923 to 1926 achieved a distance from the fountainhead of cubism—all that remains of the sense of fracture is a vague feeling that walls do not connect up as they might.[30]

Giedion's comparison of Picasso's *L'Arlésienne* and Gropius' Bauhaus in *Space, Time and Architecture* defined the terms of this distancing and conversion. The examples he chose betrayed the differences between architecture and painting while indicating the similarities. To highlight this a third item might be introduced for comparison with the other two: the Kapp-Putsch Monument (Memorial to the March Victims) at Weimar (1920–1922), also designed by Walter Gropius.[31] Solid but triangular and fractured, it exhibits those facets of cubism that never found their way into the Bauhaus building.

30 *Villa Savoye, Poissy, Le Corbusier, 1928–1931, view out of ramp window.*

31 Walter Gropius, Kapp-Putsch Monument (Memorial to the March Victims), 1920–1922, Weimar.

Humanization

In 1925 the Spanish writer Ortega y Gasset accused the cubist painters of deliberately dehumanizing art by breaking up the image,[32] but a second phase of fragmentation was soon under way in architecture, justified as a means of humanization. When Giedion inserted 40 pages on the Finnish architect Alvar Aalto in the 1949 edition of *Space, Time and Architecture*, it was, he explained, because history had advanced far enough for the next stage of modernism to be announced. Rationality had been achieved. Now it was necessary to restore a certain irrationality which he discerned in Aalto's work. Irrationality was required to draw the world back to us, because rational building, no longer the mark of the quintessentially human, had become instead the cause of our alienation from nature and from society.[33]

Aalto is a long way from cubism, but with him modern building once again began to break up. Giedion was using the words "rational" and "irrational" as if everyone knew what they meant, but Aalto, on the other hand, had recognized that the profession's interpretation of rationality caused difficulties. He proposed a broader definition: "Instead of fighting rational mentality, the newest phase of modern architecture tries to project rational methods from the technical field out to human and psychological fields."[34] This statement comes from "The Humanizing of Architecture," an essay published in 1940, in which Aalto took issue with the prevalent conception of architectural function. He thought this was, likewise, too narrow. He wanted its orbit extended from technical and economic matters to include human comfort, perception, and social needs as well. This he called "psychophysical" function. There was a distinct anaesthetic bias to it.

Aalto would often refer back to his design of the patients' rooms at the Paimio Tuberculosis Sanatorium (1928–1933), giving a detailed account of the improvements achieved by subtracting annoyances, frustrations, and discomforts, for "the duty of architecture is to eliminate all disturbing elements."[35] So at Paimio the heating was designed to reduce air movement around the bedridden patients' heads, the light to reduce glare into their upturned eyes, and the washbasins nearby were remodeled so that water from the tap would glance noiselessly across the porcelain without splashing. He chose the patient, sensitized through suffering, as an exemplar because the patient's condition was, to his mind, only an accentuation of the human condition—the condition of being discomforted and put upon. In a lecture of 1955 he said: "I have a feeling that there are many cases in life where the organization of things is experienced as too brutal. The architect's task is to make our life patterns more sympathetic."[36] Good architecture therefore alleviates; it removes impositions and smooths out the unpleasant roughness of sensation. If this orientation were pursued to its limit, building would disappear not by realizing the dream of dematerialization that enthralled so many other modern architects, but by receding into the background of perception, enabling human activity by following its varied courses without ever getting in the way. (Architecture should "function as a perfect shell for people's lives.")[37]

The Baker Dormitory at the Massachusetts Institute of Technology (1947–1948) shows how Aalto's philosophy of emollience was put into effect at the scale of an institutional building. The dormitory is situated on the south edge of the campus, overlooking the Charles River which has an arboreal strand along its bank. Between the site and this relaxing scene is a busy highway. Aalto decided that the rooms should face up- or down-river rather than straight across. Traffic noise would be reduced and the pleasant vista of water and trees would be more extensive. The winding, fractured plan allows every room this advantage. The building cranes its neck to achieve it. The rooms on the north side pull out of alignment and jut forward to get a last glimpse of the river curving round to the northwest, as if the building took the burden of physiological discomfort upon itself, so removing it from the occupants. But that which pacifies the perception of the environment from within, helped by the rising sawtooth of the exposed underside of the two longitudinal staircases, also breaks up and activates the exterior.

It is not calm. The contrast between a sinuous edge and a jagged edge in plan recalls a generation of cubist guitars. Aalto insisted on several occasions that the proper link between painting and architecture was indirect and remote.[38] Leonardo Mosso nevertheless refers to Aalto's brick buildings of 1944–1953 as his Cézanne period, "from the intense hues of the bricks he employs and his manner of handling volumes but above all from the way that light is broken up into facets in the manner of Cézanne."[39] The painterly effect seems adventitious, a side effect bringing us back more or less accidentally to the aggressive origins of modernism.

However, the north face of the MIT dormitory may also be understood in the context of emollience. Between the totality of an institution and the privacy of the individual, a full range of encounters and groupings— formal or casual, diversified or homogeneous—is possible. These may be registered in the constitution of a building, giving some hint as to the nature of its occupation. Aalto understood this too as a potential means of humanization. If buildings function as a perfect shell for people's lives, then diversified contours may be assumed to stand for diversified occupa-

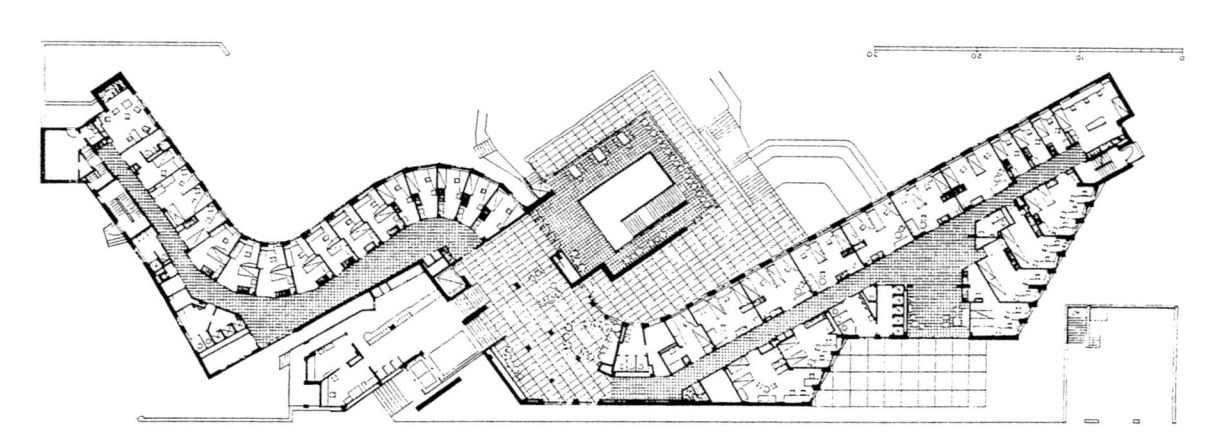

32 Baker Dormitory, MIT, Cambridge, Massachusetts, Alvar Aalto, 1947–1948, north side.

33 Baker Dormitory, plan.

tion, casual distribution for casual encounter, architectural monoliths for social monoliths, and so on. In this identification of the forms of social life with the forms of buildings lies the promise and the problem to which we shall return, pausing only to notice that most respected postwar architectural critics endorsed the program of humanization, and approved increased differentiation as its appropriate means. For Nikolaus Pevsner and Giedion it glanced off into a dubious reaction against reason only if pursued to excess.[40] For Vincent Scully and Robert Venturi it belonged within the fruitful dialectic between fragmentation and continuity.[41] For Christian Norberg-Schulz and Colin Rowe it was both the sign for, and result of, political pluralism.[42] For Lewis Mumford it was a conscious limitation of the collective power we possess over our own circumstances. For Kenneth Frampton it was a counterstroke against the homogenizing effect of corporate capital.[43] For Bruno Zevi it was the natural outcrop of functionalism and the sole means of any consequence available to modern architecture.[44] Practically everyone agreed it was the architecture of democracy. Belief in this principle remains enshrined in much that is published and taught, and still supports much that is built.

There is an emollient aspect of fragmentation that is specific to buildings. A crematorium has to operate without offending anyones's feelings. Aalto therefore, designing a crematorium, made a plan "that guaranteed that clashes could be avoided."[45] He proposed three chapels of different sizes instead of one, each with separate routes of access and egress. His design for the model primary school at Jyväskylä University (1950–1956) was also divided into several blocks, justified by the observation that

> *in general, school buildings are much too big and the system of many classes is dictated by an extreme collectivization. Instead of one large school with many classes, I attempted to design a group of many small schools. Three classrooms and a stair-landing together form a separate entity so that the illusion of a small school is created—a school that is administratively a part of the whole complex.*[46]

It might be objected that the assemblage of buildings on the Jyväskylä campus is not in any way suggestive of fragmentation, just informally arranged. However, Aalto's explanation makes it clear that something has been broken up: the perception of the total institution. We are left in some doubt whether this colloquializing of territory will be instrumental in creating the unrealized "happy medium" between individualism and collectivity that Aalto wished to foster, or whether it is an adjustment of perception permitting an avenue of escape from an unpalatable fact which, as he admits, has not been modified by the architecture. The school remains administratively one, while made several in appearance and imagination.

The condition to which Aalto was responding may be generalized thus: a holistic view of the self in a felicitous relation to the environment leads to a fracturing of the unitary enterprise of building, more or less thorough, depending on the degree to which monoliths promote psychological disquiet. Their subdivision is an emollient that softens the outline of authority, reduces the appearance of power. The architecture that performs the service is the pyramid's antithesis. But when design is brought down to the human scale it may then operate more directly on human conduct. As the scale decreases and the definition increases, it presses in on available space, eventually imposing the free impulses it was meant to release. When Louis Kahn was asked why he did not make buildings like Aalto, he replied that a building composed of designed responses to casual activity would be a monument; it would monumentalize casualness, freezing and preserving the ephemeral activity that other monuments left unspecified. Kahn was, of course, adversarial in protection of his own capacious architecture, but he had identified a genuine hazard. Architects such as Aalto were obliged to steer between the Scylla of overdeterminate differentiation at a smaller scale and the Charybdis of indeterminate differentiation at a larger scale.

At the larger scale one can usually guess whether diversified building arose from episodic construction by numerous agencies or was a coordinated orchestration of variety by a single hand. It is also easy to fall into a kind of reverie, with half-closed eyes, in which this distinction is lost. Formal diversity may thus correspond to facts about the history of design, facts about the history of construction, facts about ownership, facts about the manner of occupation, all, any, or none of these. Sometimes buildings provide sure traces of human activity; sometimes they are emblems of human activity unconnected to what they represent, except via the imagination. We are encouraged to see architecture in both ways simultaneously as if this were a special gift.

Nowhere has appreciation of the gift been nurtured more sedulously than in *Collage City* by Colin Rowe and Fred Koetter. Completed in 1973, *Collage City* was in some ways a vendetta against a triumphant modern urbanism launched at what then appeared to be its first moment of vulnerability. In fact the book is not antimodern. It is pro-eclectic and intolerant of modern architecture's suppression of diversity, which the authors immediately render into political terms: urban architectural totality is declared to be the expression of totalitarianism. The small poetic voice of modern architecture is shown silenced within a monstrous inflation of moralized functionalism that treats society and even human nature as perfectible through architecture and planning; such the authors judge to be our utopian inheritance.

Against the totalizing and reforming vision of utopia stands the charming prospect of arcadia. Rowe and Koetter realize that they have to establish an acceptable tastefulness to this counterprinciple, which could so easily degenerate into a multiplicity of reactionary hedonisms and sentimentalities. Their task is not made any easier by the provenance of the contrast, which derives, no doubt, from W. H. Auden, who treated his own proclivity for Arcadia/Eden in self-mocking fashion. In "Reading" (1962) he lists, as if confessed from a daydream, specifications that include the following: "*Weights and Measures:* Irregular and complicated. No decimal system. . . . *Size of capital:* Plato's ideal figure, 5040, about right. *Form of Government:* Absolute monarchy, elected for life by lot. . . . *Architecture:* State: Baroque. Ecclesiastical: Romanesque or Byzantine. Domestic: Eighteenth century British or American Colonial. *Domestic furniture and equipment:* Victorian except for kitchens and bathrooms which are as full of modern gadgets as possible. . . . *Public statues:* Confined to famous defunct chefs."[47] In his poem "Vespers" a serious tone is struck by Auden's meeting with his "anti-type," the utopian, with dawning realization that each requires the other to feed their prejudice ("He notes, with contempt my Aquarian belly: I note, with alarm, his Scorpion's mouth").[48] Rowe and Koetter, denied this reflective route to sobriety, obtain it through formalization.

Ernst Gombrich demonstrated that bad taste can be expunged from any image by subjecting it to a process of obfuscation. Embarrassment and repugnance reduce toward zero as the subject is hidden behind its own abstractions.[49] This is one kind of formalization. Another is found in a type of subjective perception, at once aesthetic and anaesthetic, that can disengage from the most appalling circumstances by willing an abnormal prominence to shape, light, and color, which can then be appreciated for their own sake in a flayed body or a putrescent puddle for instance. Rowe and Koetter need formalization to mask the embarrassment of copying things that are too exotic or too far removed in origin to be copied faithfully, as in the former case. But the proposals they endorse involve a peculiar mix of aestheticizing and anaesthetizing tendencies, as in the latter case.

Formalization is the technique of *Collage City*. The centralized town plan is the sign of totalitarianism; collage is the sign of pluralistic democracy.

Throughout the book both organizations are treated as the natural expression of alternative ideologies, although to what extent as emblems (that signify), as results (that evince), or as instruments (that construct) is left unclear.

Two examples will nevertheless serve to show how the authors consistently prefer emblematic interpretations. First, the *bricoleur*, whom they present as an alternative to the architect preoccupied with theoretical and stylistic unity. The term *bricoleur* was borrowed from Claude Lévi-Strauss, whose *The Savage Mind* they quote to indicate that the contemporary artist might be regarded as a relative of the anthropologist's *bricoleur*, a protoscientific inventor who puts heterogeneous things together according to what he has at hand, and according to their sensible properties, rather than according to principles or theories. Boldly, Rowe and Koetter argue that the Neolithic artisan-inventor, the modern artist, and the modern arcadian architect may all be examples of this one type.

Lévi-Strauss also sees a connection with art, but makes us aware of an enormous disparity. The prehistoric *bricoleur* was the inaugurator of civilization. The Neolithic revolution—the sudden emergence of a technologically transformed society—was predicated on pottery, weaving, agriculture, and the domestication of animals. The ingenuity behind these advances may not have been scientific in our terms, but neither was it primitive. For Lévi-Strauss the practical creativity of the prehistoric *bricoleur* was a way to explain the "Neolithic Paradox"[50] of a technological society without science. Side by side with this, the activity of any individual painter (he instances the treatment of a lace collar in a portrait by François Clouet) will appear inconsequential unless indulgently fantasized into an alchemy of the future.

Rowe and Koetter turn Lévi-Strauss's observation of related but distinct mentalities for artist and *bricoleur* into an observation of similar mentality *that will always betray itself in the same representative form:* collage. Lévi-Strauss noticed the "intermittent fashion" for collage in modern times. He thought it an etiolated reminiscence of *bricolage* as a constructive social activity.[51] The difference of opinion is significant. It is magnified as Rowe and Koetter step from art to architecture. The Facteur Cheval and Sam Rodia are reasonably described as *bricoleurs* because they make buildings with bits and pieces left over from other tasks and without a definitive plan. Compare these with *The City of Composite Presence*, a project made from a cut-and-paste assemblage of plans of great historic monuments, each ironically considered as a "vest-pocket utopia."[52] Rowe and Koetter use it to demonstrate the collage principle, but *The City of*

34 Watts Towers, Los Angeles, Sam Rodia, 1921–1945.

35 *The City of Composite Presence*, David Griffin and Hans Kolhoff, from Rowe and Koetter, Collage City, *1978.*

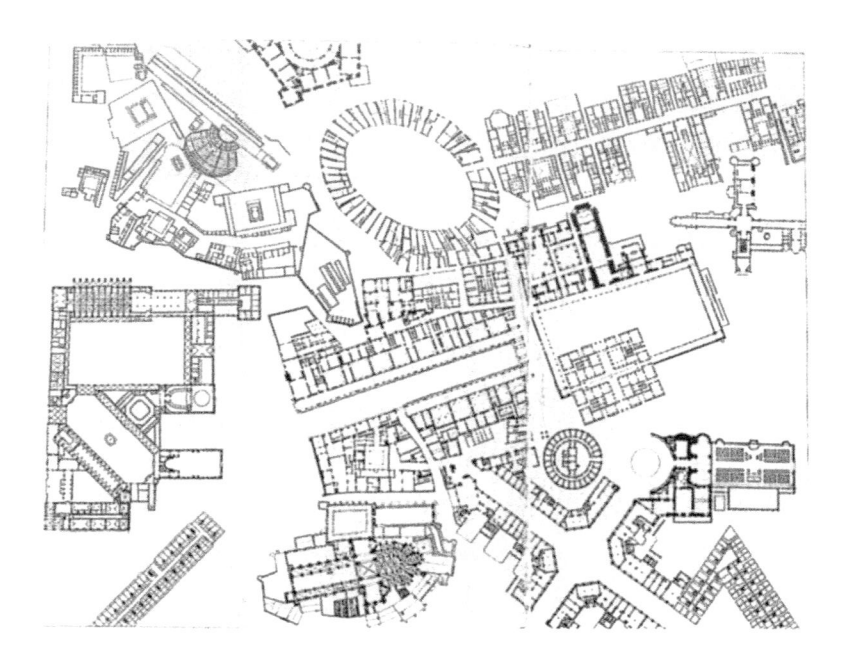

Composite Presence is an architectural master plan that permits the already formalized version of *bricolage* only as a drawing board activity. Once the glue dries, the game is up. The proposal is then as intransigent and as despotic as any urban utopia of equivalent scale. Its heterogeneity of form simulates the appearance of episodic urban development arrived at by numerous agencies, and hence with an entirely different meaning attributable to it. This misidentification seems to be the real point of the exercise.

The contrast between activity and representation is even more emphatic when Rowe and Koetter's contemporary *bricoleur* is compared to Michel de Certeau's. According to the social theorist de Certeau, writing in 1974, *bricolage* is an aspect of the everyday ruses and evasions of "making-do" that by their very nature leave little trace on architecture. He describes their "myriads of almost invisible movements, playing on the more and more refined texture of a place that is even, continuous," and then he asks: "Is this already the present or the future of the great city?"[53] The scenario involves reciprocity between the regular and the irregular, as does *Collage City,* but in de Certeau's version the visible advertisement of *bricolage* would signal its ultimate demise. For him, as for Kahn, an architecture devoted to its freezing and display could only be its mausoleum.

The second example is Rowe and Koetter's interpretation of a passage from Alexis de Tocqueville. The issue tackled is the fundamental one for them: How is architectural form related to social and political events? They admit that the argument they make is formalistic "up to a point," but defend this by proposing "the analogue of politics and perception,"[54] by which they mean that parallelism (analogy and metaphor) takes the place of determinism (causality). Yet, as the two strands lie adjacent, they may interact. They use a striking observation from de Tocqueville's *Democracy in America* to indicate how:

> *Men living in democratic ages do not readily comprehend the utility of forms. Yet this objection which the men of democracies make to forms is the very thing which renders forms so useful to freedom: for their chief merit is to serve as a barrier between the strong and the weak, the ruler and the people, to retard the one and give the other time to look about him. Forms become more necessary in proportion as the government becomes more active and powerful, while private persons are becoming more indolent and feeble. . . . This deserves most serious attention.*[55]

Elsewhere Rowe noted that de Tocqueville was not using the word "form" in an architectural context, but the caution is not repeated in *Collage City*.[56] In fact de Tocqueville was talking about protocols, politeness, social graces, accepted rules and habits affecting public behavior or institutional power. He had just been dealing with the freedom of the press and the impartiality of the judicial system. He wanted to show that the psychology of a fully democratic society would, paradoxically, ensure a more centralized state than ever, unless certain impediments, hard to distinguish from useless formalities, were put in the way to prevent it. A city filled with impediments, preventing it from running smoothly according to one principle, may indeed share a lot in common with de Tocqueville's state that is retarded by a multiplicity of protocols. If it did so, it would be because of the operative effects of these retardants, not because of their appearance. Such concrete consequences would, I think, be unacceptable to Rowe and Koetter, who continue to insist that the city is best regarded as a metaphor of society and politics, affecting consciousness in subtle ways, but to be abhorred as an instrument of social policy. They are terrified of determinism not because it is a philosophically insupportable idea but because they regard it as a real possibility. The trouble is, they express their fear by disdainfully turning their back on the problem.

Lévi-Strauss and de Tocqueville described forms as the causes and forms as the effects of activities. Rowe and Koetter, nervous about the causal relation, treat form and activity as two aspects of the same thing. From now on, because one stands for the other, the activity can slip away without being noticed, or stay at home under cover of its formal representation without being noticed, equally easily. Rowe and Koetter are convinced that it stays home but offer no further evidence than the address to prove it. The profoundly human activities of making and communicating that first gave forms their vital meaning are reaffirmed by reviving forms alone. Furthermore, Rowe and Koetter conflate the individual activity of *bricolage*, where the pattern is usually produced by one person, or a few colleagues, with the multiplied social activity of protocols, where it is produced by the interaction of many. All the accidents, additions, revisions, and chance events arising from innumerable human impulses and encounters productive of the urban environment can then be concentrated—or counterfeited—in the simulacrum of the architect's collage. An uncoordinated procedure of thinking is presented as equivalent to the uncoordinated profusion of the multitudinous historical city.

Lévi-Strauss thought that symbols are more real than what they represent, and yet in his anthropology one is constantly reminded of how effective symbols are in the construction of social practices. In *Collage City* the symbol is more real than what it represents because all the activities it can represent, except the activity of the designer at his desk, are either faint suggestions about a potential future, fake traces of a distant past, or anodyne for the present.

Foucault warned that power now operates in a proliferation of agencies, each overlapping others in a diverse network that penetrates and surrounds us all.[57] David Harvey, who thinks that art is not the mirror of nature but the mirror of money, describes two inimical forces that produce similar results. First he notices that in urban development, "the more unified the space the more important the qualities of the fragmentations become for social identity and action."[58] Then he portrays the development of recent capitalism as the dispersal and disorganization of ownership and production, so that recent disaggregations of the image in architecture reflect recent disaggregations of corporate structure. Thus the ameliorating efforts of humanization coincide disturbingly with the expressions symptomatic of exploitative capital. The latter are bound to interfere with humanization, since humanization is predicated on capital expressing itself otherwise. For such reasons we might learn to be as wary of the small and varied as we are of the large and impersonal. That would require a new sensibility, yet it seems that we may be building the conditions for its emergence as we carry on building to suppress evidence of the old forms of power that preceded those discerned by Foucault and Harvey, that continue together with them, and that still provide the recognizable evidence of alienation: size and uniformity.

The technique of military camouflage developed by Guirand de Scévola in the First World War was said to have been indebted to cubism.[59] Something camouflaged is either threatened or threatening. What then should we make of the fairly typical case of the Forum des Halles, a sizable commercial redevelopment in central Paris, backed by Jacques Chirac?[60] Most of it is sunk into an excavated pit. An undulating profile of faceted mirror glazing extends above ground, but nestles below the surrounding, decayed nineteenth-century roofscapes. The mirror fenestration gives a mixed array of reflections of the older fabric, the image in each facet deflected into waving drifts within the frame. It does not give a picture, it gives an effect; an obliterating effect that makes the prominent object withdraw into a confused, glittering, multiplied, virtual replication of the color and texture of its setting.

There is one well-known architectural precedent for the camouflage of power and size: the Picturesque of the late eighteenth century. Les

36 Forum des Halles, Paris, C. Vasconi and G. Pencreach, 1972–1979.

Halles is picturesque. Was not modern architecture, in its flight from classical monumentality, always haunted by the lower-case picturesque? Modern painting also showed itself susceptible, if the subject was architecture. Comparing a photograph of Horta de San Juan taken by Picasso in 1909 with a painting, *Houses on a Hill*, done during the same period, Edward Fry decides that Picasso's art was then in a retrograde phase, because the painting is so much like the photograph.[61] The topographical record of the irregular shapes of the vernacular buildings was uncannily similar to the disrupted ambiguities of pictorial space already discovered, but the radical effect was drained out of the picture and absorbed into the rustic object.[62]

In modern eyes the late eighteenth-century picturesque has stood condemned for supplanting social reality with a pretty scene.[63] It cooked evidence of harmoniously occupied landscapes in which social hierarchies and inequalities did not register, providing undeniable instances of

37 *Humphry Repton*, A Design to Exemplify Irregularity of Outline in Castle Gothic, *from* Fragments on the Theory and Practice of Landscape Gardening, *1816*.

ideology as style, and of aesthetics as the mask of truth. Humphry Repton's *Red Books* document the suppression. Paper overlays of projected improvements were laid over extensive views of country estates. When folded up they revealed enclosed fields, new model cottages, beggars, and other original eyesores. The picturesque architecture of Payne Knight, Nash, and Repton masks the truth. Recognized for its exploration of asymmetry, its decomposed contours and informal arrangement are nevertheless denounced as scenic shams.

Put Repton's "Design to Exemplify Irregularity of Outline in Castle Gothic" (1816)[64] next to a view of Les Halles (1979) and you will see two large building enterprises broken down and absorbed into their environment. In this respect there is very little to choose between the informality sought in the eighteenth century and that sought now. Both are ways of limiting the appearance of power, and the effects of both, to some extent, spill out beyond appearance. How far, then, and in what direction? There was a tremendous increase of effect as the scale of picturesque operations was reduced, just as there was in humanized modernism. I have argued elsewhere[65] that when the informality of the parks invaded the domestic interiors of the late eighteenth century, it became a powerful agent in the positive transformation of human relations and family structure that took place then. For a while it did not just provide the illusion of liberty, it provided the means to its attainment. Whether it will eventually be possible to make a similar claim on behalf of humanized modernism will depend on whether architecture can conspire with (rather than represent) a range of activities either unauthorized or beneath consideration.

Conservation of Momentum

A third, distinct episode of fragmentation began in the schools of architecture during the 1970s. It was a more or less immediate reaction to the historicizing tendency within postmodernism. Polemical and bellicose in tone, this reaction was also looking backward for inspiration in its fight against the past. Practically all those now designated deconstructivist[66] were involved as students or teachers in this assertive redramatization of the origins of the modern, and from it they progressed toward an ever more agitated gyration and dismantling of the architectural box. Critics, tentatively applying the portentous term deconstruction to various types of architecture in the early eighties (Anthony Vidler, for example, described Richard Meier's Hartford Seminary, which he observed to be picturesque, as deconstructed modernism in 1982),[67] converged slowly toward common usage. In 1988 the name was conferred on a movement and promulgated by the Museum of Modern Art in New York and Academy Editions in London.

The word offered itself up to appropriation by architects, as did the word's creator, Jacques Derrida, when approached by Bernard Tschumi who rang him up "out of the blue." Tschumi afterward introduced Derrida to Peter Eisenman, with whom Derrida collaborated on a proposal for a garden without vegetation in Parc de La Villette. Colleagues and commentators expressed some concern for the renowned philosopher opening himself to blatant exploitation,[68] their fears confirmed when Tschumi and Eisenman began to argue about who was the first and true heir.[69] No doubt Derridean sponsorship was an important prize in the political games at which both architects are so adept. They loved him for the name of his philosophizing, an appellation they wore (and fought over) like a royal title. It is nevertheless clear that all parties had something to gain from mutual recognition, and all were convinced that some alliance of motives already existed.

It is this alliance of motives that I want to address. For Derrida the pertinence of architecture was, indeed, alluded to in the name deconstruction. He points out that Western philosophy has been a constructive endeavor in which architectural metaphors play a large part. There is an architecture of theories.[70] A philosopher is expected to think out his philosophy the way an architect thinks out his architecture; it has a structure, a consistency, a compositional completeness, and it rests on a foundation. Deconstruction is the task of exposing the mythology of foundations, revealing the weaknesses and inconsistencies in the speculative architecture in thought (which Derrida referred to as the "last fortress of metaphysics"),[71] thus loosening its hold over us. The transparency of the constructive metaphor—its apparent self-evidence in philosophy—will be the greater in an environment of confirming examples where the tectonic schemata adopted by philosophy are exhibited in every building without exception. Derrida's argument is not weakened by the ubiquity of concrete demonstrations, because he insists that philosophy is not building; but, although he was at first suspicious, he was eventually convinced that architecture could at least provide a useful vehicle of exploration, and could perhaps show how deconstruction was practical and positive, not merely critical and negative as charged, even if it could not "question the very idea of structure,"[72] as Tschumi supposed.

Derrida wrote an appreciation of Bernard Tschumi's La Villette in 1985.[73] He did so—and this is important—before the park was built. The essay is a choice example of prolepsis, treating something as already done before it takes place. What he wrote was an appreciation of Tschumi's intentions. The extensive publications on La Villette between the competition entry of 1982 and the opening of the first phase of the park in 1988 were also prone to describe properties in anticipation of their realization. The work completed so far—very accomplished for a first commission, impressive for its thoroughness, consistency, and clarity of reference to recognizable sources, especially its solidified memorabilia of constructivism—is not the same as its foretelling in print. There was too much at stake for this to be intentional, although it was presumably recognized as probable by Tschumi.

His previous works were manifestos. In the La Villette publications manifesto and fantasy merge with and take over the more prosaic role of prospectus. A prospectus is expected to describe what will, in all likelihood, be the case. A manifesto describes what will never be the case unless, by incantation, a strong enough impulse to make it happen is maintained. The impulse may or may not persist—if not, then the manifesto will become a fantasy. If what it describes is impossible, it is already a fantasy. Pedestrianization of the architectural object, its tendency to become ordinary when built, is so common that it is taken to be natural. It is not natural (see chapters 5 and 7), but it could hardly have been otherwise in this instance because of the prominence of the manifesto.

Tschumi's manifestos, presented with unimpeachable good taste, tell of transgression. There will be an architecture if not capable of incitement to murder, then at least hospitable to it;[74] there will be a madness in architecture, which will lead us to recognize things desired yet disturbing. These intoxicating ambitions are declared against the bland background of architectural humanism. They derive from an aspect of French critical thinking that had become increasingly fascinated by the unreason on the fringes of art and literature. Georges Bataille (librarian and alleged anthropophagist), De Sade, and Artaud, relentless in pursuing the limits of humanity and the limits of writing, were gaining respect in place of their previous notoriety, acknowledged by Barthes, Foucault, and Derrida himself.[75] When Derrida reads what Tschumi writes about La Villette—that the 42 *folies* are a play on madness; that out of its "radical questioning of the concept of structure" it produces an *incoherence;* that it is *decentered*—he does not see these as claims yet to be proven on a plot of land in Paris. He recognizes, of course, that Tschumi flirts with varieties of French poststructuralism, right away notices the inconsistencies, but prefers to treat the claim as adequately demonstrated with words and pictures, insofar as it is a claim about what is possible in architecture. Interspersed with brilliant passages on the weight of architecture, in thought as in stone, are accepting recapitulations of Tschumi, like this: "These *folies* destabilize meaning, the meaning of meaning, the signifying ensemble of this powerful architectonics. They put in question, dislocate, destabilize or deconstruct the edifice of this configuration. It will be said that they are 'madness' in this."[76] Madness is evinced in the word *folies*, for which Derrida provides an imaginative etymology deriving *folie* from *feuillie*—foliage, hence folios, leaves, pages, and thence back to the haven of writing. There is a similar recoil from saying what he sees unwritten, a similar recourse to writing about what he reads, in his collaboration with Peter Eisenman (whose name, Derrida observes, means the stone man of steel).[77]

Derrida has been indicted for the extravagance of his claim that "nothing exists outside the text." Diane Ghirardo responds with the taunt that, in the context of architecture, the problem is that Derrida has never risked finding out whether it is true.[78] Were it so, then everything presentable would be legible. Print would have no special status. There would be no reason why a picture of a park, or the caption to a picture of a park, or a word in the caption to a picture of a park, should *always* receive more attention than the park, unless, of course, one is still possessed of a thoroughly hierarchical notion about where the best and most concentrated meaning lies. Derrida's apologetics to one side, his treatment of architecture, painting, and geometry[79] have all been affected by a refusal. He will only deal with what has been turned toward or turned into print already,[80] and is disposed to treat of art already dedicated to deconstructive reading, as was the case with Gérard Titus-Carmel's *Pocket Size Tlingit Coffin*, Valerio Adami's *Glas*,[81] and Tschumi's La Villette.

It is arguable that his work gains intensity and sophistication from the cautious determination not to stray; it is also arguable that the resulting concentration makes the proposition that nothing exists outside the text seem plausible only by expelling everything that does not adopt the form of printed matter to a remote outland of the Other. Accordingly, there are things outside of the text (in the wider sense of Derrida's term), but their existence can only be inferred from deflections and discontinuities within the text.

Derrida and Tschumi support each other's new prejudice. The new prejudice is obtained by overturning an old one: if speech is now judged to be neither prior to nor more authentic than writing (as previously supposed), then privilege writing instead; and if buildings are now held to be no more real and authentic than representations of buildings (as previously supposed), then privilege drawing instead. Such reversals of fortune have practical consequences. In the case of architecture it fortifies an already ensconced Platonism, whereby ideas lose vitality as they put on weight (see chapter 1).

The representation effected by the *folies* (I mean the ones in the park, not those on the page) is adequate to the assignment and no more. I see someone spin on a pivoted seat along the canal; it reminds me of the instrument used to dispel the insanity of men and women in the early nineteenth century. I see through a perforated floor; it reminds me of the opacity challenged by modern architecture. A spiral stair leads nowhere, failing to reach a destination at either end; I think it illustrates the suspension of normal functions. A wall leans; it reminds me of the idea of instability. I know that is what it is intended to signify, although it looks quite firm, just as the reinforcement bars extending from the top of a column represent incompleteness, though neatly finished. They convey meaning, they do not produce it except in contradiction of announced ambitions. I would be lying if I said that the *folies'* redness recalled the century of mechanized slaughter in the old La Villette abattoirs. The color conjures up neither a murderous past nor a libidinous future. Toys and buses parked in a pleasant pasture come more readily to mind than thoughts of madness or violation. Here is a revealing irony: Tschumi's La Villette is far less intensely transgressive than Haussmann's planned butchery which it replaces. Marguerite Duras, visiting the area in 1957, noticed a sign outside the gates, "THE FUTURE Inc.—WHOLESALE OFFAL." It set the tone, she wrote, for a place where truth really was stranger than fiction:

> *In one café a painting depicts a sow sitting next to a piglet in a cradle. The sow is knitting from a pile of tripe. From the entryway the cries of pigs having their throats slit can be heard. Across the way stands a factory that manufactures knives and poleaxes, with a sign that says "We make tenderizers". We laugh . . . and we feel pity. Then pity degenerates into literature.*[82]

The slaughterhouses were the setting for the anarchist film *Sang des Bêtes* and the photographs for an article on abattoirs by Georges Bataille.[83] Bataille was a major source for Tschumi, whose work is about transgression but never oversteps the mark.

Tschumi and Derrida share a fascination with what lies outside. They both express a desire to be working at the edge—the cutting edge for Tschumi, the margins and outworks for Derrida. Is it not likely that their radicalism issues from the need to push at the narrowed boundaries of the safer, printed territory they occupy with such assurance, but within which they remain confined?

38 Parc de La Villette, Paris, Bernard Tschumi, 1982–1989, folie *staircase*.

39 Parc de La Villette, Paris, folie in front of J. L. Baltard's Slaughterhouse Hall (1863–1867).

Wherever we look we will find similar professions of radical intent; similar examples of the disjointed, overturned object in architecture, and similar battles against the notion of formal completeness and classical order in critical writing. Foucault coined the term heterotopia to suggest a prescriptive model and described contemporary knowledge as a dispersed archipelago.[84] Deleuze and Guattari multiply new ways of thinking about relations in spatial terms: the rhizome instead of the tree; the Riemannian manifold; the thousand plateaus; assemblages and segmentarity.[85] Michel Serres charts the diffusion of the forms of thought into a thermodynamic chaos.[86] Lyotard declares war against totality. He also reminds us that Freud treated discontinuities as the most significant sites of all in speech and dreams: interruptions opened up by the unconscious to protect its contents (in some cases Freud interpreted the gaps themselves as symbolic of female genitalia).[87] For Lacan individual consciousness was no longer a single site, but multiplied and uncoordinated. Following him, Fredric Jameson and Victor Burgin develop their models of collective perception on the schizophrenic and the paranoiac.[88] Guattari, in a recent paper on architecture, suggested a polyphony of space and asked architects to "envisage chaos along new lines," as pictured for example in the physicists' strange attractor where an orbiting particle never repeats the same circuit twice.[89] Peter Dews has labeled these trends in thinking the logics of disintegration.[90]

What Derrida hoped for is already taking place: architects and philosophers intent on remodeling the model, dismantling totalities, writing about and building heterodox forms, baptizing new configurations, designing organizations without names, attacking the last fortress from both sides. What escapes no one, but what no one wishes to emphasize or even mention if they can help it, unless making jokes, and what the more elevated discussions are obliged to mention only to discount, is that fracture is the most obvious common feature. The translation seems a bit too easy for comfort. Derrida indicates that brokenness is a crass illustration of what is misconstrued to be the philosophical idea of deconstruction. Still, his collaboration with Eisenman produced a little fractured landscape of collaged pieces.[91]

Mark Wigley made the same point in the catalogue of the MoMA deconstructivist exhibition of 1988, which was full of items one can only imagine to have been chosen because they look like what the word sounds as if it means. So he was obliged to deny it. Faithful to the Derridean script, he distinguished between projects where structures are actually taken apart—using as illustration Gordon Matta-Clark's *Splitting: Four Corners* (1974), a tract house cut in two with a band saw—which he deemed not properly deconstructive, and projects, such as those in the exhibition, that challenge ideas, revealing intrinsic flaws in the constitution of architecture without destroying it.[92]

Matta-Clark once had to leave New York under threat of imprisonment because of his nefarious cutting activities. Son of the surrealist painter Matta-Echaurren, who had worked in Le Corbusier's office for a short while in 1934–1935, Matta-Clark studied architecture at Cornell, where he learned a lasting disrespect for architects. He worked with Peter Eisenman for a very short while in 1976. He was asked to contribute to the exhibition *Idea as Model* at Eisenman's Institute of Architecture and Urban Studies. During the installation he turned up with some photographs of vandalized windows in Bronx housing schemes and an air rifle. He wanted to blow out some of the Institute's windows and hang the pictures in the holes. The curator agreed that he could take out a few panes that were already cracked, but Matta-Clark went ahead and shot out the lot. Next day the glaziers were called and his contribution was repaired into nonexistence before the opening. Eisenman said it was like *Kristallnacht*, missing the point.[93]

Matta-Clark's destructive antagonism toward architects and architecture had already led to one sublime achievement, *Conical Intersect* (1975), which challenged architectural conventions in a way that no other recent work has done. It challenged them not by provocation, but by producing a miraculous result on site. A house due for demolition to make way for the Pompidou Center was used as a kind of raw material. A slanted cone

40 Gordon Matta-Clark, one of the photographs in his contribution to the Idea as Model *exhibition, Institute of Architecture and Urban Studies, New York, 1976.*

was made visible by cutting its elliptical traces through the masonry. The cone, a relatively sophisticated form in unsophisticated surroundings, a space subtracted from other spaces, exposed the rough solidity of walls and floors, whose thicknesses then became markers for a skewed, bodyless object around which the rest of the house seemed to have crystallized afterward.[94]

It is second nature for sedentary critics to lionize risky personal adventures of this sort. In 1988 Matta-Clark's second exclusion from architecture caused a stir. Several commentators saw it as indicative of curatorial conservatism, and beyond that, indicative of an intrinsic conservatism in deconstructivist architecture as defined by the MoMA exhibition.[95] Jürgen Habermas's barbed description of those whom he calls new conservatives (including Foucault and Derrida) who "recapitulate the basic experience of aesthetic modernity" comes to mind: "They claim as their own the revelations of a decentered subjectivity, emancipated from the imperatives of work and usefulness."[96] Insofar as the third phase of fragmentation is the same in the guise of the Other, the recognizable in the guise of the inconceivable, insofar as it seeks to confine architecture to words and pictures, the accusation stands. But there is another way to look at it. A certain conservatism is necessarily exhibited in the pursuit of anything that takes a long time. Unfortunately, arguments for and against new, or ostensibly new, trends presume that conservatism is always regressive. The case is stitched up in accord with that erroneous blanket presumption.

During the afternoon of December 13, 1954, Picasso, at the age of 73, began his 12 variations on Delacroix's *Femmes d'Alger*.[97] In a meticulous review of these variations Leo Steinberg showed why unfinished business was resumed almost 50 years after its annunciation and 40 years after its alleged resolution in synthetic cubism. Picasso wanted to display the whole woman without dismembering her. His earlier cubist figures had been desiccated for the purpose of fracture. Steinberg wanted to demonstrate that, in a last attempt at possession, the many-sided body, more pliable, less brittle, was finally reinfused with an erotic sensuality.

Writing in 1972, Steinberg saw the aging Picasso as a man on his own.[98] Now, 20 years on, the enterprise seems less idiosyncratic. Nor was Picasso's quest entirely unappreciated then. The previous year Jean-François Lyotard had published *Discours, figure*, wanting to show that some aspects of visual experience were beyond the scope of writing. Lyotard argued that the inaccessibility of visual art to an overstructured and exhausted discourse of words was its merit, not its defect. The de facto role of visual art was to provide a volatile, unordered, unpoliceable communication that would always outwit the judicial domination of language.[99] As an example he used a pencil drawing of a nude, sleeping woman made by Picasso in 1941—a body not quite ductile, not quite broken, but severally parted and pulled, that shows everything at once.[100]

A reversion to that unfulfilled promise has since been enjoined from several quarters, including philosophy, literature, and architecture.[101] All sorts of mutually incompatible reasons have been given for this, but one is left with the strong impression that collage and simultaneous presentation were founding axioms not just of modernism but of everything that will follow; that by tearing, spreading, cutting, and busting a decisive historical break was made, on which we continue to build from ground zero. Remember the pedigree sketched for the Atheneum, a building that ended up more like a neat Corbusian shipwreck than a Corbusian ship. Think of the brilliant reminiscence of cubist effects inside the Borges and Irmão bank at Vila do Conde (1979–1986) by Alvaro Siza, who has described himself, disarmingly, as quite conservative and traditional.[102]

The distinguishing feature of deconstructive architecture is its exaggeration of a property already recurrent in twentieth-century architecture. Reviving the disruptions presaged in cubist and constructivist painting and sculpture, it raises, once again, specific technical problems of realization ignored in its public presentation, or quite deliberately flouted as insignificant by its promoters. In order to overcome them architects will either have to become conservative, in the sense of having to undertake a long, maturing travail, or they will have to retreat into Derridean graphology from which increasingly violent representations will produce decreasing effects, or the effects achieved will be achieved through the labor, skill, and ingenuity of others.

In an interview with Edward Robbins, the late Peter Rice, an engineer who acted as consultant to many projects of the new tendency, said he preferred working with designers who had no pretensions to structural knowledge, because the real challenge was left entirely in his hands. That is why he had enjoyed collaboration with Zaha Hadid.[103] What follows from the architect's emancipation from structure is the architect's release from it, not the building's. Even when the new borrows its structure from what exists, the increasingly complicated object defies technical description by architects who do not have the training to define the connections between freely oriented components so they can be assembled together. With this task handed over to technicians, engineers, and consultants,

41 Rooftop Lawyers' Office, Vienna,

Coop Himmelblau, 1983–1988.

there is a bifurcation of design into uninhibited expression and laborious technicality. At the same time it becomes harder to say where the art lies: in the immediacy of the first expression which is made easier, or in the subsequent description and materialization which is made considerably more difficult. The fiasco of the Sydney Opera House is rerun hundreds of times on a smaller scale.[104]

Wolf Prix, spokesman for Coop Himmelblau, gives an indication of this state of affairs. They will do anything to preserve the calligraphic exuberance of the first sketches, sometimes done with eyes closed. They continue, after 20 years, to treat their work as an expletive directed at the establishment. They talk like adolescents about their motives. Their work is a giant psychogram. Subjectivity must be preserved at all costs.[105] Yet a building that *looks as if* it were thrown together or blown up, such as the Rooftop Lawyers' Office in Vienna (1983–1988), where the overall effect of apparent damage is reliant on the complete control of every junction, must be more carefully fabricated than one that does not. Prix even said that the engineer "is the only real Deconstructionist in our group" because he can take things apart piece by piece and show how they go back together again.[106]

This is to some extent an issue of structure and construction, but fundamentally it is an issue of representation—of descriptive drawing, of projective geometry—which is hidden from professional view almost as completely as it is hidden from public view. This subordinated reliance is not new to architecture (see part two of this book), nor is the tendency to glorify the first moments while the most demanding tasks come later (see chapter seven). But let us appreciate the irony in the situation: it was the technique of orthographic projection, supposed to be within the competence of the architect, the technique used to explain the nature of cubist painting at its inception, that was destined to become the blind spot inhibiting architectural realization of fragmented form in three dimensions. That is why examples are still so extremely rare.

It has taken Coop Himmelblau 20 years. It took Hans Scharoun 50 years. Their work is not similar in feeling; the scale, the materials, the nature of the achievement are different; but both found ways to project the chaotic dislocations from the virtual space of cubist painting back out into real space, this time without rectification.

Scharoun took so long that his work extends into all three historical phases of fragmentation. His late accomplishments were prefigured in his youth. At the age of 17, in 1910, he had made a drawing titled *Kirche als Fels* (Church as Mountain).[107] It was a precocious but otherwise unexceptional effort; up-to-the-minute cubist composition applied to a thoroughly picturesque and sentimental scenography. He built the Philharmonie Concert Hall between 1956 and 1963. The entrance aspect of the building (it does not have a facade) looks like a squat recollection of the volcanic, faceted outcrop from 1910.[108] The technical imagination required for the sketch of 1910 was modest; that needed for the completion of the Philharmonie in 1963 was vast. Such was the time it took. In 1987, 15 years after the architect's death, Wim Wenders chose the interior of Scharoun and Wisniewski's Berlin Staatsbibliothek, just across from the Philharmonie, as a set for his film *Wings of Desire*. Then David Harvey singled this out as an exemplary portrayal of the incessantly fragmented space of the postmodern era.[109]

In between times Scharoun developed a sleek modernism, often dubbed expressionist, subject to increasing fraction, and much affected by the work and writings of Hugo Häring, who wished to develop architectural forms that were neither defined by geometry nor copied from nature.[110] It was function, more specifically the movement and repose of human bodies, for whose benefit architecture was made, that would help generate these as yet undetermined shapes. Giedion could easily have intro-

42   Hans Scharoun, Kirche als Fels,

1910, Akademie der Künste, Berlin.

duced Scharoun and Häring as originators of the new humanism, rather than Aalto, were it not that he had already dismissed them as reactionary.[111]

There is no single explanation for the ever increasing irregularity in Scharoun's architecture, which reached a climax after the Second World War. The jagged perimeter of the Philharmonie, for example, is certainly not responding to context or function in any normal sense. The extended, meandering foyer of the original competition design, sited behind an existing neoclassical entrance shared with another institution, was made redundant when the concert hall was moved to an open area on what is now the Kultur-forum, yet it was retained in modified form, wound tightly round and under the body of the auditorium stranded on

43 Philharmonie, Berlin, Hans Scharoun, 1956–1963, plans.

the vast and empty site, thus contributing to the mountainous profile externally, and creating tremendous three-dimensional complexity internally. There is also an odd contextual double take in the resiting of this exemplary destruction of the architectural box. It was put in line with the grand triumphal avenue planned by Hitler and Speer, and was also the first new building on a bomb site among the flattened remnants of buildings wrecked by other means.[112]

Scharoun himself argued the Philharmonie from the inside out. He claimed that the architecture was responding to the requirement of audition. The epicenter is the orchestra, and the epicenter of the orchestra is the conductor. As Wilfried Wang wryly observes, the conductor in this case was the notoriously authoritarian figure Herbert von Karajan.[113] Functionally, the Philharmonie is a centralized building, a focusing of 2,218 points of perceptual attention. In order to unify the sound at these 2,218 points of reception, which are distributed all around the orchestra—to get rid of audible echo—the coherent reflective properties of conventional architectural geometry had to be broken up. Scharoun exploited a well-known exchange whereby the unification of what is heard demolishes the regularity of what is seen. He used this to support his own vision of something he described as like a terraced valley inside a tent, while insisting that the result was "not motivated by formal aesthetics."[114]

44   Philharmonie, exterior from west.

45   Philharmonie, auditorium.

PERSISTENT BREAKAGE **97**

The foyer and its multitudinous connections to the auditorium, producing what Julius Posener called the mysterious Piranesi effect of consciously created confusion,[115] is surely amongst the most impressive works of twentieth-century architecture. Conspicuously lacking in good taste, the living-room light fittings, the polychrome tiles, the colored lens glass blocks, and the standard doors and windows hang like reminders of normal arrangements where that most fundamental of normal arrangements, the right angle, indispensable cue to judgments of distance and shape, is no longer much in evidence. The extent of its eradication in all three dimensions of so large a complex is one measure of Scharoun's success. Very little has happened. Architecture is still a composition of intersecting planes with the occasional curve; it is just that fewer of the intersections are rectangular.

46 Philharmonie, foyer.

47   Philharmonie, foyer.

48   Philharmonie, foyer.

Posener recognized that the Piranesi effect was no mere building from pictures. It had, after all, been demonstrated that Piranesi's *Carceri* etchings were inconsistent projections, therefore not constructible in three dimensions.[116] Scharoun, with the indispensable aid of his associate Werner Weber, had nevertheless "succeeded in actually building rooms which it seems impossible to reduce to the projections of plan and section though, naturally, they have been drawn as such—which, incidentally, makes the Philharmonie one of the most difficult works of architecture ever realised."[117] And Posener was right. The difficulty of the enterprise lay in the difficulty of representing it. The means of representation will always set limits on what is representable. Since the Philharmonie extends beyond the normal range of achievement, I want to return to the question of how it was drawn in the next chapter. For the moment I want only to point out a paradox: architecture as we know it can only escape the flatness imposed by drawing through drawing.

Steinberg indicates that even Picasso, genius of the instant, needed to take his time, but the performance of fragmentation in architecture was beset with difficulties not encountered in painting. The resistance of the architectural medium is greater, whether the medium in question is building material, space, or projective drawing. The physical weight of the material, the perceptual unambiguity of the space, and the opacity to imagination of standard orthographic projection had all to be overcome again in new ways.

Scharoun was striving for a unity with man at the center; he said so in 1919 and he said so again in 1963.[118] This is still the architecture of humanism, where potential conflict between the Wagnerian architect orchestrating the complex object under his command, like the musical conductor who stands at the center, and the polymorphous needs of a diverse audience is resolved into coincidence.[119] Only the traumatic endeavors of individual genius can overcome the tyranny of industrial repetition, but the results can be enjoyed by all, as a kind of holiday. Things are explained differently now. The situation has changed. What has not changed is the urge to fragment, which stays more or less constant independent of its explanations. For this same property now signifies not humanism but its abolition.

Posthumanism is a condition extrapolated from a pregnant statement at the end of Foucault's *The Order of Things* (1966), where he predicts that man, as a subject of human knowledge, will be erased "like a face drawn in sand at the edge of the sea."[120] Tschumi says that posthumanist architecture uses the decentered subject as a conceptual instrument to attack traditional architectural shibboleths such as coherence and unity.[121] Eisenman says that posthumanist architecture is also postfunctionalist, and exhibits modern man's inability to grasp rationality or perfection.[122] Vidler is disturbed by the posthumanist body he discerns in the work of Coop Himmelblau, Libeskind, and Tschumi, a morselated, fragmented body, "contorted, deliberately torn apart and mutilated almost beyond recognition."[123]

Fragmentation induces anaesthesia or thoughts of pain. It indicates humanism or the end of humanism. It has been described as a critical instrument, and it has been described as irrational. Half the time it is a way to soften things, half the time a way to sharpen them. It is a means of liberation, a means of intimate control. It is the escape from capitalism, and capitalism's destiny. It is a way of responding to function and a way to impede function; an escape from geometry or geometry's apotheosis; a method of unification or a triumph over unity; a way to enhance or destroy individuality. A fracture may be the scene of spent anger or unrealized desire. It may represent a political state, a state of consciousness, a state of the world, a state of the body. Or it may represent ruptures in perception, memory, knowledge, self, substance, space, or time. Tracing the fortunes of this activity from the beginning, one must conclude that it has been said to mean almost anything about almost everything. Is that because fragmentation is always a side effect and therefore incidental? Or is it because the explanations are always a side effect? The very fact of its persistent recurrence being confined to the twentieth century would suggest the latter. Some explanations make a great deal of local sense but do not account for the recurrence.

As time passes, what was once an unpredictable step sideways becomes a definite step forward for art. Repetition of the step may then be urged in recognition of the advance. The repetition may produce another step forward into an uncharted future, or it may produce a sign that a step was once taken that way. In the latter case fragmentation would be a sentimental recollection of an anarchic gesture that paid off, a gesture now loaded with meanings it could not have carried at the time of its first performance, when no one could have predicted that in 1926 Roman Jakobson's distinction between synchronic and diachronic, essential to the development of structuralism, would be inspired by Picasso's art as well as Saussure's linguistics, or that in 1927 an abiding interest in cubism would coax Niels Bohr toward the principle of complementarity, its coexistent wave pictures and particle pictures essential to the advance of quantum physics.[124]

The accretion of meaning, coupled with a decrease of effect, may help us to understand why the graphic energy of dissociated and disrupted form has got to the pitch it has. As the origins of modernism drift further into the past, and as representations become more visible than the buildings represented, the manifest energy must increase. It is a conservation of momentum: what is diminished by time is augmented by the pencil. As the idea fades, the agitation around it has to increase to keep it alive; and as the scope of the architectural profession also reduces, the designer's visible impact on the object increases, as if to recall lost powers.

There are some recurrent themes in the explanations of fragmentation. Usually it is directed against legitimated authority. Perspective, classical conventions of composition, dictatorial government, corporate capitalism, the father/superego, and mechanical production have all been ritually violated in architectural effigy. The histories intimated by apparent fracture work in favor of antiauthoritarian interpretation. Something has been broken, so something has been attacked, metaphorically. But witnessing so many skirmishes against so diverse an opposition, we would be forced to conclude that fragmentation is a destructive urge that has been sublimated into representation and thus kept manageable.

Centrality has been among its most conspicuous targets. The cubist and constructivist artists wanted to dispatch central-point perspective. Rowe and Koetter repeatedly contrast their collaged urbanism with centralized town plans from the Renaissance. They open with Tom Schumacher's startling photograph of the Pantheon's oculus, together with its fractured sunlight projection on the dome coffers below, indicating that the prototype gives rise to its antitype, as Auden had said.[125] Mark Wigley's deconstructivist catalogue essay begins with a résumé of qualities the new architecture will disrupt: pure forms arranged so that all conflict between them is resolved, producing a serene effect of harmony, unity, and stability. These were exactly the terms used by Wölfflin to describe the Renaissance centralized churches.[126] Vidler makes a similar contrast between Renaissance anthropomorphism based on the encircled Vitruvian figure and current dismemberments.[127] Foucault's heterotopias were set against his description of Bentham's centralized Panopticon, an institution devised to enforce the unified, focal authority represented in the centralized churches. Finally there is the decentered self, symbolized in dreams, according to Lacan, "by a fortress, or a stadium—its inner arena and enclosure, surrounded by marshes and rubbish-tips, dividing it into two opposed fields of contest where the subject flounders in quest of the lofty, remote inner castle whose form . . . symbolises the id."[128]

And all these need to be set in the context of the continuing accusation that modern art is an agent of disintegration, a complaint made by Ortega y Gasset, Nicholas Berdyaev, Hans Sedlmayr, and Karsten Harries that still commands widespread support.[129] At the end of the Second World War Sedlmayr's *Art in Crisis: The Lost Centre* (1944) gave, in negative form, practically every argument encountered so far in support of the third phase of fragmentation. The keynote was W. B. Yeats's line: "Things fall apart; the centre cannot hold." Sedlmayr was concerned with the appearance of fragmentation in the object as a portent of wider destruction, especially the shattering of man's self-image suggested by it.

49 Pablo Picasso, Weeping Woman,

1937, Tate Gallery, London.

In our age, he wrote, art "has in a very definite sense become eccentric."[130] It is the symptom of a mental disorder to which the critic must react as a psychopathologist. When the center is canceled, concepts of person and God will be threatened by that cancellation. In architecture the first signs of morbidity were discernible in eighteenth-century English landscape gardens, when ruins became popular, when the house lost its priority in the park, and when Ledoux forced domestic accommodation into a sphere, destroying the coordination between form, use, and signification. But it is in the twentieth century that the morbidity becomes chronic. Sedlmayr sees Picasso's spirit, through Ivanov's eyes, flying out like a raven from his own subjective center and pecking at "the dead husks of a world whose ordered semblance lies in ruins.... And so humanism dies." We are the first civilization ever to reach what he called the "trans-humanist" stage.[131]

Sedlmayr and the advocates of deconstruction are in general agreement on all major points about the tendencies of art and architecture. Their only disagreement is whether or not to denounce them as pathological. As with the concordance between Wölfflin and Wittkower, it is here, where there seems nothing to argue about, that we need to be most discerning. Note how neatly into this scheme of things dovetails Wölfflin's and Wittkower's shared conception of the serene, stable unity of Renaissance centralized churches as reflective of Renaissance thought and feeling. Sedlmayr accepted it as given,[132] just as Wigley does now. Wittkower's historical research in *Architectural Principles in the Age of Humanism* has been assimilated within an ideology common to modernism, antimodernism, postmodernism, poststructuralism, and deconstructivism, all of which accept loss of center as an article of faith.

But the idea that human consciousness suddenly fell out of a world with a center and an edge, as in an uncomfortable awakening, is a demonstrable fiction. There was no such age with such a state of mind. There was only art which gave that impression from certain points of view, so that the imagination could feed on what the intellect could not accept. Suppose that fragmentation operates now as centrality did in the Renaissance. It would neither reform nor reflect the world as it is. It would instead, rather surprisingly, be offering a presentiment of what the world lacks. So is it the consoling vision of our times?

The Consolations of Architecture

The imprisoned Boethius awaiting execution in the early sixth century was brought consolation by Philosophy, a large, imperturbable woman, shining her beacon in the tempestuous night. Throughout his last work he kept reciting his fears, expressing them in metaphors of darkness, being at sea, drowning, and storms. Philosophy brought tranquillity of spirit by bringing order to the chaos of experience, thus arming the soul against assault.[133] It is easy to see that Boethius' philosophy satisfied an emotional need. The emotional tenor of contemporary critical philosophy in general and deconstruction is particular is incomparably different, but it does have an emotional tenor, and it also satisfies an emotional need. So does the architecture to which the name deconstruction has stuck. Explanation may be sought here rather than in the battle with a mythical past. If the past is mythical, the battle must be symbolic. The emblems of classical wholeness and unity may be under attack, but something else for which they stand will be the source of agitation.

Protagonists and antagonists agree that our age is one of enormous dislocation and uncertainty. How can they tell? Because everything signifies that it is so, everything that is within the chain of signification, from the news on television to the history of twentieth-century philosophy; but the experiences of everyday life do not always confirm it. From the standpoint of ordinary existence, as led by many of us in the privileged haven whence all these ideas of dislocation come, the world is much safer, much more consistent, and far more certain than it ever has been. The truly haunting Other is not what lies outside the text but what lies outside the picturesque garden of the Western world. Beyond are the vast majority for whom dislocation needs no representation, and whose visions of chaos are more likely vengeful and in preparation of a new order.

It may all change tomorrow, but our experience of the world today is of its reliability and its resistance to catastrophic change. It is surely obvious that, the more is known about uncertainty, chaos, and dislocation, and the more widely and quickly news of disturbance is disseminated, the greater the potential and actual control of events will be. This is the situation we find ourselves in vis-à-vis monetary control, military control, the control of deviance, and perhaps also self-control. Ian Hacking argues eloquently that while certainty retreats in the face of probability, the superior grasp of probability has led to far more accurate prediction, and so "the greater the level of indeterminism in our conception of the world and of people, the more we expect control."[134]

The accidental collapse of buildings, once quite frequent, is now rare. Fire risk has been reduced by orders of magnitude. Under the wildest architectural singularity will be the ubiquitous standardized provisions of electrical cable, water and gas pipes; around it will be a network of metaled roads and communication links, national and international; and within, an armory of security devices. All one has to do is read something by a deconstructive architect in conjunction with a contemporary work on environmental control systems or building safety requirements to see the former as a difficult compensation paid for the latter. Twentieth-century art, which denies escapism, still thrives on dreams of escape that must stay as dreams. The maintenance of the dream is what counts, not the realization of the dream's content. A wish kept alive is a wish unfulfilled; a realized wish extinguishes the desire, either for the time being or forever. It is an interesting feature of our times (just ours, not the others) that we need desire more than we need satisfaction.

It is not historical conventions that define architecture's real stability (as we shall see, they never did), but the host of codes, rules, and practices that surround construction now. In this respect construction is typical of a wider pattern. Episodes of fragmentation have been aimed at an illusory historical stability as a stand-in for the blind predictability of the present, which cannot be fully accepted without conceding that architectural form, playing its frenzied, dazzling counterpoint against the predictable base, cannot materially alter that base without destroying its own raison d'être. That is why the landscape of more complete predictability could turn out to be a landscape of apparent chaos. Another reason has already presented itself with equal force; the landscape of more complete control could be a landscape of intimate, particularized adjustments and responses, as complex as the circumstances of its occupation. Getting rid of repetitive and therefore predictable forms does not necessarily get rid of predictability in human affairs or predictability of material performance, which brings us to the third reason. We have already seen that the more complicated the architecture, the greater the control required at all stages of its design and manufacture. A designed environment of apparent chaos has to be predicated on precise technical control of the object. The greater the control, the greater the phantasm of release produced by it, an old formula for cathartic art that still applies.

In the most recent episode of fragmentation, larger bursts of energy seem to have been directed at building so as to force a multiplied identity upon it that it would otherwise not accept. Thus the sins of the profession are expiated ("With property developers around we cannot have too much collage")[135] and, far from following the ectopic contours of a world that already exists, a scenery is invented to provide an imaginary complement to the world that exists for us—the world experienced day after day. The complementary scene can mean almost anything so long as it takes a disjunctive form in the face of expectation, just as, prior to the Enlightenment, philosophy could reach almost any conclusion so long as it provided relief in the face of experienced uncertainty.

The sustaining condition of modernity's convulsions is stability.

Part Two   *Chapter Three*   **Seeing through Paper**

A history of architecture that dealt with the impact of drawing would need to explain two things: how architectural spaces arose out of the deployment of depthless designs, and how architectural space was drawn into depthless designs. With this in mind, the distance between Raphael and Scharoun might once again be measured.

In his pioneering study "The Rendering of the Interior in Architectural Drawings of the Renaissance" (1956)[1] Wolfgang Lotz concluded that a decisive change of practice occurred in the first decades of the sixteenth century, and that the painter-architect Raphael helped to bring it about.[2] Lotz showed that centralized buildings in general and circular interiors in particular had posed an intractable problem; they could not be drawn so as to provide useful information, such as scaled dimensions. At that time most notable Italian architects, trained as painters, were prone to make perspectives, but perspectives gave only partial views of the inside of a building, and a particular viewpoint would inevitably result in distorted proportions. The answer grasped by Raphael, and recorded in his letter to Pope Leo X of 1519, was that if such drawings "do not diminish at the extremities, not even in round buildings,"[3] they would be better adapted to the purpose of the architect.

Lotz presented this as a discovery made by stepping backward. The imagined viewpoint of the observer was withdrawn further and further from the building; foreshortening was reduced and eventually eradicated. The result is more detached than a perspectival picture—less visual and more professional, is the way Lotz put it.[4] In this same letter Raphael describes plans, external elevations, and internal elevations (which I presume to include a sectional slice through the building's shell). It could be said that bringing the three types of drawing together was the revolutionary step. However, what really matters is what holds them together. Raphael spells out that corresponding parts are joined by parallel lines, from plan to elevation, or from one elevation to another. These parallels—conservers of true measure—are most readily understood to be representations of light paths, and understood thus they emphasize remoteness, one way or another, because either the light source projecting the information or the eye receiving it has to be imagined at an infinite distance. These parallels are also the agency through which the space outside the surface of the drawing is brought into it. Proclus held space to be "nothing other than the finest light."[5] In architectural projection space is nothing other than pictures of light. Images drawn as if transmitted to a surface by light are pushed around and explored within the surface by simulacra of light that have been flattened into a comb of drafted lines.

Few things have had greater historical significance for architecture than the introduction of consistent, coherent parallel projection into architectural drawing, and few things have been more transparent to critical attention than its effects.

Were it not for light's straightness, our vision would be muffled, smeared, and stunted. This is a fact of sight, the miracle of which, in all its expansive plenitude, depends on the unswerving uniformity of light's propagation. Light is, after all, the ultimate geometric instrument—attested to by contemporary laser surveys, as also by Euclid's abstruse definition of the straight line, which can only be interpreted as a line that looks like a point when seen from end on, in other words a line that is identical to a light path.[6] But this geometry immanent in vision and yet invisible crept into Euclid's definition unannounced. The geometry of images propagated by light was developed as a historical postscript to the geometry of the surveyed plot, which seems already to have been intimated by John Dee in his 1570 *Mathematical Praeface to the Elements of Euclid*. To his mind geometry's second coming would release it from the terrestrial surface. Euclid's *Elements* only taught us how to measure land, said Dee:

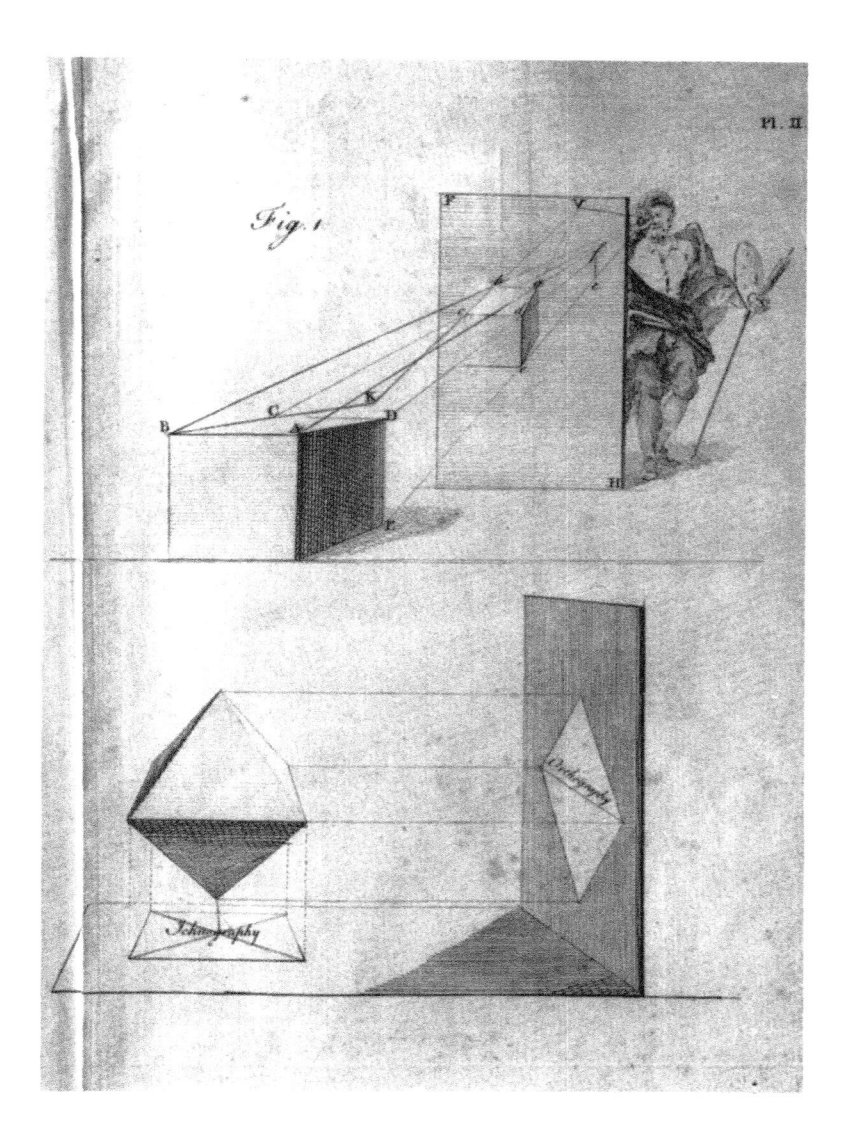

50 The Principles of Parallel Projection, *from Daniel Fournier,* A Treatise on the Theory of Perspective, *1761.*

An other name, therefore, must needs be had, for our Mathematical Science of Magnitudes: which regardeth neither clod, nor turf: neither hill, nor dale: neither earth nor heaven: but is absolute Megethologia: *not creeping on the ground, and dasseling the eye, with pole perche, rod or lyne: but lifting the heart above the heavens, by invisible lines and immortall beames.*[7]

So too in architectural drawing, light showed up late. And what it showed up when it arrived was space. It was a great conquest, but it was neither complete nor sudden; architectural space would remain, one way or another, limited by and bonded to the pictures that normally gave access to it.

The perspective projections of the painter and the parallel (orthographic) projections of the architect both bring space into pictures.[8] The differences between post-Renaissance painting and post-Renaissance architecture are subsumed within this similarity; both arts had conquered the same territory with similar equipment at much the same time. Alberti and Raphael emphasized the differences between painters' and architects' drawing, but the two kinds of projection were allied in all sorts of ways too.[9] So, in practice, how great was the difference? Lotz believed that two successive revolutions of spatial sensibility had been prompted, first by Bramante's perspective method of designing, and then by Raphael's orthographic method. He was convinced that since the introduction of the latter, "the architect's manner of seeing and representing has deviated from that of the painter, and the unity between the two arts ... which Bramante had put into practice, has dissolved."[10]

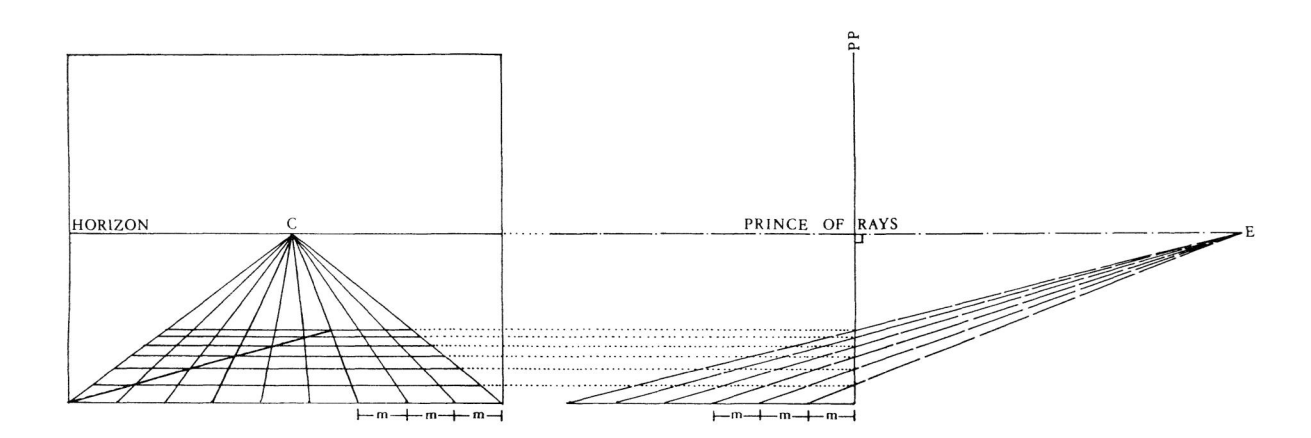

51 Alberti's perspective construction (drawn by the author). E = eye; C = centric point; PP = picture plane; m = tile module.

I believe that Lotz exaggerated the practical differences between perspective and parallel projection. Certain tectonic properties were fostered by both. For the most part, painters and architects have favored easy methods of setting out; the tectonic properties in question derive from these easy methods. It is important to realize that they arise out of specific constructions normally employed in studio and atelier and are not intrinsic to all projection. In 1435, or thereabouts, Alberti advertised the perspective method described in his book *On Painting* as an easy, practical method. It was so, and proved immensely influential. A line, called "the prince of rays," commands everything in its path.[11] It travels from the observer's eye, parallel to the ground. On it is hung the picture plane which it penetrates, perpendicular and in the middle. Alberti's method

52 Raphael Sanzio, The School of Athens, *1511, Stanza della Segnatura, Vatican.*

gives a perspective of a squared pavement behind the picture plane. The tiles are aligned with the picture plane and aligned, therefore, with the prince of rays. The point of its penetration is the point of convergence of the foreshortened images of the pavement lines coursing toward the horizon. It is the prince of rays, not perspective, that is the organizing force here, because it is the prince of rays that orchestrates the rectangular constituents of the picture as frontal, bilaterally symmetrical, and axial. Using this method, it is easy to make a picture with these properties and hard to do otherwise. Who could ask for a more impressive demonstration of their combined effect than Raphael's fresco *The School of Athens* (1511)? And who could ask for a better demonstration of the same properties in architecture than Raphael's project for Pope Clement VII, the Villa Madama on Monte Mario in Rome (1517–1521)? An

53 Watercolor of Villa Madama (anon.), eighteenth century, Victoria and Albert Museum, London.

54 Villa Madama, Rome, Raphael Sanzio, 1517–1521, plan (redrawn by the author from Uffizi 314 A).

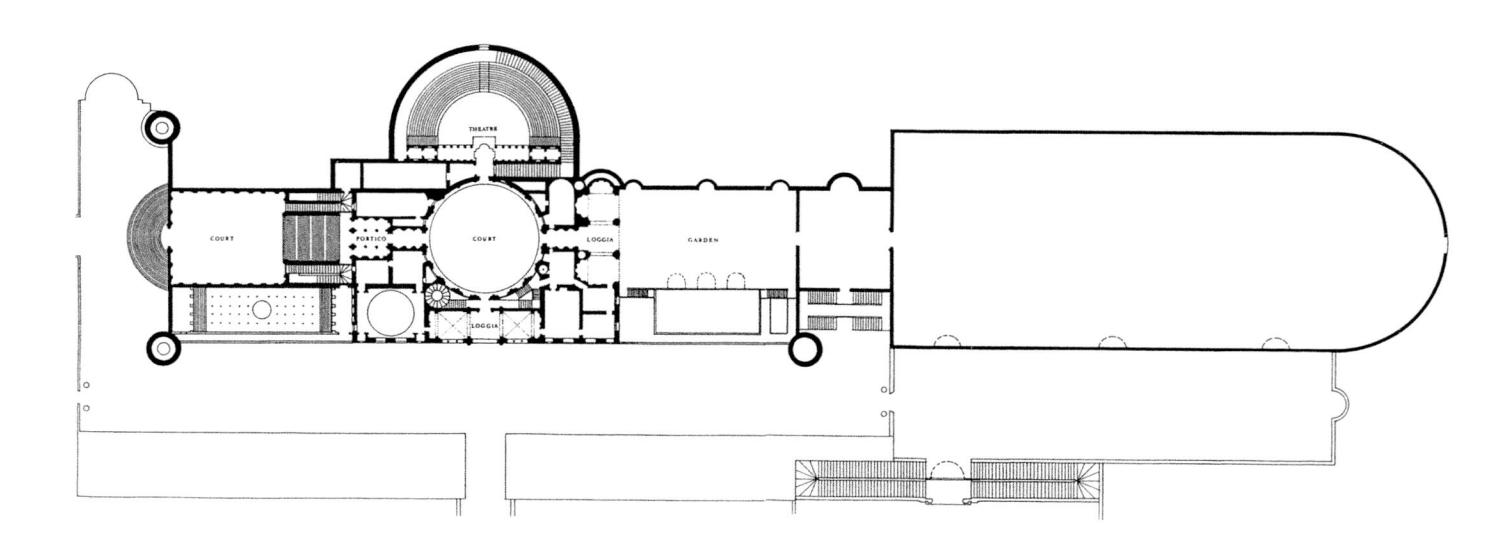

eighteenth-century watercolor of the partially constructed villa, looking from the garden, through the open loggia, onward through the telescoping tribune toward the Vatican, which can be seen in the distance, could be mistaken for the fresco's setting.[12] In his letter to the pope, Raphael maintained that all his architectural work had been produced with plan, section, and elevation,[13] and yet the villa is eminently perspectival.

The introduction of linear perspective was not as abrupt as was once thought. Erwin Panofsky and John White have traced its evolution from workshop ploys adapted and improved by generations of painters better to suggest pictorial depth.[14] These ploys used geometric elements but they were developed in an intuitive, piecemeal way, so the celebrated rationalization effected by Brunelleschi and Alberti was more like the last word than the first.

Again, the same is true of parallel projection. Plan, section, and elevation, considered independently, are almost prehistoric. They can exist, and even coexist, without invoking projection (or indeed light) at all. A plan need not be regarded as a picture; it can just as well be thought of as a set of geometric operations on a flat sheet that can be repeated at a larger scale on a flat plot. The geometric layout of primitive plans and of their leveled sites are similar in the exact, Euclidean sense of the word. The constructability of the one demonstrates the constructability of the other, but the plan is no more a picture of the site than a triangle is a picture of a similar triangle. *Grundriss, pianta, assiette,* ground plan, and the antiquated English "platform" give reminders of their earthbound origin; so does Raphael's anthropomorphic simile of the plan as footprint (which is not a picture of a man but his trace on the ground).[15] The principal plan would later evolve into a horizontal slice of the lower part of the major rooms of a building. Its pictorial surface, now raised above the ground with which it was once identified, rains down nominal projectors, but they rarely capture much indication of depth in their shallow drop to the floor.

With the principal elevation we move from a primitive nonpictorial equivalence to a primitive pictorial equivalence that can be just as depthless. Vitruvius called it "a picture of the front of a building, set upright and properly drawn in the proportions of the contemplated work."[16] Alberti and Raphael generalized this to include all sides. They were as preoccupied with maintaining proportional equivalence as Vitruvius was, but Raphael tells us explicitly that this was to be maintained despite the obliqueness of the surface represented,[17] whereas the Vitruvian definition might suggest that it would be maintained because the

55 Christ Disputing with the Elders in the Temple, *fourteenth century, Dragonelli Chapel, San Domenico, Arezzo, detail of tiled floor. The receding geometric tiles are depicted with no diminution of breadth. They diminish in depth, but according to no predetermined ratio or construction.*

56 Sant'Andrea, Mantua, Leone Battista Alberti, designed 1470, detail of facade.

57 The Arch of Augustus and the church of San Michele, Fano. On the facade of the church, by Bernardino di Pietro da Carona, 1513, is the carved perspective of the Arch of Augustus.

58   Andrea Palladio, final facade study for San Petronio, Bologna, 1579, Worcester College, Oxford.

elevation could be marked on an upright wall like an inscription on a board, or like ink on paper. The close resemblance between wall surface and paper surface has never been entirely overcome in architecture, any more than has the geometrical equivalence of plan and floor. It is far too convenient. Even Alberti, patron of projection, kept it that way. For instance, the facade of Sant'Andrea, Mantua, designed by him in 1470 and erected after his death, is as flat as a board. In photographs it looks more highly modeled than it really is, because in photographs graphic depth cues tend to become more pronounced. Egg and dart moldings, applied like piped icing to the church's stucco frontispiece, outline a giant order of pilasters that cut through the indications of a subordinate structure, and so look as if they are in front of it. But because the molding, which has a quadrant profile, gets embedded in the surrounding stucco, the pilaster surface is in fact recessed further than the rest of the facade, though discernible as such only at close quarters.

It was probably an accident, but this reversal highlights a feature exploited ceaselessly by architects. Ways of implying depth in two-dimensional pictures are applied to the almost flat surfaces of three-dimensional objects, but then interact with the perception of real depth to give more or less of a flutter between the real and the imaginary. In these circumstances design and object, as well as perspective and ortho-

*59   San Giorgio Maggiore, Venice,
Andrea Palladio, 1564–1580, detail
of facade.*

graphic drawing, become so similar, so entangled, that when Bernardino di Pietro da Carona carved a frontal perspective of the Arch of Augustus on the facade of San Michele, Fano, in 1513, the result was hardly distinguishable from the architecture on which it was carved.[18] The layering of thin parallel planes, already evident in early Egyptian architecture, soon became the governing artifice of the elevation, drawn and built, as can be appreciated either from Palladio's last, ingenious design to complete the unadorned brick front of San Petronio, Bologna (1579), or from the facades of his Venetian churches. The effects were obtained by drawing lines on a flat sheet that looked as if they represented something with considerable depth, then building it with limited depth. The result was compaction—architecture as a low relief of its own pictures. The similarity between wall and paper was emphasized, moreover, by the common practice of using a close-textured, monochrome facing material such as marble or stucco for monumental buildings, the very color of which resembled paper. Under these conditions, projectors need not extend very far from the picture to reach the thing pictured, and the imagination of the designer need travel no further than the projectors to envisage what has been designed. Thus limited, architects can cope with space because of a safety factor built into their drawing that protects them from the unimaginable.

Sections, too, can be depthless and nonpictorial when they take the form of templates or profiles, but that is certainly not what Raphael meant by "internal elevation." As Lotz understood, and as Jacques Guillerme and Hélène Vérin also recognize, the architectural section breaks open contained space in order to show it as an elevation, forcing entry and revealing the interior to the distanced eye of the architect. The immediate antecedents of projective sections were pictures of half-demolished ancient ruins, and comparable developments occurred in the performance and illustration of anatomical dissection.[19] In sixteenth-century engravings the cut fabric is often indicated as a mechanical fracture rather than a smooth conceptual severing. The typical orthographic section therefore ends up as two kinds of drawing synthesized into one: a profile of a cut, which need not involve projection, and an elevation of what lies beyond, which not only involves projection but tends to open up a deeper space for it to survey.

Thus projection was a late, extra ingredient grasping more or less cautiously at the imaginary space behind the three drawings. What, though, is the simplest, most effective relation among projective plan, elevation, and section? The answer is dependent on what is being depicted. Yet if we take a representative sample of architectural drawings from the sixteenth to the nineteenth centuries a strong pattern quickly emerges: one of each type does maximum service. Monumental architecture, before and after the institution of parallel projection, was, typically, a masonry carcass round a large ceremonial space. For reasons unrelated to drawing, the vaulting or roofing would be highest around midspan. Theoretically, a section can be taken through any part at any angle, but in practice, sections are taken vertically through the crown of a vault or the apex of a truss, because only then is the full extent of one side of the interior exposed to view. If the side you see is the mirror image of the side you do not see—if, that is, the building is symmetrical about the sectional plane—you see it all through one cut. So vertical, bilateral symmetry is economical within the confines of the technique.

A plan, which can be any shape, may tend to be rectangular for a host of reasons, but the internal logic of parallel projection will push it that way too. Five minutes at a drawing board will convince anyone unfamiliar with the technique that this is the way things have to be set out. The instruments at your disposal will lead you to produce frontal pictures of the several sides of boxes as soon as you have gained the slightest idea of what you are doing. It is easiest to deal with the three types of drawing if they are perpendicular to each other, and it is easiest to align the principal surfaces of an object with the surfaces on which it is drawn; in consequence, a building will be a box in a box of pictures. So planar, rectangular form is economical too, within the confines of the technique.[20]

In his brief account, Raphael says that to make an elevation, you first make a base line the full width of the edifice, then divide it in half with a vertical line called the "centre line of the building."[21] This line is not hard to explain, since architecture, according to the classical ideal, was bilaterally symmetrical, but it is also the line of the most advantageous sectional cut, which has to slice through a cavity beyond the facade, otherwise it will not show much. A center line projected through a cavity easily converts into a processional axis. Then the axial route will show up on the principal elevation as a principal entrance, thereby converting the simple, binary equality of left and right sides (a-a) into a tripartite, therefore hierarchical, centralized symmetry (a-b-a).[22] Things begin to add up. The three drawings are not just plan, elevation, and section, but ground plan, front elevation, and axial section. That is why in most classical architecture, design and building are in a near perfect accord. Maximum descriptive power is obtained at minimum price—a good bargain, so long as what is required is frontal, symmetrical, axial, and predominantly orthogonal.

But what happens when it is not? Some modern architects, deprecating the use of the classical orders, wanted to retain elements of classical order; others, wishing to be more radical, rejected that too. Yet, strange to say, though there are plenty of jibes at the paper architecture of the academies, plenty of warnings about the illusions created by pretty pictures, I have been unable to find any modern architect or critic expressing serious opposition to architectural projection as evolved in tandem with classical architecture. The triumvirate of plan, elevation, and section were not identified as part of the problem and so continued as working desiderata of the profession. On the other hand, as we have already seen, modern painters attacked perspective vociferously, dubbed it a mere convention, removed evidence of it from their work, and claimed that vision itself was not perspectival. For them the dominant means of representation was an issue, for modern architects apparently not.

Scharoun's Philharmonie has none of the properties that are bolstered by classical projective representation; it has no front face, it is not rectangular. There is a residual axis along the length of the auditorium, but it does not divide the space into exactly mirrored halves, nor does it correspond to a processional route. In postwar Germany, Jean Gebser's zealous eulogizing of aperspectival consciousness had some influence, and it has been

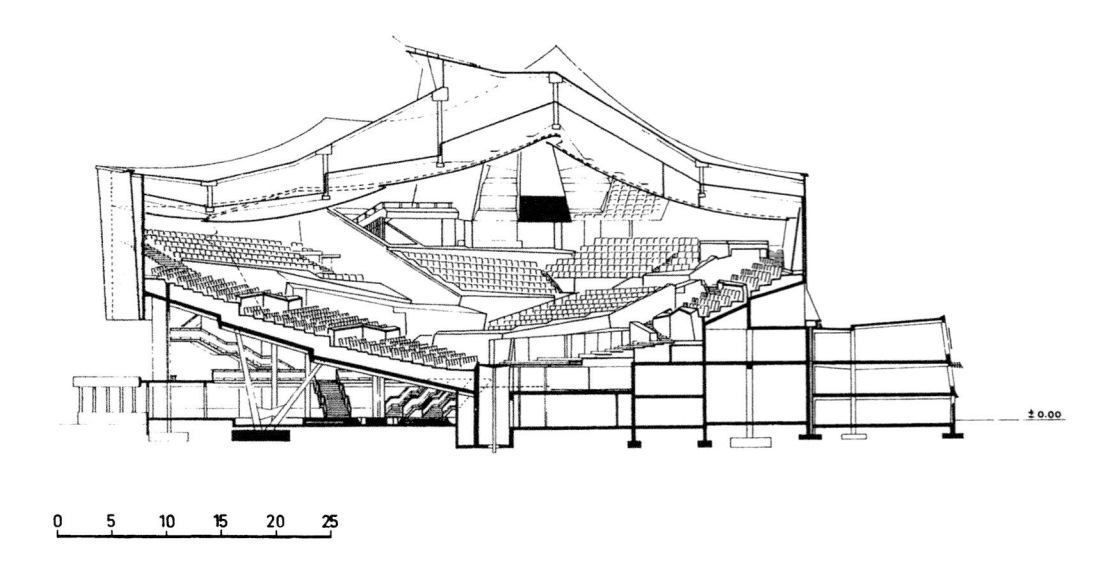

60   Philharmonie, Berlin, Hans Scharoun, 1956–1963, published sections, Akademie der Künste, Berlin.

suggested that Scharoun's architecture is an attempt to produce Gebser's notion of aperspectival space.[23] If this is true—and I think it makes sense to call the Philharmonie aperspectival, not in contrast to optical perspective, but against the tendency of Alberti's perspective construction—then it emphasizes the difficulty of describing the building by means of conventional architectural drawings, since we have also seen that they would be apt to assert the same schematic formation as Alberti's perspective invariably did.

There came a time when Scharoun and Weber found their drawings of the project to be inadequate. Scharoun had worked from sketches to models in the design stage, but now the complex was being built. There had been irremediable errors in setting out the foundations. At this point they felt it necessary to supplement the standard set of drawings by taking large-scale sections at closely spaced intervals across the entire breadth of the building, ensuring its surveyability to regain control of the work.[24] The two published sections through the residual axis of the auditorium gave a fair account of its interior but were insufficient to define the geometry of its canted and curved surfaces, while neither they nor any other single or multiple sections through the building ever hinted at the space of the foyer. Some grasp of the totality was achieved by compressing the classical section until it became in effect a profile and then chopping the building into thin slices with it. Nevertheless, it was a myopic way of seeing things, less closely tied to experience than the classical section had been, and anything but economical. Nor was it particularly well suited to describing the metric properties of the Philharmonie fabric.[25]

±0.00

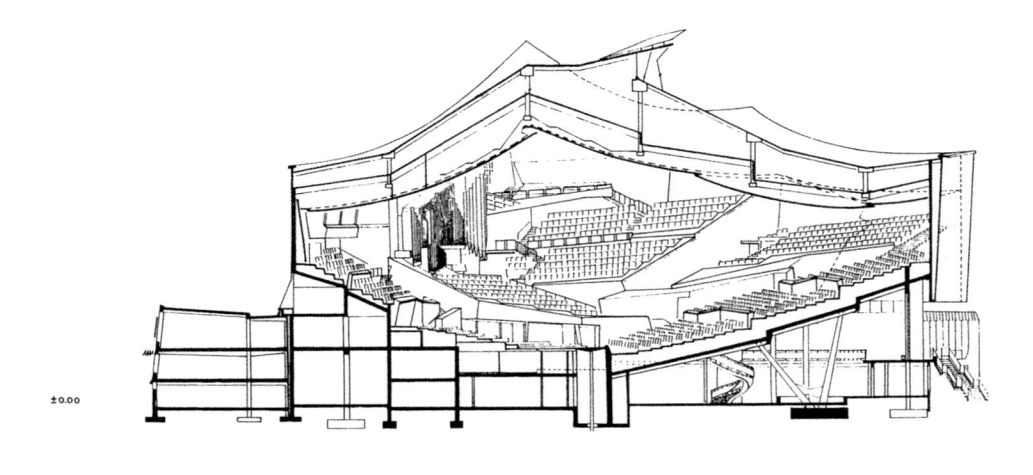

0 5 10 15 20 25

To the extent that modern architecture relinquished the underlying order of frontality, symmetry, planarity, rectangularity, and axiality, it was no longer in easy accord with its drawing techniques. Buildings, though not wholly determined by the means of their production (which is to say, from the architect's point of view, the means of their representation), are mightily influenced by them. Demands on the architect's imagination are far greater when the object has to be envisaged despite its drawing, and that is why it is remarkable that the Philharmonie interiors, especially the foyer, turned out so coherent in their unpicturable diversity. Toward the end of chapter five I will explain how a similar situation arose in the making of certain late Gothic buildings.

Classical order was a tendency, not an imperative. Throughout what I am calling the classical epoch—from the Renaissance to the end of the nineteenth century—it was constantly undergoing modification and adaptation. The happy marriage of style and drawing was a strong consolidating force, but the hold of projective drawing could be loosened or transformed. It could be loosened by reducing its power and influence over form. I think that Francesco Borromini worked this way on the church of San Carlo alle Quattro Fontane in Rome (1638–1641). The assumption that the building corresponds to the geometry shown in Borromini's "definitive" plans is wrong, and I have found it impossible to account for the actual shape of the dome, or for the coffering, by any projective means.[26] Or the hold of the drawing could be transformed by using it differently. That is not a speculation but a demonstrable truth which I now want to demonstrate twice, first with an infrequently encountered technique of perspective construction described by Piero della Francesca, then with a supplementary mode of architectural drawing called stereotomy, or *l'art du trait*.

Chapter Four **Piero's Heads**

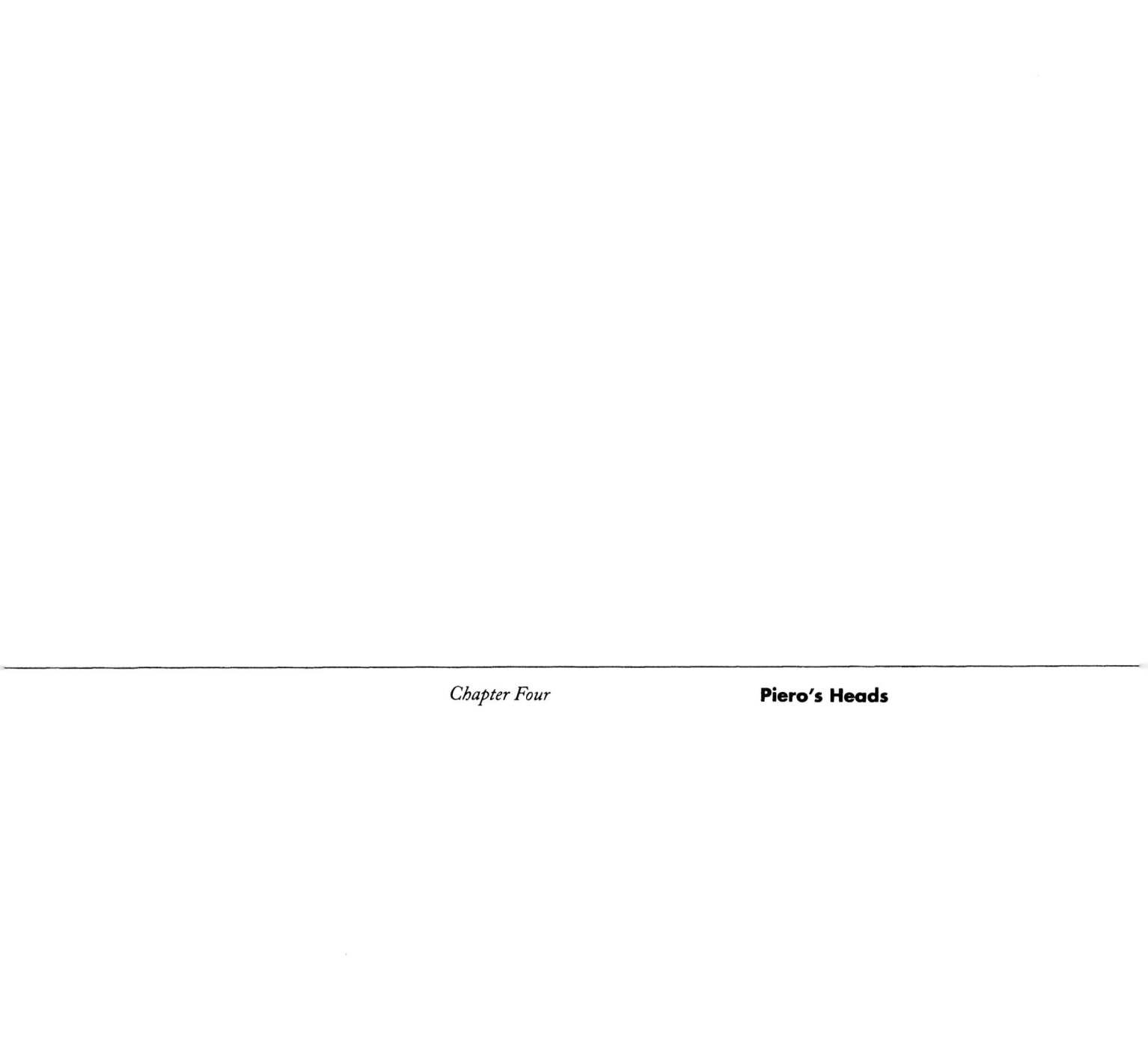

We have come to believe that perspective is a cultural artifact, a set of conventions through which we have defined our own vision. This is not quite right. Perspective is a convention, but varieties of convention are involved, and not all are equally arbitrary in respect of what is depicted. It is easy to recognize the difference once it is pointed out. Compare, for instance, any amateur snapshot taken in the open air with a typical Renaissance painting. The perspective format of the painting will not be duplicated in the photograph. The same difference can be observed in two early Renaissance perspective techniques; there was a perspective construction for architecture and a perspective construction for heads. The former is well known, the latter hardly recognized; the former has vanishing points, the latter does not.

Perspective's Reputation

During the twentieth century, perspective has been subjected to intense and often hostile interrogation by philosophers and critics as well as artists. They conclude, in the main, that although it was once vaunted as the exclusive road to visible truth, perspective gives no unique or privileged access to reality. Some have pushed one step further to convict perspective of obscuring precisely what it pretends to reveal. According to one credible line of argument, perspective is the least credible way of seeing because it has convinced us so completely that it is the only true

61 Bernardino Pinturicchio, one of the series of scenes from the life of Aeneas Silvius Piccolomini (Pope Pius II), Piccolomini Library, Siena cathedral, 1505–1507.

62 Snapshot (author).

way to see. As faith in perspective's capacity to convey the truth erodes, reaction against its partisan rule spreads wider. Its hegemony over vision has been linked to various other kinds of dominion and power, hence the increasing resort to terms such as "the domain of vision," "the empire of the gaze," and "scopic regime," drawing politically charged metaphors into the vocabulary used to describe the very thing that is accused of being politically charged. But how do we tell whether this verbal branding is justified? Could it be another episode of iconoclasm?[1]

Perspective has for ages been regarded as deceitful.[2] What is so remarkable about the twentieth-century revision of this idea, as developed in German and French criticism, is the reversal of an earlier understanding of its shortcomings. In the eyes of Renaissance commentators, perspective was a deception because it distorted true measure; because, that is, it departed from the inalienable truths of Euclidean geometry. In the eyes of many of its twentieth-century detractors perspective is suspect because it imposes Euclid on the way we see. So, what was once insufficiently Euclidean is now too Euclidean. For example, Maurice Merleau-Ponty, in his working notes for *The Visible and the Invisible*, treated perspective geometry and Euclidean geometry as, to all intents and purposes, identical. He spoke out against their conspiracy: "I say that Renaissance perspective is a cultural fact, that perception itself is polymorphic and that if it becomes Euclidean, this is because it allows itself to be oriented by the system."[3] A little earlier he had reminded himself to "take topological space as a model for being," noting that "Euclidean space is the model for perspectival being." Presumably, then, perspectival being is a special case of being—a rather questionable one—and Euclidean space a particular type of space, both masquerading as universal. Merleau-Ponty joined perspective to Euclid and then joined these to Cartesian geometry because of the reticulated network of straight lines he thought characteristic of all three. Later, Jacques Lacan continued to insist that, from the point of view of vision, the three kinds of geometry were in an unholy alliance.[4] These two writers maintained that all classical geometry was implicated in an attempt to capture and colonize the way we see. Merleau-Ponty at least suggested the possibility of unmasking vision by removing the perspective tracery that had been imposed on it. Lacan was less hopeful. He extended the accusation beyond perspective, beyond geometry, to vision as a whole, which for most of us, most of the time, must remain irredeemably bound up with the process of domination. The conquered eye becomes the organ of conquest. Surely this is one very striking instance of the tendency, well observed by Martin Jay, to treat vision as a dangerous weapon within us and others, against which we must constantly be on our guard.[5]

Foucault's account of *panoptisme* is another.[6] The gathering of lines of sight into a point, like the gathering of reins by a charioteer, is a symbol of control because it is an instance of control. In the late eighteenth century Jeremy Bentham, the utilitarian philosopher who wanted to be a jailer, proposed building an institution that would expose its inmates' every move by making the architecture conform to the lines of sight emanating from a central kiosk. The governor's eye would be all-pervasive in this position. Social power would be focused by turning architecture into the lines of one man's sight. Likewise, according to Lacan, the seventeenth-century philosopher-mathematician René Descartes modeled his idea of the perceiving subject on the infinitesimal point where the lines of perspective construction meet. So it would seem that on this one characteristic of perspective geometry—the convergence of lines into a point—two tyrannies of Western thought are balanced: social determinism and mind/body dualism.[7] They make a heavy burden. Many instances could be cited of more recent authors who say that the geometry of perspective has to be imposed on vision, and that the imposition is pregnant with political and philosophical implications.

There is a related argument, prosecuted with only slightly less indignation, that holds perspective to be one convention among many for the picturing of reality. The best-known contemporary exposition of this relativist view is in the philosopher Nelson Goodman's *Languages of Art* (1976). The earliest thematic exposition came from the art historian Erwin Panofsky in his essay "Die Perspektive als 'symbolische Form'" (1924),[8] which deals with ancient art, Byzantine, medieval, and trecento painting, as well as what was going on in Flanders and Germany during the fifteenth century. It is an attempt to begin a historical anthropology of pictorial practice. Each distinct culture has its distinct symbolic methods of mapping the world into pictures, and each method can divulge to the attentive historian a fund of intelligence about the mentality of its period or people. A generous, egalitarian motive lay behind the essay, the thrust of which was to demolish linear perspective's special claim to truth, a claim also attacked by Goodman.

There are two kinds of rejoinder to perspective's claims to truth: that it gives an incomplete picture, and that it gives a different picture. The first is the least contentious, pointing to all those aspects of visual experience that are excluded from perspective images, such as binocular vision, movement of the eye, focusing in depth, the full gradient of light and shade, and so on. In the face of these criticisms, the advocates of perspective have to retreat, relinquishing more and more of the complete experience of vision as they go. There is no doubt that perspective fails to tell the whole truth. The question is whether there is *any* truth in it. It is a vexed question. Critical philosophy has devised some pretty caustic agents for detaching perception, representation, and reality from one another. Goodman is in full possession of the chemistry set. Even so, he has some trouble demonstrating that there is no necessary connection between what perspective draws and what we see.

At first sight this seems the easiest task of all. Obviously a constructed perspective view on a flat sheet of paper is not the same as the full-bodied view in three dimensions. It is now well understood that the normal experience of perspective is an experience of a distorted image.[9] Unless we view the picture with one eye from a fixed point that corresponds exactly with the center of projection assumed by the artist when he made the picture, it will appear to a greater or lesser degree anamorphic. Its contour as perceived will not match the contour of what it represents. The advocates of perspective's truth have to retreat to the rock of fixed, immobilized, monocular vision. Goodman, having chased them there, says it is a simple matter to dispense with the one remaining defense but does not do so. In fact, as Michael Kubovy explains, Goodman's attempt to dislodge this "futile" claim reverts to discussion of the normal way we look at such perspective pictures as opposed to the way they should be looked at if we want them to provide an exactly corresponding image of the thing represented.[10] So why does Goodman fumble the *coup de grâce?*

The last stand of perspective takes place on the following proposition: A perspective picture is an instantaneous section of the straight lines that join a point in space with the salient features of the things pictured. The fidelity of geometrical perspective results from this being similar to the way that light enters the eye. The key word here is similar. Compare the two well-known illustrations of perspective machines in Dürer's *Underweysung der Messung.*[11] The behavior of the taut strand drawn from the nail in the wall to the contour of the lute, and the behavior of the line of sight that passes from the draftsman's eye through the framed grid and onward to the contour of the voluptuous woman he is studying, are sufficiently alike to make the geometrical description of the one event the same as that of the other.

The only straight lines are those that are idealized in geometry. Neither light nor bits of string travel in exactly straight lines. String is bent by gravitation; light is also bent by its passage through varying densities of atmosphere.[12] The only points are those in geometry. Neither an eye nor a nail will collect these not quite straight lines into a precise point. So Goodman is absolutely right when he says that perspective projection is a set of conventions (anything that is formulated geometrically must be), but this tells us nothing of perspective's capacity to render likeness. The behavior of these three analogical systems—the optical (Dürer's artist looking at his model through the gridded veil), the mechanical (Dürer's contraption for drawing a lute, which could be operated by a blind man), and the geometrical (linear perspective)—will not be identical. But for practical purposes they will be so similar that no appreciable difference will be noticed in the results obtained. And if that is so, then pictures abstracted by means of geometrical perspective will be sufficiently similar to substitute for direct vision, albeit arrested and radically reduced in scope.

Are there other means of representation that can claim to be closer approximations to what we see? Panofsky thought he had found several, and gave particular prominence to curvilinear perspective. However, he emphasized that even curvilinear perspective was necessarily imperfect.

63 Albrecht Dürer, perspective machine, *from* Underweysung der Messung, 1527.

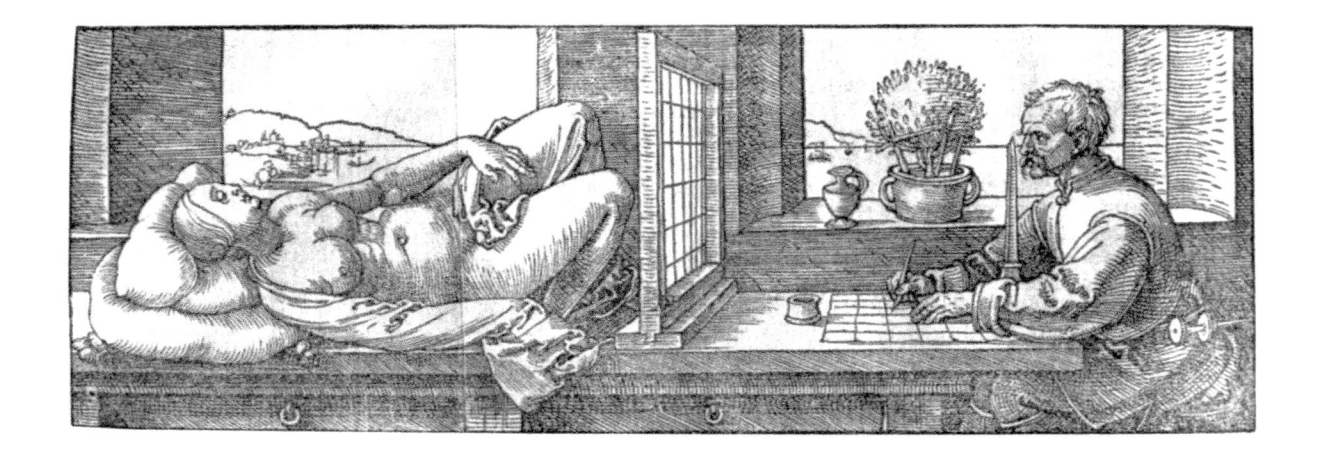

64 Albrecht Dürer, aids to draw the perspective of a reclining woman, *from* Underweysung der Messung.

Panofsky imagined that linear perspective does not give a true image because it is projected onto a plane surface, whereas the image that we see is projected onto the curved surface of the retina inside our eye. Here we have an instance of the strange tricks we can play on ourselves when we try to picture our own sight. Panofsky imagined the shape of the retinal surface as if this were really what he saw. He became a phantom looking at his own organ with another, interred eye. A "true" image would be an enlargement of what this second sight saw. But because a sphere cannot map into a plane surface without distortion, he thought no flat picture could be the final solution.[13]

The reasoning is false because perspective projection has nothing directly to do with either the physiology of vision or the psychology of vision. It deals only with how the pattern of light may be intercepted and copied before it gets into the eye, not with what the eye does with the light when it gets in there. It is blissfully simple: verisimilitude is judged by the degree of observed correspondence between two overlaid patterns. The observing eyes suck the image into our consciousness, but the correspondence takes place independent of our eyes and without reference to the shape of the "image" inside them. Dürer's two perspective machines make this absolutely clear. Anyway, if you think about it, the experience of vision is no more spherical than it is flat.

I do not wish to appear unsympathetic. Like Goodman, Merleau-Ponty, Lacan, and Foucault, like so many twentieth-century painters and architects, Panofsky wanted to find a means of escape from the tyranny of linear perspective. All I have done so far is close a few bolt holes without giving any reassurance that perspective is not the monster it is made out to be.

Why offer reassurance? Despite all the efforts to dislodge it from thought and art, perspective is more pervasive than ever because of its continuing power to provide striking likenesses in photographs, films, and videos, which are produced from electronic or mechanical analogues of perspective construction, just like Dürer's nail, string, and frame. Art and philosophy may wish to be excused, but we are still under perspective's sway, and will presumably remain so for some time. Neither is it obvious that perspective is operating its alleged tyranny through these media in the time-honored way.

Is perspective really the evil in our eye? Is it so hopelessly bound up with the philosophy and politics of domination? Is there not something indiscriminate, something of the witch hunt in the accusations against it?

How can anyone take seriously the lumping together of so many geometries—projective, Euclidean, and Cartesian—as accessories to the crime of perspective, unless as a symptom of distrust and animosity? Every kind of classical geometry has been condemned already for murdering intuition by substituting quantative for qualitative judgment. For the most part the case rests on this assignment of collective guilt. There is some truth in it (let us nevertheless stay wary of intuition). There is some truth in the complaint that perspective dominates and shapes vision; that it puts perception in the service of authority. There is some truth in all this, but it is not the whole truth.

The task, as I see it, is to discriminate. It is not all of geometry, nor even all of perspective, but only one general feature of perspective, plus one particular feature of one well-known perspective construction, that give perspective its bad reputation.

A culprit has already been named. The general feature of perspective construction just referred to is none other than the convergence of sight lines into a point. That, however, is not a mask imposed on vision by perspective, but a good approximation of the way light gets to our eyes. It led an independent career, as for example in Euclid's *Optics*,[14] before being coopted into perspective projection. Thus, while Descartes's *cogito* and Bentham's Panopticon were dependent on the convergence of light into the eye, they were conceivable without prior knowledge of Renaissance perspective. Whatever the relation, it was not one of concomitance.

The other culprit is the vanishing point, which, by contrast, is unthinkable without prior knowledge of Renaissance perspective.

The Vanishing Point

Perspective plays a large part in two histories: the history of painting and the history of geometry. In both histories the same property is singled out and considered to be of fundamental importance: the convergence of parallel lines into a vanishing point. For those unfamiliar with perspective construction it might seem that our second culprit is impossible to distinguish from our first. The pattern is the same—straight lines radiating from a point—and this has led to countless confusions and countless conceits from the early days of perspective to the present. In fact, though identical as images, they are very different functionally and conceptually.

To such an extent is the "crucial discovery"[15] of the vanishing point identified with linear perspective that the author of a sophisticated article in a learned journal can write: "An image constructed in accordance with the rules of linear perspective always has a centric point, that is, a point at which the orthogonals converge."[16] Thomas Frangenberg's assertion does not sound contentious, but it only remains true if we believe in fictions without functions. If there were no orthogonals (lines at right angles to the picture plane) either in the object mapped into perspective or in the construction used for mapping it, where would the centric point be? Well, we could say, it is there but you do not need it and you cannot see it. Likewise, if there were no parallels in the object mapped nor any converging parallels in the construction of the picture, where would the other vanishing points be (the centric point is a special case of the vanishing point)? If we say that these must be invisibly present too, because if there were parallels they would all converge in a point, then we might as well say that anything that need not be there and is not there could be there.

The method described by the painter Piero della Francesca in the last part of his treatise, the method he referred to as the "Other Method" *(altro modo)*, has no vanishing points. I intend to trace some of the implications of this straightforward but at the same time astonishing fact. In order to do this I will have to say something more about how the vanishing point affects the making of perspectives and how its prominence in discussions of the subject affects our understanding of perspective.

On the one hand the vanishing point is the source of numerous peculiar conceptual paradoxes. These recall the investigation of centrality in chapter one. Perspective did what theology aspired to, and what painting and architecture simulated, effortlessly amalgamating irreconcilable opposites. In some treatises the centric point is indicated by an eye drawn in the middle of the picture, signifying that it is the projected position of the private and particular organ of sight. Piero himself calls this point the eye *(l'occhio)*. In Jacques Androuet du Cerceau's *Leçons de perspective positive* (1576) an incongruous projection of the viewer's entire body is repeatedly thrown on top of the perspective construction to help indicate how the system works. In du Cerceau's hybrid projections the diverging lines from the eye are perfectly coincident with the converging parallels behind the picture plane.[17] The self-same centric point with the self-same radial lines indicates a localized point in the eye and an infinite distance on the horizon. The ambiguity is easily resolved by pointing out that the lines from the eye exist on the near side of the picture while the lines receding to infinity exist on the far side, but this does not explain why the lines on the far side should duplicate the cone of vision the way they do. Why should infinity be represented in exactly the same way as

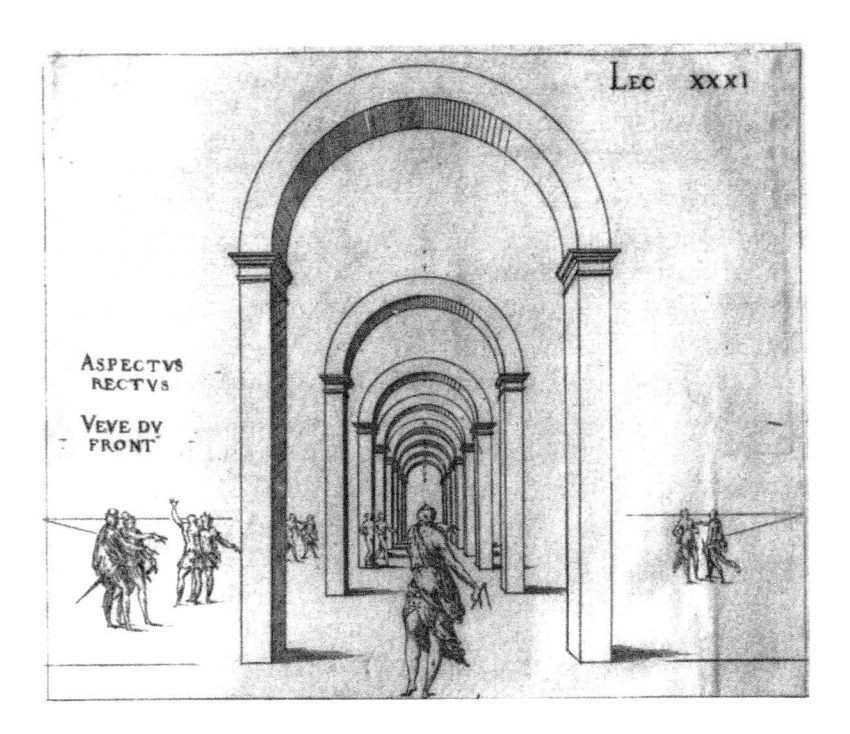

65 Jacques Androuet du Cerceau, perspective construction, from Leçons de perspective positive, *1576.*

a focus or center? It is not what one expects. It is contrary to intuition. Alberti skips over it with a few hedged words ("these lines drawn, as if to infinity . . .")[18] and proceeds to base his entire system of construction on it, whereas we now dwell on what he passed over. Listen to Norman Bryson on Raphael's *Marriage of the Virgin:*

> *All of the orthogonal lines across windows, doors and pavements converge there at the vanishing point where, par excellence, the viewer is not. The lines of the piazza race away towards this drain or black hole of otherness placed at the horizon, in a decentring that destroys the subject's unitary self-possession. The viewpoint and the vanishing point are inseparable: there is no viewpoint without vanishing point, and no vanishing point without viewing point. The self-possession of the viewing subject has built into it, therefore, the principle of its own abolition.*[19]

Only after 1600 did the vanishing point receive mathematical attention, various proofs being adduced by Guidobaldo del Monte (1600), Simon Stevin (1605), and Girard Desargues (1634),[20] but for nearly two centuries it had been impressed on the consciousness by visual means alone, pulling infinite alleys of virtual space behind innumerable pictures. What Renaissance artists accomplished in silence we look back on and celebrate with words. And it seems right that we should do so. Bryson's, Foucault's, and Damisch's eloquence on this same point helps us understand why.[21]

Whoever first recognized the convergence of parallels into a vanishing point (possibly Agatharcus in ancient times,[22] perhaps Brunelleschi in modern times—the evidence is unreliable) was making a momentous discovery. This property conflated the objective and the subjective, the universal and the idiocentric, the private and the public, expansion and contraction, truth and falsehood, and, as already noted, it undermined two fundamentals of Euclidean geometry (that parallels never meet, and that congruent figures stay the same shape when moved). In these intellectual senses, the vanishing point construction of perspective could be disturbing in its implications. On the other hand, while it may have pushed certain ideas awry, it marshaled the image of the world and shoved it into line. It turns out that Bryson's existentially perplexed subject has to occupy, in imagination at least, a very neat and orderly space; a space in which tidily arranged bits of architecture (walls, windows, doors, and pavements) provide all the pointers from the decentred subject to the black hole of otherness, which is ironic.

It is also important. In Renaissance painting the relation between human presence and perspective scenery is polarized in other ways too. What else can be drawn in perspective except architecture, furniture, and a few other man-made objects; what else, in other words, but things already defined and made according to the regulative lines and planes of classical geometry? The answer is everything and nothing. Nothing else can be *projected* by Alberti's construction, unless previously reduced to a reticulated armature. But everything else can be sketched in the boxes thus provided. Hubert Damisch asks us to consider clouds, an extreme case.[23] There is a good deal to fill the gap between Damisch's diaphanous clouds and Alberti's pavement, such as: clothing, rocks, vegetation, animals, and of course men and women.

Are these not the real subjects of painting? Does Alberti not anyway make this clear by spending more time telling painters how to show figures expressing emotions in their faces and actions in their bodies than on perspective? Also, if you look at Florentine artists' sketchbooks from the fifteenth century, you find few perspective constructions. Their studies of figures and drapery were accomplished without it.[24] Nevertheless, Alberti begins with perspective because one has to build the house before people can live in it. Alberti's perspective is a preparation for figure compositions, and that is how it was treated by the many painters who used

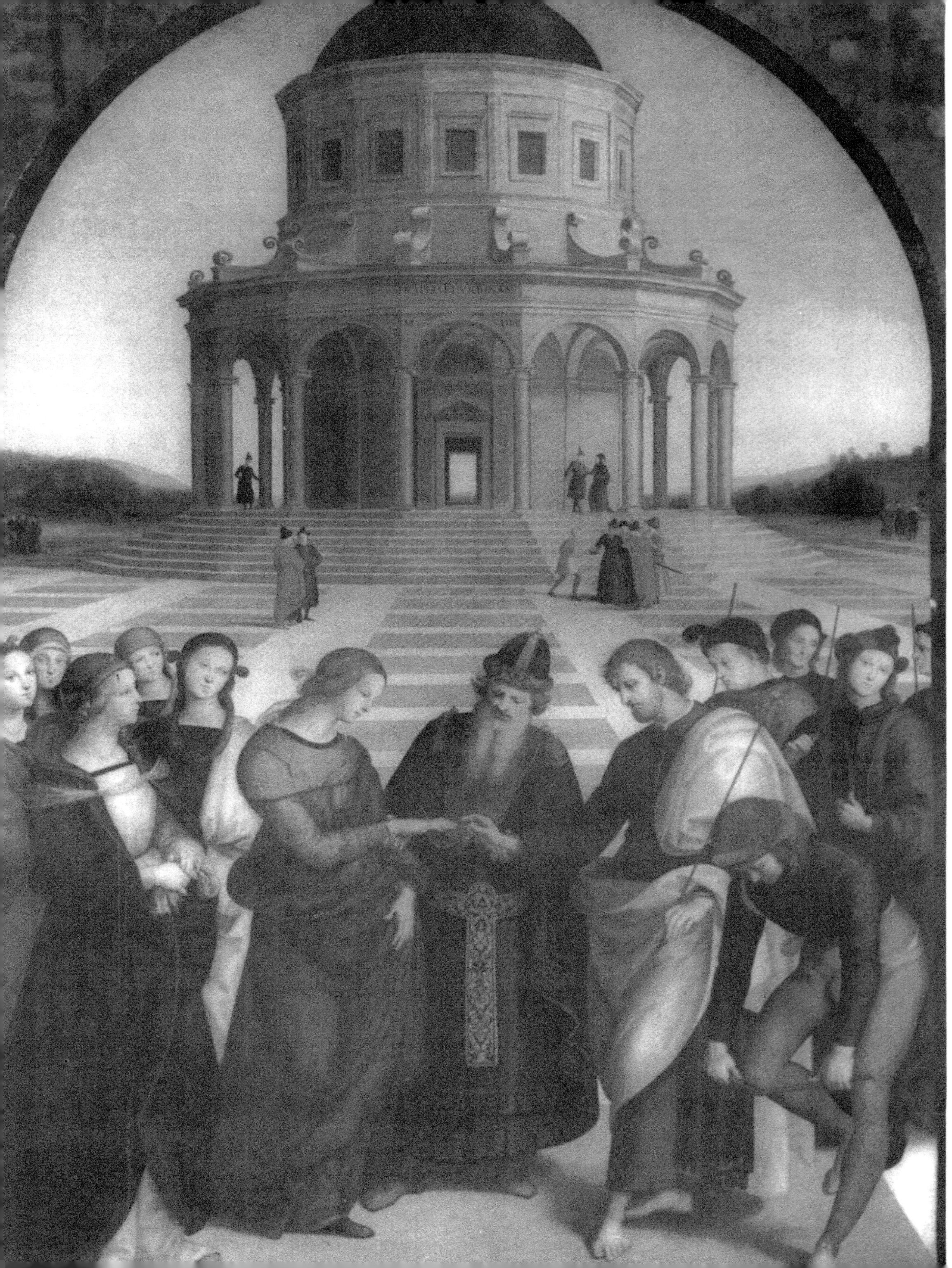

66 Raphael Sanzio, Marriage of the Virgin, *1504, Brera, Milan.*

it, although it took some time to get the occupants to move into the accommodation that perspective had prepared for them. Often, as indeed in Raphael's *Marriage of the Virgin* (an early work), the figures queue up to do their business just behind the frame of the picture, leaving the architecture as a backdrop largely empty of human drama.[25] With a slice of commotion in front and a telescope of vacant buildings behind, the story and the setting could be treated independently. Perspective might just as well have been added after as prepared before, and such must have been the case in at least one of Perugino's two paintings of the Pietà, which have different perspective backgrounds but identical figure compositions.[26]

But people eventually began to move in. A more consequential relation between perspective setting and figure composition is evident in Leonardo's drawing for, and incomplete painting of, the *Adoration of the Magi* (1481). The tessellated pavement is precisely and accurately inscribed in the drawing. Firmly outlined stairs and part of an arcade rise from the pavement. Then come the figures in faint wisps and splashes of sepia ink, over the floor, up trees, on the stairs and in the galleries, born into perspective space but not themselves determined by means of perspective projection; the quick rising from the dead. By the time Leonardo starts to paint, the balance has completely altered. The figures are now dominant; figures solidified but still mobile and vivacious, making full use of the space around them, "with a variety of gesture and emotion that is quite new to Italian art,"[27] to quote a recent assessment. The tessellated pavement has disappeared under a covering of earth. The only remaining clues to the originally pervasive central projection are the ruined vestiges of architecture in the background, which still tell us the way to the obscured vanishing point. Recall Leonardo's map of Imola, where the shape of the system of measurement was clearly distinguished from the shapes of the things it measured. Twenty years before he made the map, Leonardo was already making good use of a similar distinction in perspective painting.

The construction grid was an enabling format hardly discernible in the finished work, from which we might conclude that central projection, in practice, was a relentless ordering of one thing that helped emancipate another. Artists think of themselves as free but they still work on flat surfaces. They borrow their freedom from within a geometrical imprisonment. Alberti's perspective gave them a second order of freedom within a second order of confinement. Inside the rigid cage of central perspective they developed unbridled, graceful bodies, conceived without any recourse to geometry but directly dependent on it for the intuition

67 Leonardo da Vinci, Adoration of the Magi, *drawing, 1481, Uffizi, Florence.*

of their liberated form. The later painting of Leonardo, Raphael, and Michelangelo, what Vasari called the *terza maniera*,[28] was therefore, at one and the same time, a relinquishing of perspective and perspective's most profound triumph over the image.

Renaissance perspective enabled the release of the human figure, previously immured in the surface of the picture, but what did it do for the setting? Unless evasive action was taken, this turned into architecture, because it was a virtual architecture already, even before it was a thing. What could be easier than to turn a grid into a pavement? What could be easier than to allow the line that Alberti called "the prince of rays,"[29] coursing from the eye of the beholder to the centric point, to career on through the pictorial space as an axis of symmetry or to open up an avenue in its path, since this focus-cum-vista was already intimated in the preparatory construction?

In a premonition of central perspective, Lucretius, in the first century B.C., observed the apparent convergence of parallels when looking down a straight colonnade.[30] Architecture led to the vanishing point which led to architecture. But was the painted architecture of the Renaissance taken from the real world? Obviously not. After Brunelleschi's celebrated demonstrations, made sometime between 1413 and 1425, show-

68 *Leonardo da Vinci*, Adoration of the
Magi, *c. 1481, Uffizi, Florence.*

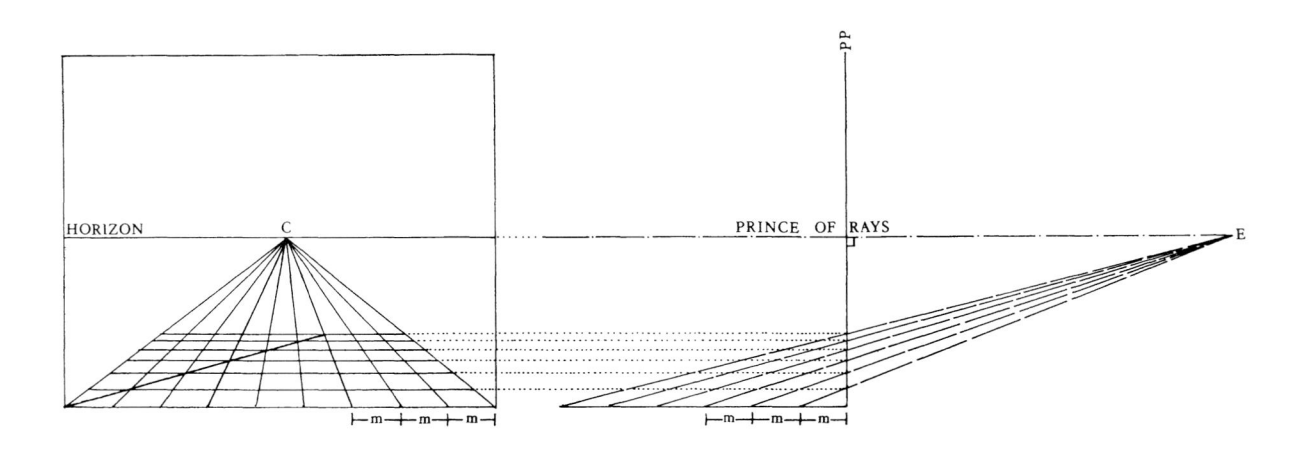

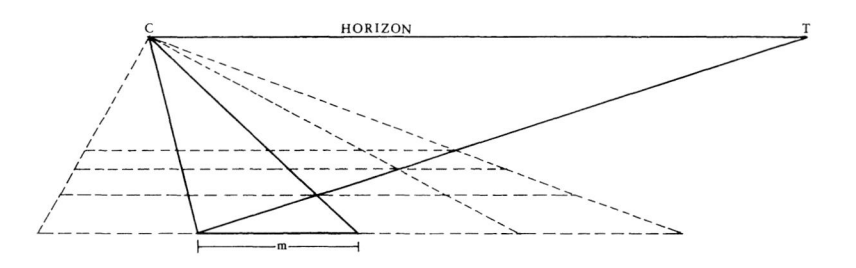

ing the Florence baptistery and the Piazza della Signoria, in which the perspective panels were held in front of one eye to substitute for the actual views,[31] we might expect a spate of topographical paintings to exhibit the verisimilitude of the new technique. If such there were, they have disappeared without trace. No, instead we have inventions and fantasies. In one important respect the architecture was less well connected to the existing world behind the picture than were the figures, because the figures, though largely inventions, were based on closely observed studies from life, whereas most of the architecture was not. So if the architecture did not come from the world at large, where did it come from? From the painters' imagination? To some extent yes, but also from the technique of perspective construction.

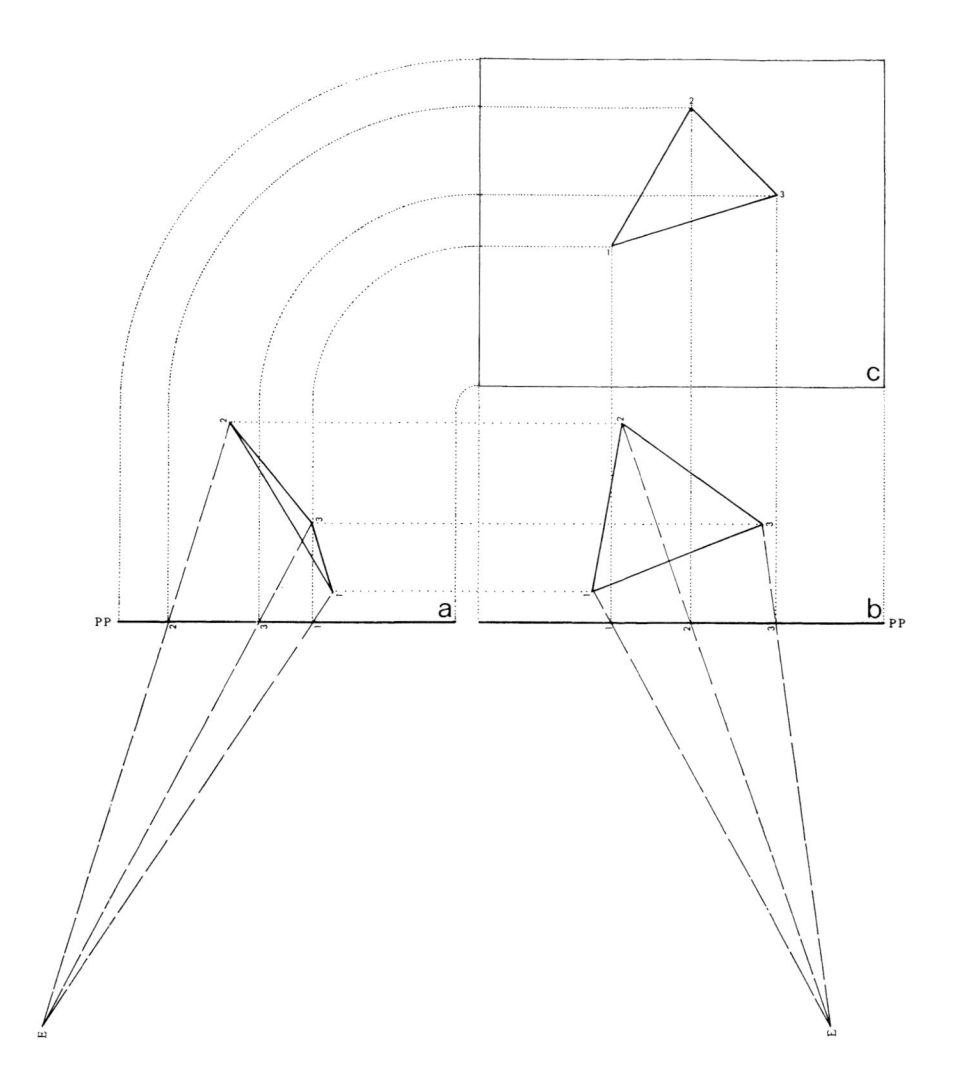

69 Three early methods of perspective construction.

E = eye; C = centric point; PP = picture plane; m = tile module. Diagram 1 is Alberti's method for making a perspective of a square tiled floor. Depths of tile rows are determined by mapping the intersections of sight lines with the picture plane in the equivalent of a side elevation, then transferring the heights of the intersections onto the perspective picture. The result is verified by drawing a diagonal continuing through several tiles; it should be straight. Diagram 2 is Pélerin's method, which also depicts a square tiled floor. T = the diagonal vanishing point (or tiers point), which can be put anywhere on the horizon line. The length CT then defines the distance of the eye (or center of projection) perpendicular to C. Diagram 3, Piero's Other Method, makes a perspective of points in space, in this instance a plane triangle 1, 2, 3. Here a is a side elevation of the triangle, picture plane, and eye, b is a plan, and c is a front elevation of the picture plane with the perspective image of the triangle mapped onto it. (Drawn by the author.)

Practically every published demonstration of perspective shows the image traveling from behind the picture to impinge on its rear surface, whereas I am saying that the image usually came from everywhere *except* that direction. The perspective treatises contain propaganda to persuade us that the technique records the real. They show pictures of objects with an independent existence captured in the searchlight of perspective rays. Perspective can indeed report direct from reality. That is what was going on in Dürer's machines and in Brunelleschi's demonstrations, but that is not often what was going on in Renaissance painting.

One early treatise drew attention to this. The intriguing thing about Jean Pélerin's construction in *De artificiali perspectiva* (1505) was that he did not bother to maintain the illusion that the lines from the eye had to extend through the back of the picture.[32] Pélerin's construction (called the tiers point method, a modification of Alberti's) requires no information from outside the picture plane. Using his method, the world, or rather, something like the world, is dreamt up in a sheet of paper. And looking at his plates we notice that it is an unsurprising world composed of tiled floors, lots of buildings, and some conventionalized outlines of figures and trees.[33]

Pélerin's construction renders the world on both sides of the picture as an inference to be drawn from within it, showing conclusively that whatever an artist using measured perspective is tied to, it is not the objective data from the real world. It shows with equal force that the artist is tied to something, namely the method of construction. Everything else points this way too: it is not perspective as such but the particular means of its construction that gives bias to its contents.

Perspective, as made in Dürer's machines, or as witnessed in Brunelleschi's deceptive substitutes for actual scenes, comprises a set of conventions that have a necessary relation to some reality existing beyond. Alberti's construction is a set of conventions *without* any necessary relation to reality. Using it, we can go fishing for real objects to catch within its net. We can also use the net to make imaginary objects that exist only in the picture. These phantoms, so very easy to produce as long as they stay in line and follow directions, are contrivances of the technique. Once the phantoms have been formed in the image of their construction, it is possible to make real things in the image of the phantoms. Alberti's perspective has the power to make reality in its image precisely because it is not like reality. If reality were already arranged that way, his perspective could have had no impact, and the issue of its power would not arise. In this sense the real world has latterly recorded the image implicit in Albertian perspective, standing the accepted relation on its head.

When critics and historians say that Renaissance architecture was perspectival, they usually mean that the axial vistas of Renaissance architecture were derived from Renaissance perspective compositions. A recent exposition of this idea by Lise Bek maintains that our modern conceptions of architectural space come from Alberti's framed, orthogonal, telescopic vision. She instances the axial vistas in Alberti's own architecture, the new development of axiality in domestic buildings, such as Bernardo Rossellino's Palazzo Piccolomini at Pienza, the Villa Reale at Poggio a Caiano, Raphael's Villa Madama and Vignola's Villa Giulia in Rome, and then Palladio's villas. She notices that the insistent vistas through these buildings and landscapes usually were identical to lines of human movement such as avenues, colonnades, and enfilades of doors, and that the scenery to either side of these optical channels was treated as, ideally, symmetrical.[34] All are pertinent observations. She realizes that the architects who instituted these arrangements soon began to play with their mitigation, as had painters. In numerous perspective paintings the extended view was diverted into two oblique, lateral vistas, while the center was blocked with an aedicule, a wall, or a figure. So too in architecture, paths of movement sometimes parted from an axis of vision, weaving in and out of correspondence with it. The superb garden architecture of the Villa Giulia is a case in point.

But when it comes down to it, the new synthesis of movement, vista, and symmetry is under suspicion as a political strategem projected from canvases into the cities and palaces of Europe. Perspective became a potential effectuator of power. We could perhaps work this out for ourselves, but Alberti gives us a little help. In the *Ten Books on Architecture* he compares the city layouts appropriate for a republic and a tyranny. A tyrant should divide the districts of his city with great impenetrable walls overlooking all the houses, walls that are also thoroughfares to convey his army from his fortress; while a republic is best provided with a continuous network of winding streets in which all the citizens' houses have extended local views. The two kinds of government require different terrains, and the differences reside in the way surveyability and accessibility are centralized and unified in the one, or multiplied and mutually linked in the other.[35] When any vista cuts through the lives of others it is a political instrument of a certain complexion. And when architecture petrifies its passage and forces movement and vision into the same privileged paths, even more so is architecture politicized.

These observations are reminiscent of the accusations cited earlier, except that they are now more definitely linked to the practice of a specific construction: Alberti's. For only Alberti's construction, and its variants, emphasize the axial vista; only they are dominated by "the prince of rays." The title conferred on this line is certainly appropriate.

That is how I see it. Lise Bek sees it differently. At first she blames Alberti, but things got worse. Alberti, she thinks, was not so bad after all when compared to Piero della Francesca: "What had been a mere base for Alberti becomes essential in Piero's theory and the Albertian storia with its humanistic didactic rhetoric must here submit to geometry."[36] She believes that the more geometry there is in perspective the more it will dominate the picture and then deliver its domination to the world outside the picture. Think of how immobile everything became in Piero's painting. Even his figures are like statues, like columns, like architecture, like geometry.

Piero's Reputation[37]

Piero's reputation has waxed while perspective's has waned. The twentieth century is the period of his critical apotheosis as it has been the period of perspective's critical demise. He used to be regarded as a skillful but not greatly inspired painter, a perception that extends back to Vasari, who put Piero first in his list of offenders guilty of the "dry, hard, harsh style" that came from the excessive study of perspective: "These artists forced themselves to try and do the impossible through their exertions, especially in their ugly foreshortenings and perspectives which were as disagreeable to look at as they were difficult to do."[38] Their work was inimical to liveliness and expression, qualities later perfected by Leonardo and Raphael according to Vasari, as we have just seen. Piero's fortunes have changed, but the same reservation lingers on. Even his greatest admirers admit that there are times when Piero's art is overborne with mathematics and geometry. "Unfortunately," wrote Bernard Berenson, "he did not always avail himself of his highest gifts. At times you feel him to have been clogged by his science."[39] Kenneth Clark tells of the painter's decline into mathematics during the last years of his life, so that one is no longer sure whether Piero's failure of vision was due to blindness, as Vasari records, or geometry.[40]

There is an equally persistent contrary strain of opinion from those who see science as the guide of art,[41] but any argument between those for and against its influence is contained within a common presumption about the way the mathematical interest expresses itself. What Piero does, according to this common presumption, is depict an imaginary world, more ideal than the one we occupy, in which all things reveal their resemblance to regular geometric solids. Piero, "whose name," it has been said, "is synonymous with the study of regular bodies,"[42] insinuates their geometry into everything, and perspective is the medium of insinuation. Piero's painting is seen therefore as the illustration of one kind of geometry by means of another:

> *Perspective . . . causes Piero to organize everything within the most simple and monumental outlines. It is as if every object must become one of the five regular bodies in order to serve the perspective construction. This construction is central, arranged on planes converging on an ideal axis.*[43]

So said Roberto Longhi in 1913, and many have since said that they see the same.[44]

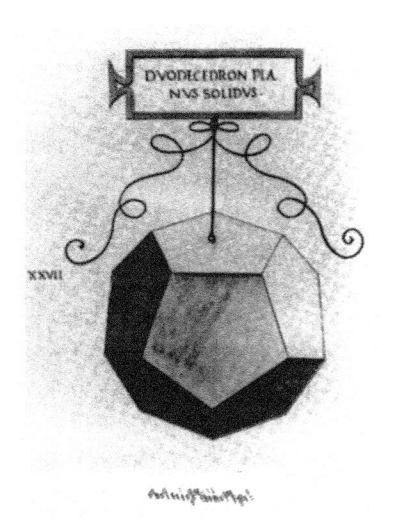

I do not see these regular bodies inside the human bodies painted by Piero. I see bodies that are less complicated, less animated, certainly more stolid, but not more regular. Nor do I recognize an ideal axis in all Piero's paintings. Sometimes I am made more conscious of the absence of any such line. I would suggest that Longhi's perceptions, turned into critical clichés, have been imposed on the paintings because of what we think we know about Piero's mathematical studies. We know that Piero produced the most rigorous treatise on linear perspective from the Renaissance period, *De prospectiva pingendi*, as well as an accomplished work on the geometry of the regular solids, *Libellus de quinque corporibus regularibus*,[45] so the properties assumed to belong to these studies are projected onto the paintings, which seem to affirm their origin by a kind of sympathy that exists between the suspended stillness of the geometrical diagrams and the motionless human figures in the paintings.

70 *Solid dodecahedron, from Luca Pacioli,* De divina proportione, *1496.*

This persistent idea about the geometry inherent in Piero's work cannot account for our more appreciative reception of it. Berenson's marvelous passages on impersonality in art, from *The Italian Painters of the Renaissance* (1897),[46] began the approach from a different direction. He noticed the peculiar, inverse relation between expression and expressiveness in the paintings. Piero's figures show no emotion, yet they arouse strong responses in us. Berenson sought to describe and explain these feelings, as have succeeding generations of critics. Michael Baxandall, for example, was intrigued by the way in which Piero's figures, which look so static even when portrayed in movement, evoke tenderness, perhaps because their postures are reminiscent of children learning to walk, always seeking stability. This suggestion is of course Baxandall's own, and he describes very well how these paintings, apparently conceived as virtuoso performances without expressive intent, induce speculative emotion in the observer seeking to account for what is absent, or maybe suppressed: "And perhaps this is the final paradox of Piero; that what we describe as rich technical sophistication on his side becomes on our side the matter of such rich emotional experience. His means somehow become our ends, and no one can say quite how."[47]

Nowhere has this dumb solicitation provoked more fervent response than in the recent discussion of the *Flagellation*. In this painting the scourging of Christ takes place in the background, diminished by perspective as if incidental, while three figures paying no heed to the event, and little heed to each other, stand large in the foreground. The painting looks perfectly proportionate in its perspective and vastly disproportionate in the narration of its subject, because what we expect to be prominent is not, and what we expect to be the gruesome center of attention is not, while the incidental scenery (the architecture that Creighton Gilbert declared to be the real subject of the painting),[48] and some bystanders, take the stage. The bystanders, who have to be acknowledged as the principal figures, look elsewhere, their mind on something else. This narrative anamorphosis is presumed to be explicable, so critics and historians of art have taken upon themselves the task of writing a story worthy of it.

Marilyn Lavin's monograph *The Flagellation* (1972) can be read in this light. Since its publication a war has raged around it. Historical evidence has been refuted, sources reassessed, judgments impugned, and motives questioned, but, it seems to me, everything that has followed is heavily dependent on Lavin's original way of making sense of the picture, however questionable.[49] She sees the painting as setting the seal on a reconciliation between two powerful political figures whose sons had suffered death and disfigurement. The painting alludes to the shared grief of the fathers, who stand in the foreground with an angelic youth. The flagellation is an allegory of the cruelties visited on the sons. The poignancy of Lavin's interpretation is bound up with the indirectness of its expression in the painting, bound up, that is, with the passivity of those portrayed. The problem has since been reduced to identifying who was who amongst the foreground figures. But whoever was who, what extraordinary state of consciousness could account for their combination of self-possession and distraction, their inward concentration and remoteness, as a harrowing torture commences? The "enigma" of Piero is called forth by the painting itself, not by lacunae in its documentation as so many of Lavin's critics seem to believe.[50]

It has recently been claimed that Piero is not an arid formalist but a great narrative painter.[51] In a way this is true; his paintings enlist us as their

71 Piero della Francesca, **The Flagellation of Christ**, *Galleria Nazionale delle Marche, Urbino.*

narrators. Something in the figures, who refuse to act out their assigned parts, stimulates us to make sense of what is plausible but mystifying.

I am emphasizing more recent appreciation of the emotive power in Piero's art not only because it is at variance with Vasari's judgment, but because it underlines the question posed by Baxandall. To what extent is emotion *transmitted via* Piero's technique, and to what extent *created by* the technique? The question will fall like either a brick or a gauntlet. Everyone knows that art is more than technique; but art must involve technique as more than an obedient instrument of intention, otherwise the accomplishments of art will not transcend intentions either, in which case there would be little point in venturing further than intentions.

Art often conspires to cheat us of our expectations. We wait expectantly for the conspiracies to hatch between artists and the things artists use. These manufactured things that artists use already have art in them: red pigment, straight lines, sheets of canvas, musical instruments, words, sentence structures, reinforced concrete, and perspective construction, for example. The aspects of art that are not subject to predetermination are those that arise out of the interplay between fabricator and fabric. By itself, the "idea" of the artist is nothing.

Does this perhaps explain the improvidence of so much that is still written on Piero, where we are offered either verbal or geometric ideas as sufficient explanations? Iconographic treasure hunters search for an original idea that will reveal what the paintings really meant. To make sense of the hunt, they have to assume that the meaning was there before the painting began. The dispute over the identity of the three figures in the *Flagellation* has become like this. Like this also is the endless wrangling over the egg that hangs suspended in the Brera altarpiece. Whatever its meaning, it could make little impact on our understanding of the painting in which it is so small an incident. The problem is not that these items receive attention, but that receiving so much, they substitute for the paintings and become the content of the work.

Then there are the lines superimposed on the paintings to advertise the presence of rational proportions, golden ratios, hidden rectangles, and central perspective constructions—always *central* perspective. For every contribution on the identity of the three persons in the foreground of the *Flagellation* there is another on the perspective construction and the real-measure proportions of its architecture.[52] Thus intelligence is divided, passing to either side of the shape and deportment of Piero's figures, which tell a different story.

72 Luca Cambiaso, pen and ink figure study, Uffizi, Florence.

So, back to the figures and their moving immobility. How were they made? There are two sources of evidence for this, indicating different methods: Piero's perspective treatise and Vasari's biography. It is a platitude that Piero's figures are statuesque, and so Vasari's brief notice in passing that Piero "was . . . very fond of making clay models which he would drape with wet cloths arranged in innumerable folds, and then use for drawing and similar purposes"[53] may be suggested as the cause of this effect. If Vasari's information was reliable, then Piero's repeated choice of so lumpish and unleavened a medium must have been fueled by the desire to achieve gravity in the results. Alberti thought that painting should "show the movements of the soul by the movements of the body,"[54] but Piero showed the movements of the soul by the stillness of the body.

The evidence from Piero's perspective treatise, *De prospectiva pingendi*, is his study of a human head—an unusual subject, though not unique. Human figures are recalcitrant to direct survey by Alberti's linear perspective, but that did not prevent occasional attempts to render them surveyable. Alberti himself spoke of the body as an assembly of faceted planes, which, by implication, could be recorded with the same perspective precision as pavements.[55] Several artists, including Albrecht Dürer, Erhard Schön, and Luca Cambiaso, tried to realize such figures, though more in emulation of measured projection than as examples of it.[56] Piero went much further. To find out how much further we only have to look at the series of 25 drawings toward the end of *De prospectiva pingendi*.

Piero's Perspective

Momentous things are contained in this treatise "written for idiots," to quote Daniele Barbaro (who nevertheless borrowed a good portion of it for his own book on the same subject), but the only way of telling is from the treatise itself, which has never received its due. Indeed, the degree to which it has been dismissed, ignored, and misunderstood is nothing short of scandalous. Passed over in the history of projective geometry, discoveries contained within it attributed to later figures,[57] passed over in the vast literature on perspective in painting, passed over in studies of Piero's art, it is generally considered as a long-winded codification of limited theoretical or practical interest, impressive only for its labored thoroughness. Even Boskovits, a perceptive and informed writer, concludes that while it is a remarkable document full of interest for the historian of projective geometry, it "had only very loose ties with artistic work," and complains that because Piero wrote nothing about composition or beauty, we cannot understand how they relate to his perspective methods.[58]

I maintain, on the contrary, that we can understand its significance if we look at the drawings a bit more, and spend less time on the admittedly tendentious text. I shall try to show, moreover, that Piero's development of perspective geometry affected the nature of his painting, not in the ways normally supposed, but in ways we can only appreciate if we understand the difference between Piero's perspective and Alberti's.

In his introduction Piero says perspective is composed of five elements: the eye, the form of the object seen, the distance between the eye and the object, the lines that join the eye to the outermost edges of the object, and finally the surface of the picture.[59] Everything within this system is confined between eye and object. There is no reference to, no intimation of, that staggering, metaphysical void that opens up beyond Alberti's construction. This is not a matter of words. Alberti said almost nothing about the vanishing point and then used it. Piero avoids saying anything about it and avoids using it too. At first he gingerly acknowledges the convergence of parallels, then he dispenses with it altogether.

There are no sure dates for the compilation or completion of *De prospectiva pingendi*. Like Piero's exclusively mathematical works, the *Trattato d'abaco* and the *Libellus de quinque corporibus regularibus*, it was done late in his career, although before the *Libellus*, in which it is mentioned. In all probability Piero was working on it during the 1470s or the very beginning of the 1480s. The earliest dating of *De prospectiva pingendi* would still place it after almost all of Piero's known paintings, so we cannot assume that any of the techniques and insights that set it apart from Alberti's construction were available to Piero as a painter, since the techniques and insights may only have come to light during his late study of perspective geometry. We are obliged therefore to work with the gift of hindsight, but without presuming the fullness of Piero's prior knowledge. His lucid visual exposition of the "Other Method" allows us to see how radically its wholesale use would have altered the character, composition, and space of Renaissance art. Once this potential difference is established, we will know what to look out for. It will then be possible to surmise what use Piero made of the Other Method in his own painting and, furthermore, to suggest that the same construction without vanishing points had been employed by Uccello.

De prospectiva pingendi is divided into three books. The first describes the projection of horizontal plane figures using an adaption of Alberti's central perspective. The second describes how these may be extended vertically to represent three-dimensional bodies. The third describes the Other Method.

In practical terms the main difference between Piero's central perspective and Alberti's is that Piero instructs the reader on how to twist the items he wishes to draw out of alignment with the picture plane, whereas Alberti does not. Alberti treats his perspective grid as an endless pavement, whereas Piero treats his like a doormat folded up behind the picture plane. In theoretical terms the main difference is that Piero, even when dealing with central perspective, concentrates on the local relations among eye, picture plane, and object. He persists throughout in describing all orthogonal foreshortening in terms of proportional ratios, not in terms of converging parallels. He employs convergence when convenient, but refuses to acknowledge it as anything but a shortcut.[60]

Still, the culminating examples of central perspective in *De prospectiva pingendi* are buildings aligned with the picture plane—the kind of archi-

73 Piero della Francesca, De prospectiva pingendi, *fig. xxix, plan-and-vanishing-point construction of an unaligned octagon.*

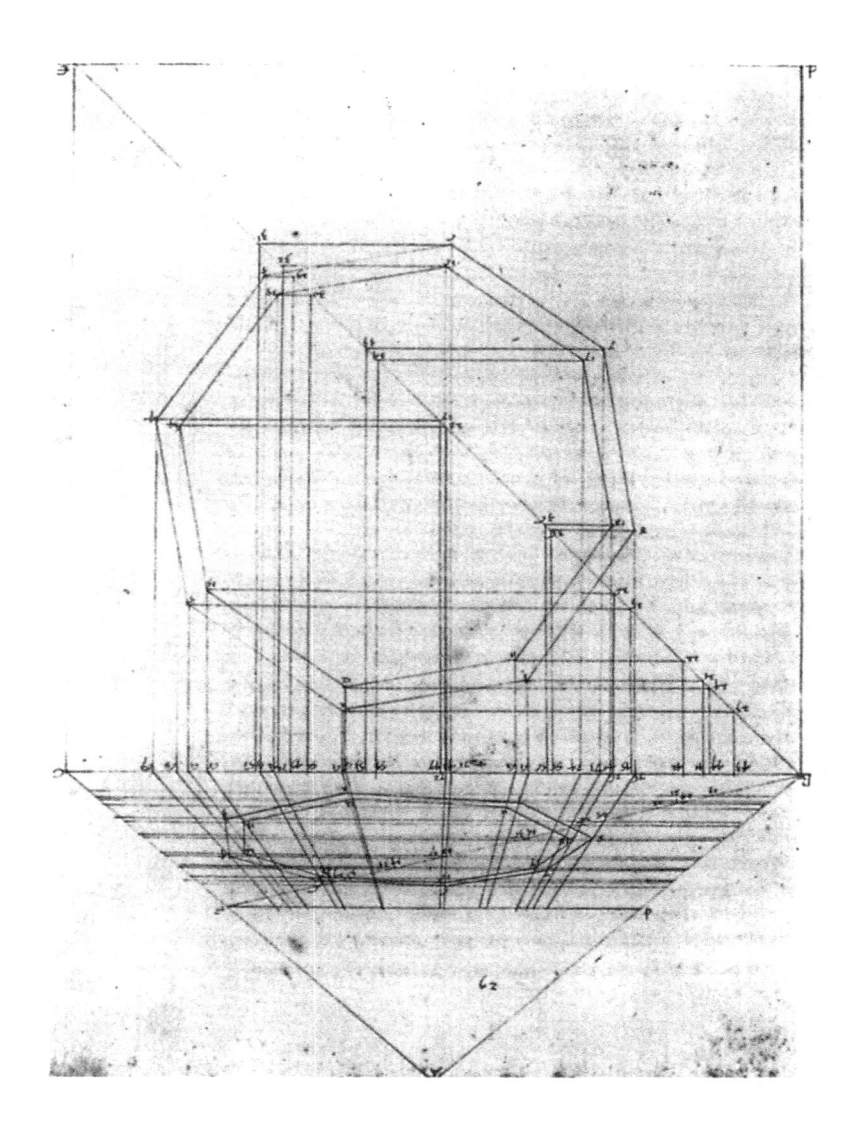

tecture that springs easily into being from the oriented Albertian void. Using Piero's version, items of more varied geometry can be projected, but a practical limit is soon reached. Piero then explains why it has become necessary to introduce the Other Method:

> *Now in this third [book] I intend to treat of the foreshortening of bodies composed of diverse surfaces variously positioned, but having to deal with more difficult things, I shall follow another route and use an Other Method for their foreshortening, which I did not use in the preceding demonstrations; but in effect it is the same thing, and what results from the one [method], results from the other. But for two reasons I shall change the foregoing rules; the one is because things will be easier to demonstrate and to understand; the other is because of the great multi-*

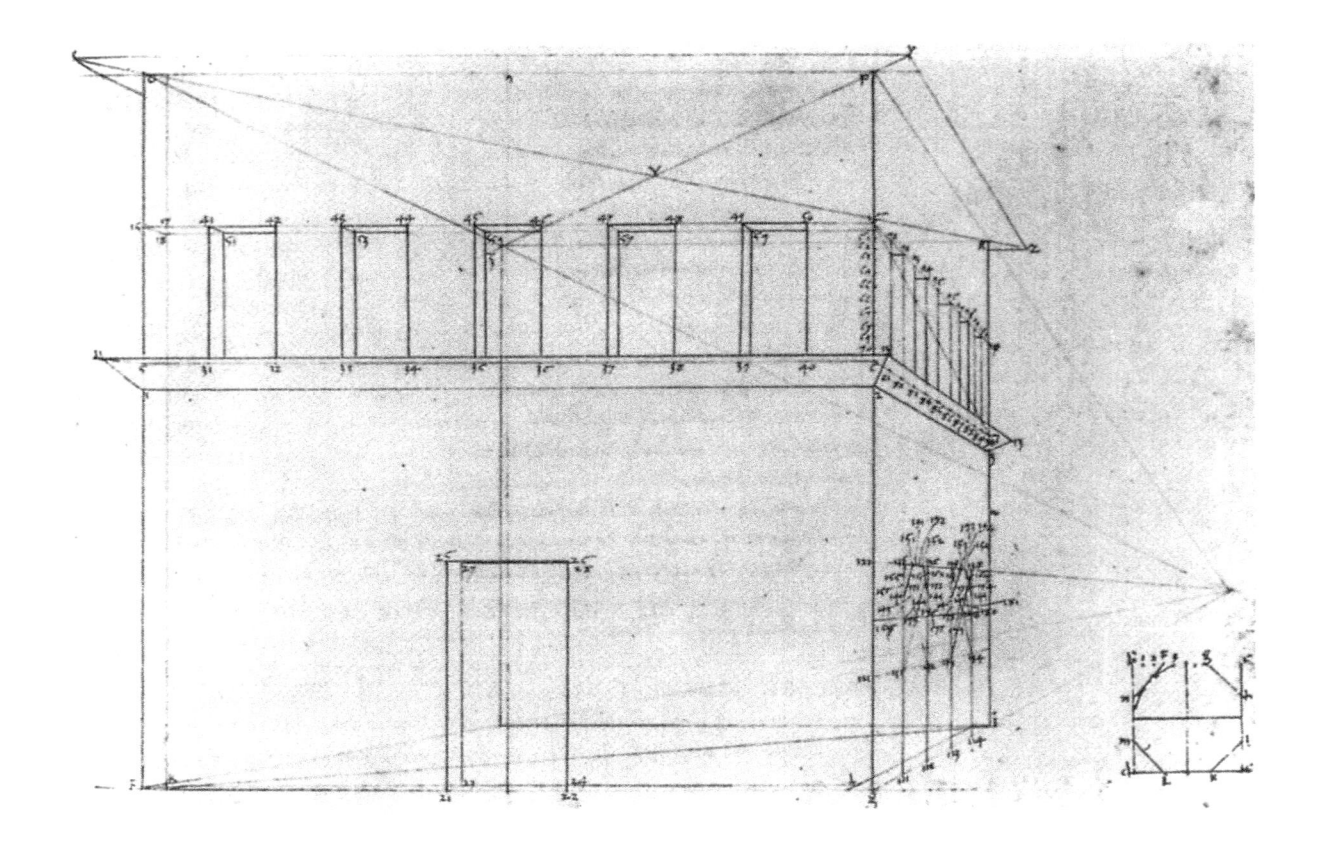

74 Piero, **De prospectiva pingendi**, *fig. xlii, a building projected with the aid of a vanishing point.*

tude of lines that would be necessary to make these bodies following the first method, so that the eye and the mind would be confused by these lines without which these bodies could not be foreshortened properly, nor without great difficulty.[61]

The easiest way to explain the Other Method is with the aid of Dürer's machine for making a perspective of a lute; it imitates with drawings what Dürer's machine did with weighted line, nail, frames, and cursors. Piero begins with a plan and a side elevation of what he wishes to draw in perspective. He includes the picture plane and the position of the eye in both. On these same two drawings, sight lines are taken from the eye to corresponding points on the object (lute, cube, or whatever). These will intersect with the two lines representing the top and side of the picture. To obviate the necessity of tracing innumerable lines to represent the rays of light into the eye, Piero recommends stretching long filaments, such as hairs from the tail of a horse, from nails fixed in the eye positions, then marking intersections with the picture plane using numbers to correlate them. The vertical and horizontal coordinates obtained from the plan and side elevation of the picture plane are then mapped onto a front elevation, which is, of course, the surface of the picture. This produces an array of dots, as did the cursors in Dürer's machine. Join up the dots and there is the image. The whole system is conceived in terms of plan, side elevation, and front elevation.[62]

Everything about the Other Method is so straightforward, so directly related to local relations between tangible things. All the paradoxes fly out of the window, including the vanishing point. But all this straightforwardness has been bought with two other giant abstractions: orthographic projection and the dissolution of surfaces into constellations of dots.

Piero's treatment of orthographic projection is extraordinary. His is the earliest recorded account of it, and also a remarkably sophisticated account that stands comparison with much later expositions, presaging, it has been said, the descriptive geometry of Gaspard Monge, devised at the end of the eighteenth century.[63] On the other hand, all this is incidental and taken for granted in the text, where Piero refers to objects drawn that way as "put in true form" *(ponne in propria forma)*, without saying how they might be put in it. It is seldom observed that there is no perspective *projection* in Piero's Other Method for perspective. There is a perspective *result* that is achieved entirely by orthographic means—just like architecture.[64] Many are aware that perspective was first described by architects; few are aware that architectural drawing was first described by a painter.

Piero's Other Method makes pictures of light paths between points in exactly the same way that architects make pictures of buildings. A statement such as this may lead us to think that we have found a handy explanation for the monumental aspect of Piero's painting, so reminiscent of architecture. Perhaps we have, in part, but the matter needs further elucidation, because in one important respect his method diverges from normal architectural usage.

Mention has been made already of the alliance between Alberti's perspective and architectural drawing. Rectangular, frontal, symmetrical, and axial forms are easily engendered from either. None of these attributes arrives so automatically using Piero's Other Method, except frontal orientation, and even this Piero soon instructs us on how to remove. Piero's achievement was to separate the form of the object from the form of its projection. The separation was not a complete divorce. The manner of projection still leaves its traces, but a certain zone of liberty was found. Here again the comparison with Alberti's method is instructive. Painters

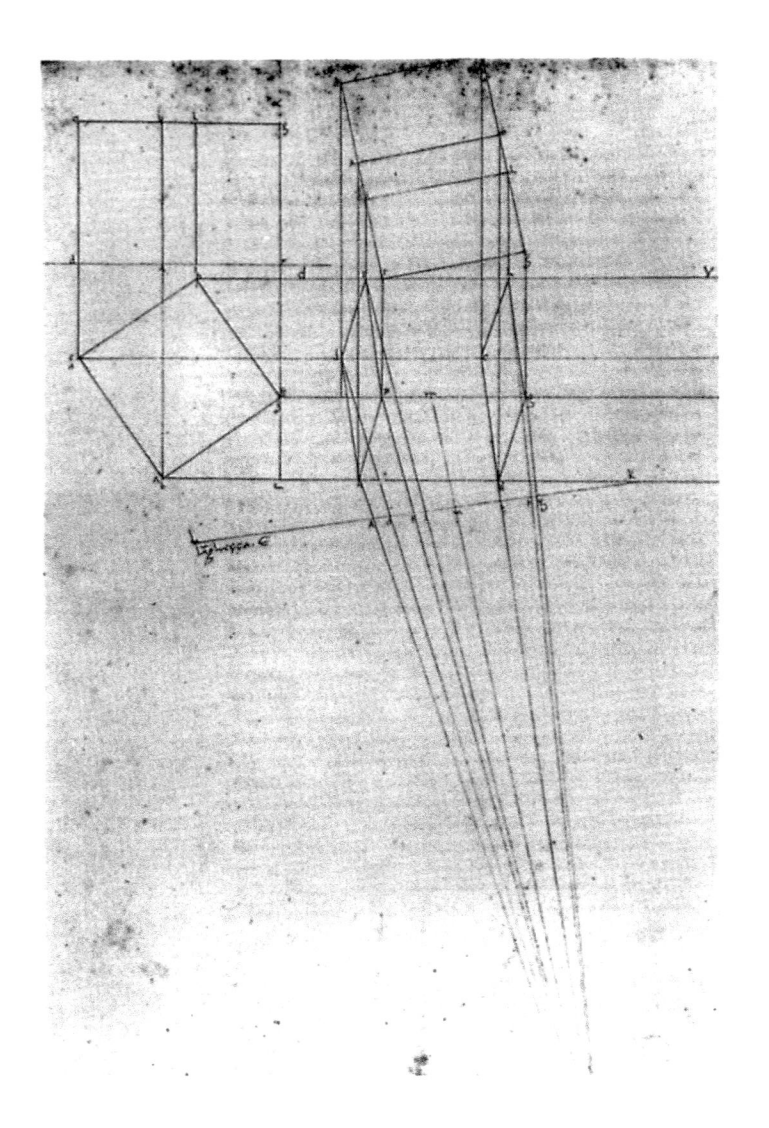

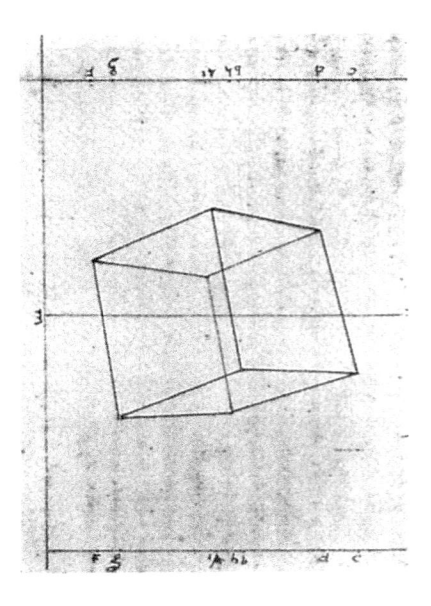

75 Piero, De prospectiva pingendi, *fig. lii, rotated orthographic projection of a cube.*

76 Piero, De prospectiva pingendi, *fig. liv, perspective of a rotated cube.*

THE PROJECTIVE CAST **152**

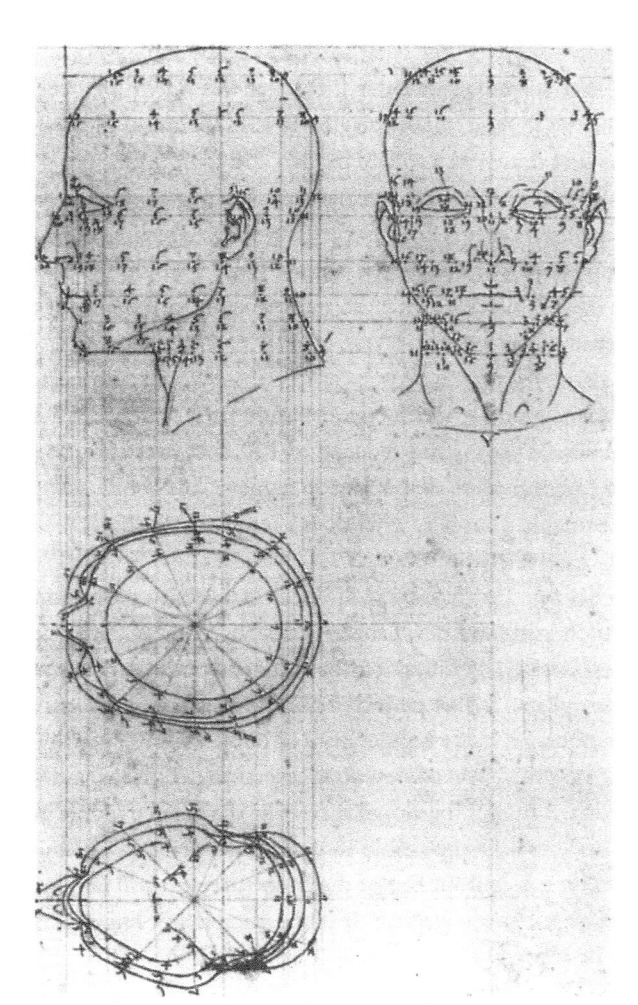

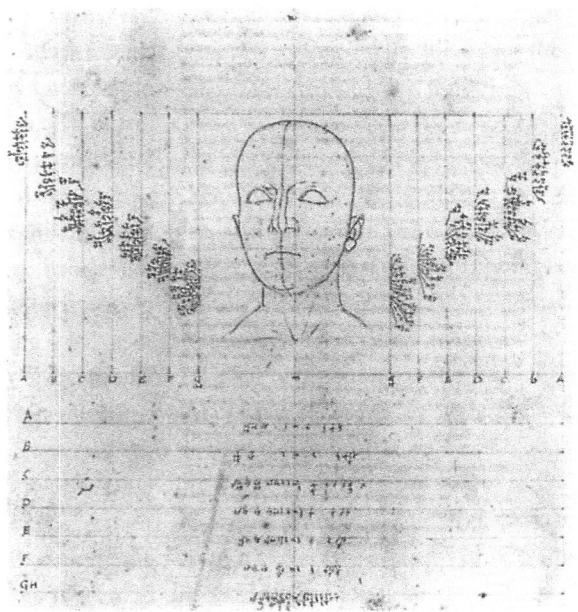

77 Piero, De prospectiva pingendi, *fig. lxiv, orthographic projections of a head.*

78 Piero, De prospectiva pingendi, *fig. lxvii, perspective of a head. Each of the numbered measure lines, left, right, and below, carries information about one sectional ring of the head.*

like Botticelli and Raphael found a different type of liberty in Alberti's central perspective. Architecture (plotted) is marshaled by the system of projection, while figures (intuited) move freely in its measured space. In Piero's art people and things seem to be more alike. Such an effect would be the natural consequence of using the Other Method for both figures and objects, as happens in *De prospectiva pingendi.*

De prospectiva pingendi proceeds from the simple to the complicated; it also proceeds from things that have a defined geometric relation to the picture plane to things that do not. Their progressive release is demonstrated in Piero's projection of rotated cubes, which also gives a marvellous insight into his ability as a draftsman and as a geometer. The problem was: "To proportionally foreshorten the given cube put at an angle and with no one of its sides parallel to the picture plane."[65] To get the cube in this position Piero uses what is now called auxiliary projection, a technique that according to Peter J. Booker was only developed in the latter half of the nineteenth century by British technical illustrators.[66] The cube is first drawn as a square in plan and then prized out of line by successive rotations within the projection. It is amazing enough to find this in a fifteenth-century manuscript, but more amazing still to find the same rotations performed on the human head. I would say that the turned head was the ultimate virtuoso performance in Renaissance perspective projection.[67]

The head is not planed into a schematic armature of facets, as in the examples cited earlier, but is instead reduced to 130 dots, thus resisting the geometric reformation of its shape. But to achieve this gentle scanning of so unconformable a surface, what does Piero have to do?

The human body does not easily submit to linear perspective projection because it has features without lines. Linear perspective deals with lines, as Piero himself emphasized, defining it as "profiles and outlines put proportionately in their place."[68] The projected outline of any continuously modeled, closed surface, such as a human head or body, will alter continuously as it rotates or as the point of view changes, and the contour of the outline will change position on the closed surface as the surface moves in relation to the eye. Now it is impossible to determine, prior to making the perspective, where the contour line of the body surveyed will lie. But this is what has to be put in perspective. If its position is not known, its projection cannot be drawn.

Dürer's machine shows why the problem does not arise while surveying a real body. Imagine the voluptuous model laid out on the table in place of the lute. One operative of the machine would pull the stretched strand so that it glanced tangentially across the surface of her skin, just touching her. The other operative would then adjust the cursors, mark the point of intersection on the mounted sheet, repeating the performance till they got sufficient points to represent the woman's contour as "seen" from the nail.

But in the treatise Piero is restricted to using drawings as the starting point. That is why he has no way of knowing where the contour will be. He can only find out by performing an act of conceptual surgery on the head. He does not carve it into facets; he slices it into sections. To slice it he has to have measured it in such a way as to enable drawing of the slices. A person has to stand still to be measured. We shall see that, because of the way Piero uses the technique, the immobility is not absolute.

Nevertheless, it is a calming influence that would tend to produce figures with the same sort of distracted arrest one notices in people being measured for a dress or a suit.

The Other Method starts with plans and elevations. It is easy to draw a cube that way because we already know what shape it is. How, though, can a human head be drawn in plan and elevation? Piero does not say. The evidence is not in the text but in the drawings. The 130 points on the geometrical surface of the head are distributed quite methodically. Sixteen radial points at $22^1/_2$ degree intervals are located in each of eight horizontal rings, giving 128 points, plus two extras for the end of the nose. How were these precise points obtained and what were they obtained from? The pattern of their distribution is all that we have to go on.

In his book *De statua*, Alberti describes several instruments that the sculptor may use to survey the human body in its various postures. One of these, the *finitorum*, is a disk divided into radial segments that sits on the top of the head. A fulcrum extends from its center, and from the

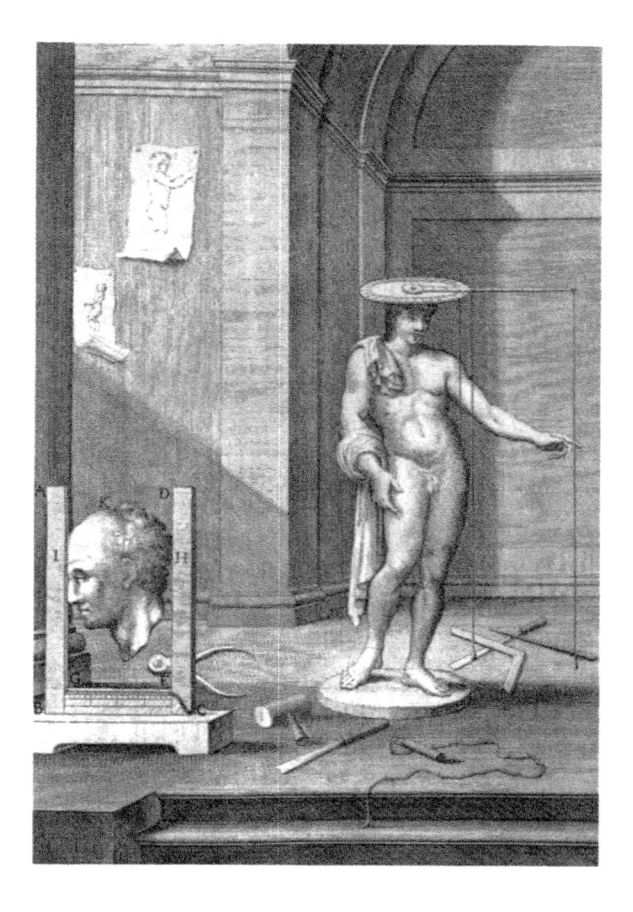

79   L. B. Alberti's finitorum, from De statua, *Bartoli ed., 1782.*

fulcrum an adjustable plumb line is dropped. The *finitorum* is, he says, "the means whereby we record with sure and accurate method the extensions and curves of lines, and the size and shape and position of all angles, protuberances and recesses."[69] An instrument of this sort would explain the radial distribution of Piero's survey points. Alberti claimed to have successfully measured several living persons with the *finitorum*, but he also describes a more certain way to take the same measurements from a statue embedded in a vertical cylinder of wax.[70] We may therefore choose to imagine a person keeping stock still to avoid disturbing the swaying instrument, or we may imagine statues, perhaps the clay models, measured in a cylinder of wax, which could be what Vasari meant when he wrote of the models being used for drawing "and other purposes" (although this is pure speculation).

Auxiliary projection means that the front, top, and side views of the head can be turned in any combination of directions within the system of drawing. Of course, through all these manipulations in picture space the head stays rigid, as measured. But human bodies are not rigid in movement; they are pliable. One little modification, easy to miss, indicates that Piero tried to overcome this limitation in the technique. A second series of orthographic projections shows the head tilted upward. The lower rings of surveyed points are no longer plane sections.[71] The fifth ring begins to curve up behind the ear. The eighth, round the neck, is in a full arc. As the head tilts up, the flesh on the neck and at the back of the head compresses and distends, pulling the superficial points with it. Thus a man lifts his head, like Frankenstein's creation, in the imaginary, constructed space of orthographic projection.

What else did Piero project in the Other Method? A ring with 96 facets, the base and capital of a composite column, and a domical apse with 28 coffers. Later demonstrations of this same method confirm that it sorts out its own subject matter from a middle region between rectangular architectonic objects and unsurveyable forms. Nobody after Piero took on the human figure in the way that he did. There was somewhat less concern with complex architectural details. The emphasis moved to items such as the *mazzocchio*, a faceted ring used as formwork for stylish headgear, and the lute, a body of amphibious definition, part in, part out of geometry.[72]

Piero's exposition of the Other Method is enough to change our idea of the way in which Renaissance perspective shapes the world that it portrays, enough to dislodge the tenacious idea that Alberti's perspective is Renaissance perspective. And it demonstrates that the vanishing point is

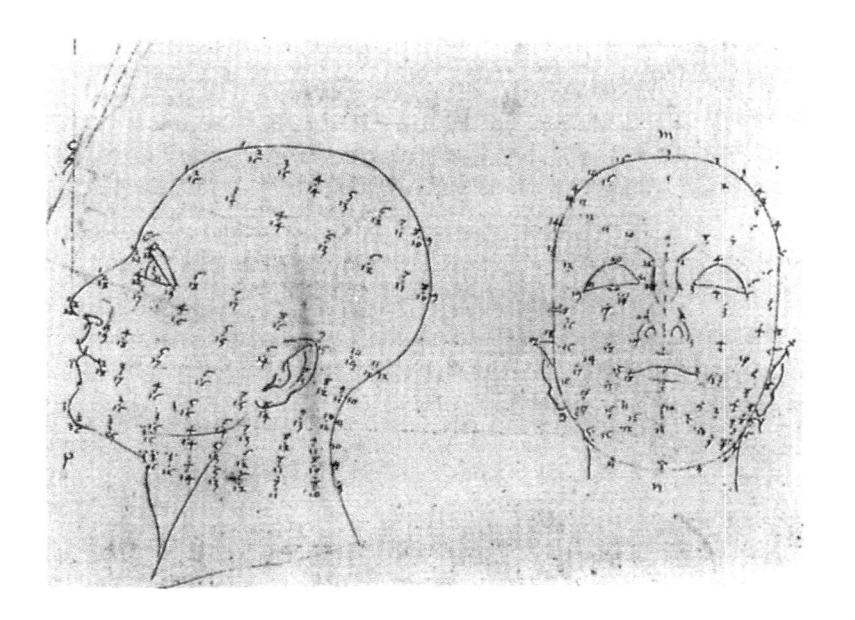

80   Piero, De prospectiva pingendi, *fig. lxviii*, orthographic projections of a tilted head (elevations).

81   Piero, De prospectiva pingendi, *fig. lxx*, orthographic projections of a tilted head (plans).

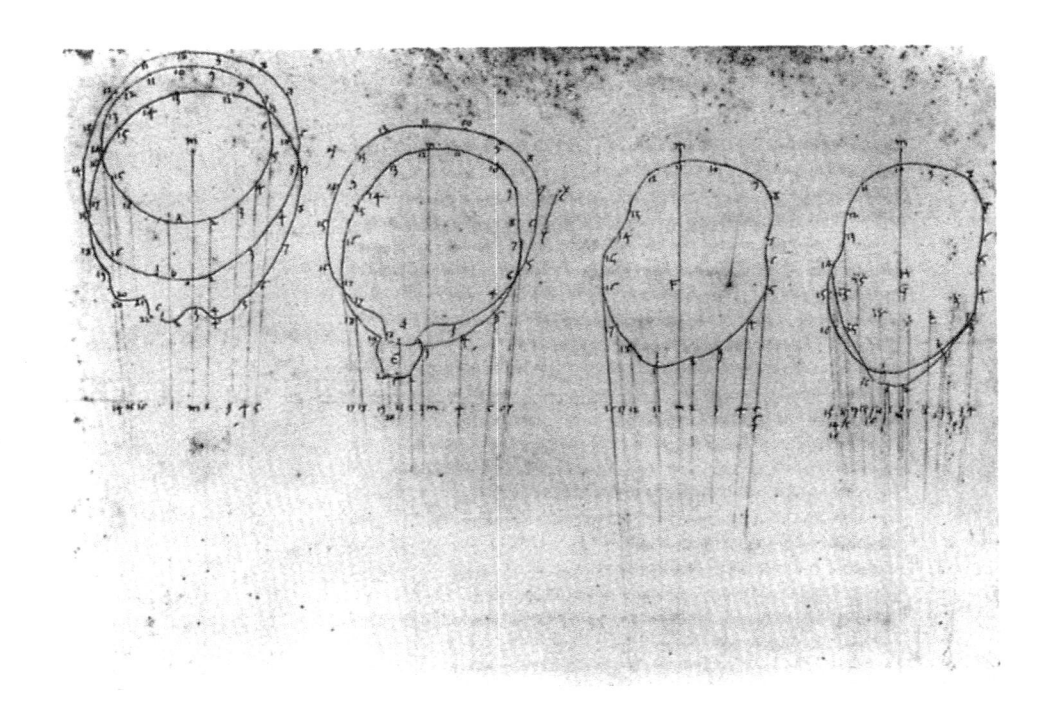

not inherent to perspective construction. But did the Other Method make any impact on the practice of Renaissance painting? Is it not a dead end? Giedion had a resounding phrase to dispatch things like this: "transitory phenomena." [73] Condemned to inconsequence, they could safely be erased from the annals of progressive history. What matters is what develops. Even were we to agree, we would notice that things do not develop by continuous extrusion through time like toothpaste squeezed from a tube. Any set of events will appear discontinuous at a small enough temporal scale. Some appear so at a large scale. Eudoxus's astronomy was a dead end until Copernicus came along; classical architecture was a dead end until Brunelleschi and Alberti came along. Dead ends may be of interest because they define where things were given up, suspended, neglected, or left incomplete. They may also reveal a good deal about what was pursued by contrast with what was not. Perhaps by the time of its appearance in *De prospectiva pingendi* the Other Method was a dead end in relation to painting, not because it had no effect on it but because it had already had an effect.

Painting with the Other Method

What does *De prospectiva pingendi* tell us about Piero's painting? First of all it tells us that we can no longer rely on clichés about perspective axiality and human figures that look like the five regular solids. If this is the impression conveyed by the paintings it has nothing to do with the perspective methods described in his treatise.

Forgetting Piero's paintings for a while, we could speculate on how the Other Method might affect the composition of paintings in general. It would encourage the mapping of discrete objects with no preordained metrical relation to one another. They would tend to tumble into pictorial space through a sequence of rotations from the picture plane. In its simplest use, things might look as if they were stuck to the surface of the picture. Nothing could extend very deeply behind the surface of the picture, because everything has to be connected to the observer's eye by the length of a cord or strand of a horse's tail. The things in the picture may occupy a space as consistent and unified as that defined by central perspective, but this space would have no underlying direction or bias. Once items were defined in perspective, alteration of their position within the picture would result in an accumulation of objects in incompatible spaces, producing an effect not unlike that of modern photomontage. The illusion of spatial depth would arise from the assemblage of objects, rather than from the telescopically foreshortened grid. The contents would tend to float unless held down by some means. The Other

Method would favor complex, quasi-geometrical shapes if rigid; if flesh they would have to adopt a stable, surveyable posture, for only a modicum of movement could be imparted within such a system of projection. Finally, things would tend to look flatter, less fully rounded, than we have come to expect, in part because conceived in terms of contour, in part because there is a very strong countertendency to exaggerate depth when sketching things into a vanishing point construction. By now there should be no need to repeat that if this method were pursued in painting the way that Piero pursued it in *De prospectiva pingendi*, all the well-known signs by which we recognize the presence of perspective would be absent. There would be no need for parallel lines, no need of vanishing points, no principal axis, and no biasing of space prior to matter.

How much of this applies to Piero's collected works? Probably about half. Human figures are less mobile, there is a tremendous emphasis on subtly modulated contour, and there is a noticeable flattening that Thomas Martone refers to as "spatial understatement."[74] On the other hand, Piero's compositions maintained a high degree of frontality and a fair degree of axiality throughout his career. The Other Method does not supply a master key to unlock the essential secret of Piero's paintings, but it does, I believe, provide a very useful means of elucidating some of their less explicable features.

It is very unlikely that the Other Method, so laborious and time-consuming, was used by Piero to draw each figure in every painting; it is far more likely that sufficient practice would impress its effects on the painter's intuition of form. This may seem a rather vague notion, but there is some definite evidence to support it.

Decio Gioseffi drew attention to the uncanny similarity between the head projected from various aspects in *De prospectiva pingendi* and certain groups of women's heads in Piero's fresco cycle of the Golden Legend in Arezzo. Gioseffi went on to compare both to the modern technique of photogrammetry. The women in the fresco look very much alike, as if from the same model, as were the numerous views of one man in the treatise, and the 24 systematic shots of one man from different angles in the photogrammetric survey.[75] Let us follow in Gioseffi's footsteps and look more closely at the women in Arezzo, and at one group in particular: five witnesses to *The Proving of the True Cross*, each seen from a different angle, whose faces are hard to tell apart. Surely these five austere sisters should be able to indicate something about the cause of their likeness, and, thereby, something about the manner of their creation.

82 Piero della Francesca, five women from The Proving of the True Cross, *San Francesco, Arezzo.*

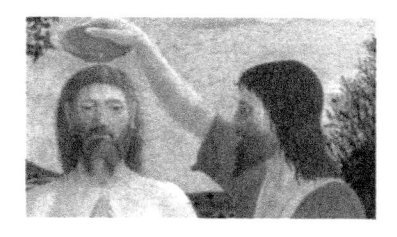

83 Piero della Francesca, The Baptism of Christ, *detail, National Gallery, London.*

The faces are alike but not identical. There are small differences in shaping and expression. It should be possible to tell if the similarity is genetic and the differences cosmetic, if, in other words, all five originate from the same source. And then, if the likeness is genetic, it should be possible to tell whether it is attributable to the projective rotation of a single figure, as suggested by Gioseffi.

If Piero's cartoons for the fresco were drawn direct from life using a single model, we might expect a variety of challenging postures with no particular geometric relations between them. If, on the other hand, he worked from one model *in propria forma*, he would work from an original pair of projections, full-face and profile, thence to single rotations, thence to combined rotations. He might wish to eradicate all geometric relations between the five resultant projections, but that would be very hard work indeed; much easier to use each successive stage of the rotation as you go.

Piero's taste for true form was displayed early in his career in the ludicrous orthographic relation between the near-identical heads of Christ and St. John in the *Baptism of Christ,* which are painted as front and side elevations of the same model. In the Arezzo cycle, profile views still

84 Five women from **The Proving of the True Cross:** *schematic rotations.*

predominate (4 out of 10 heads in the *Adoration of the True Cross;* 4 of 13 in *Solomon Meeting Sheba;* 3 of 8 in the *Beheading of Chosroes;* 8 of 12 in the *Exaltation of the True Cross;* 13 of 36 in the *Battle between Heraclius and Chosroes*).

But amongst the five women in the *Proving of the True Cross* only the central figure is in profile. We can see more easily how the other four are related to her by fitting a little muzzle over each face. The muzzle, shaped like a pyramid, joins five defined points: the corners of the eyes, the end of the nose and the corners of the mouth. The first three points are rigid in relation to one another. Lips are labile; they move on the face, but since our serene sorority are uniformly demure, we need not worry too much about the distortions that would derive from varied expressions superimposed on a straight-faced original.

The pattern is clear. Head B rotates clockwise from head A on a horizontal axis. Head D rotates clockwise from A on a vertical axis. Head E is a smaller rotation in the same direction as D, to which a counterclockwise rotation on a horizontal axis has been added. Head C also rotates clockwise on a vertical axis which is then tilted backward slightly on a horizontal axis perpendicular to the above-mentioned horizontal axis.

85 *Five women from* The Proving of the True Cross: *rotations in detail. Because perspective foreshortening would be insignificant under such a small angle of vision, all linear proportions, such as* a:b *on face A, should be almost identical throughout if these were rotations from one original. The point* x *on faces B, C, and D shows where the apexes of the pyramids should be if the ratio* a:b *held. Likewise, the ratio RS:SW on face C is very different on face D and E. (Drawn by the author.)*

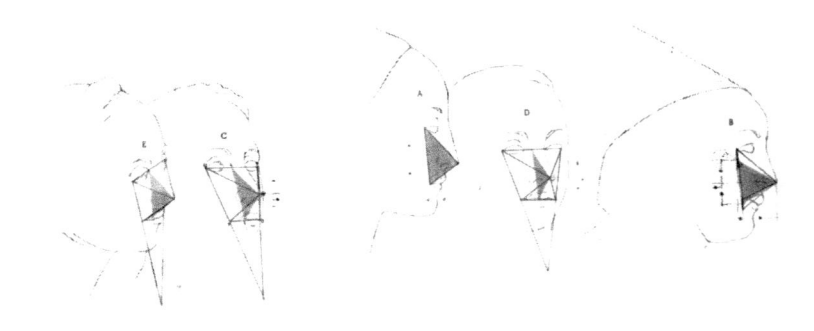

This provides substantial evidence in favor of projective determination, because the pattern of successive rotations is exhibited, as if to indicate how the heads gave rise to each other out of auxiliary projection, as Gioseffi guessed. But when we look closer the substance quickly dissolves. The pyramids ought to map into one another as projections, yet none will transform exactly into any other. There are lots of essential differences of shape in the five rigid frames. This indicates that the similarity of the five heads is not attributable to their origin in a single generic projection.

So, although the resemblances in appearance and procedure are too considerable to disregard, it is almost certain that they were not obtained by projection. It is reasonable to assume that Piero, having previously practiced such rotations, was composing his figures with that idea in mind. He would paint them as if born out of the Other Method, though in fact they were not. Thus measurement and intuition would aid one another—the antagonism between them being a theoretical fiction, not a practical fact.[76]

De prospectiva pingendi presents central perspective and the Other Method as equally viable alternatives. Likewise in his paintings neither method prevails. He seems to have used them to counterbalance each other, so that the tendencies of neither attained dominance. Such was the case in the *Flagellation*, where certain characteristics of the three foreground figures can now be cautiously attributed to the effects of the Other Method. Despite their subtle exactitude of outline, the three figures do not belong in the same perspective space as the Albertian tiled pavement on which they stand, nor are their spaces compatible with each other. Ruth Brocone noticed that the middle figure has feet about two feet long, judged by the depth of the tiles.[77] Because the viewer is to the left, the outermost figure would not be seen in profile if, as indicated by his feet, he stands parallel to the picture plane. The bearded man would make more sense to the left of the centric point than to the right, where the twist between feet and head exaggerates into contortion. So it looks likely that Piero devised the figures independently and then fitted them into the picture as well as he could; hence their rootlessness, and their air of paralyzed distraction.

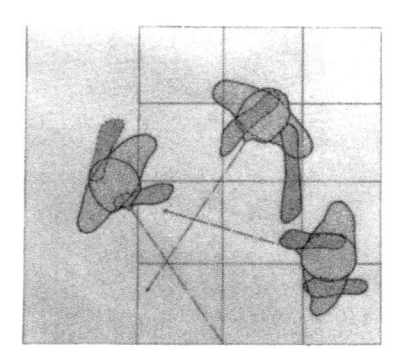

86 *Plan showing orientations of feet, bodies, and heads of the three foreground figures in Piero's* Flagellation.

87 *Piero della Francesca, Brera altarpiece, Brera, Milan.*

88 Piero della Francesca, Nativity, National Gallery, London.

89 Vanishing points of the stall roof in Piero's Nativity (drawn by the author).

Minor spatial disjunctions of this sort are very common in Piero's meticulous paintings, which range from the Brera altarpiece, a monumental homage to the central vanishing point, in which the Other Method was almost certainly used to draw the details of the floreate coffered barrel vault,[78] to the London *Nativity*, an equally extreme counterexample that makes a mockery of Alberti's construction.

The *Nativity*, probably a late work like the Brera altarpiece, has no obvious, ordaining geometry.[79] Yet the underdrawing, now visible in places because the paint surface is badly damaged, is sharp, clear, and characteristically parsimonious. Figures, lutes, and saddle, all candidates for the Other Method, are precisely contoured. During conservation, pricked outlines were discovered beneath the drawing, indicating that the shapes had been imported. Only the saddle on which Joseph sits is aligned with the picture plane. The rude and ruinous stall immediately behind the figures is slightly out of alignment—the kind of thing that happens all the time in snapshots, quite often in trecento paintings, especially those of Giotto, and only twice, so far as I am aware, in Renaissance perspective painting: here and in a work by Paolo Uccello, *The Subsiding of the Flood*.[80]

Paul Barolsky has been censured for suggesting that the *Nativity* contains an element of parody (the braying ass sings with the angels).[81] Yet he may be right. It is tempting to see the teetering plane of the stall roof as a parody of Alberti's pavement. The roof is the nearest approximation to a consistent geometrical surface in the picture. But instead of the extensive, perfectly horizontal basis of regular tiles, there is a rustic surface, inclined and skewed, with no one of its sides parallel to the picture plane,[82] hazardously propped up in the air and decaying. To emphasize its singular disorientation, the parallel sides can be extended until they meet. The sense of inversion is undoubtedly heightened by the painting of the ground as if it were as insubstantial as the sky (it is presumed incomplete). The small goldfinch on the worn mat of grass, perched just

THE PROJECTIVE CAST   **166**

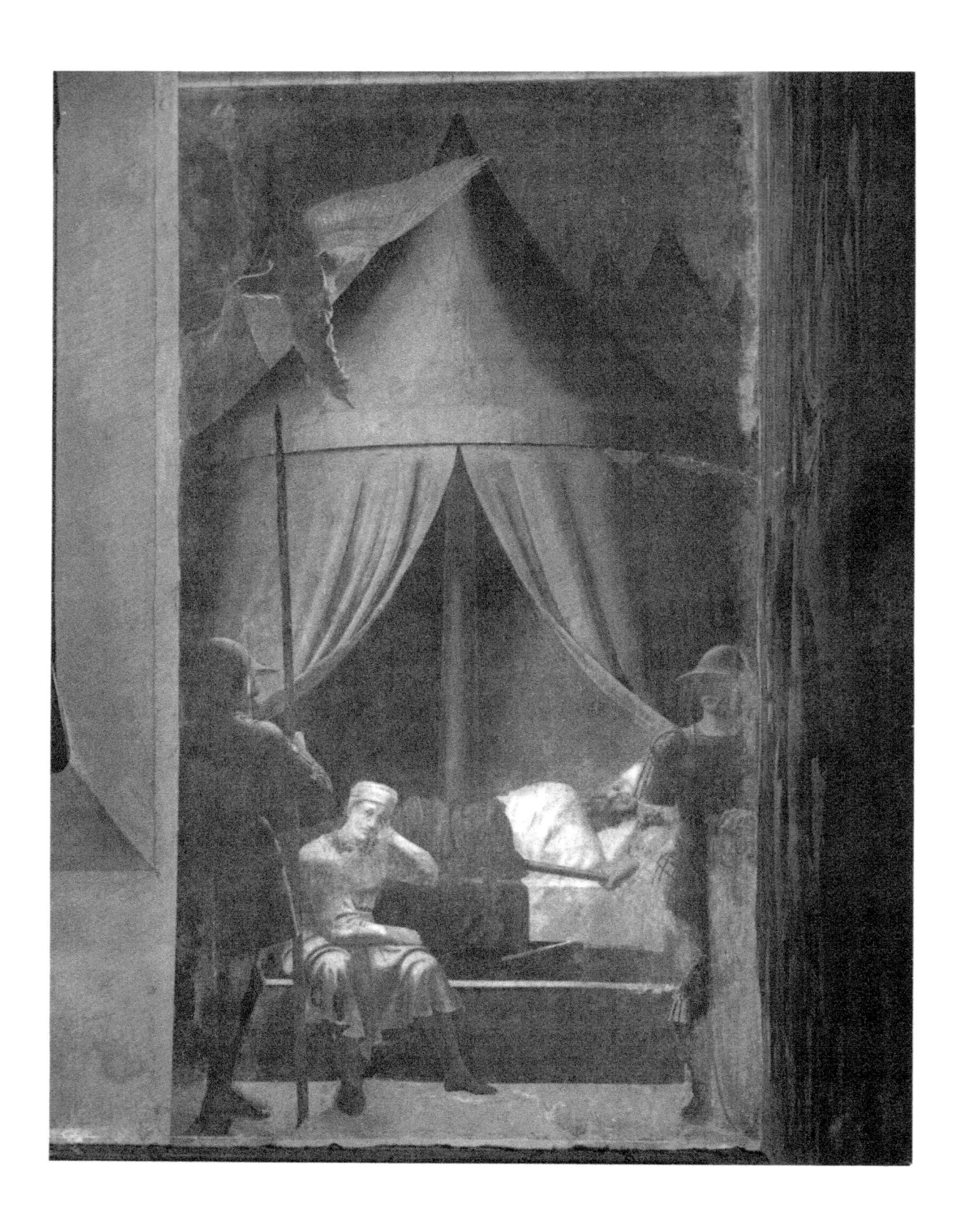

90 Piero della Francesca, Dream of Constantine, *detail*, San Francesco, Arezzo.

like the magpie on the roof, draws attention to their comparability in another way.

Mostly in Piero's painting it is a question of figures versus architecture—Alberti's division in another guise—though in one case the playoff between them is arranged differently. *The Dream of Constantine*, part of the fresco cycle at Arezzo, shows that Piero was quite capable of making Albertian compositions without the architectural directions supplied by central perspective construction. The *Dream*, which has no rectangular architecture, is frontal, axial, and symmetrical.[83] Moreover, the overtone of political potency, so often conveyed in central perspective composition, turns into an allusion to sexual potency. The prince of rays has its way with the tent, aided by the tumescent center pole rising from behind the loins of the sleeping Constantine. I do not think I am inventing this. Look at Piero's unusual treatment of Mary in the *Madonna del Parto* where the opened skirts of a similar tentlike canopy and the dress, split and drawn apart over the abdomen of a very pregnant Madonna, are explicitly linked to insemination and birth.[84] It is typical of Piero to alter the course of a composition, a tendency, a story, or an allusion like this. But then, running counter to the central thrust through Constantine's tent, connected to it only by the position of the spectator's eye, is the messenger angel, oblique to everything, coursing downward and back into the picture, statuesque, like concrete even in flight, but foreshortened in a way that would defy normal intuition (the face is completely obscured by the body). This apparition is surely a practical result of the Other Method.

Vasari was right about Piero: something was lost. Something was gained too. The quickness of life departs while the subtlety of contour increases tenfold and the commanding imperative of the vanishing point is subdued or perverted. For the same reason his figures, undressed of rhetorical gesture and expression, seem to be in a dream rather than a drama.

The Chaos of Battle

Two further episodes in the Arezzo cycle, the *Battle between Constantine and Maxentius* and the *Battle between Heraclius and Chosroes*, are composed of not just one disorienting figure, like the *Dream*, but a multitude. This is where we might reasonably expect to find the Other Method come into its own. The elements are right—armor, shields, weapons, and helmets—and the dense melee of fighting bodies scotches any lingering notion that Alberti's construction might have been of service. Neither scene was constructed as a unified perspective space; the parts must therefore have been composed in piecemeal fashion. All is contour,

91 Leonardo da Vinci, A Town Overwhelmed by a Deluge, c. 1515, The Royal Collection, Windsor.

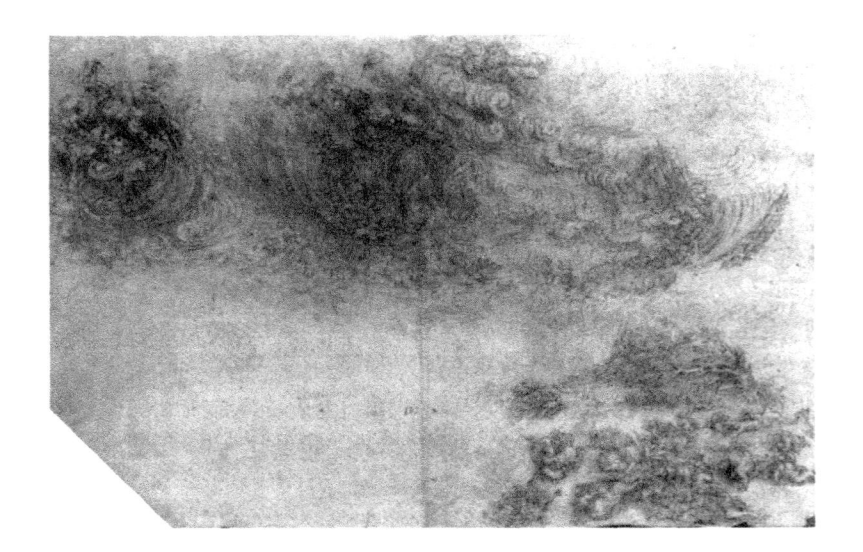

the circumstances are propitious, yet, disappointingly, there is little of the sophistication of outline attributable to the Other Method.[85]

Battle scenes are, nevertheless, an interesting genre. We see the whole range of visible phenomena in perspectival cones, and we can record just about anything we can see in plane perspective; a camera has no trouble with riotous assemblies, but the idea of chaos or tumult cannot easily be conveyed through the conventional schema of Renaissance vanishing point construction, which is military in the sense of the parade ground rather than the battleground. All the same, a number of painters attempted to elicit chaotic patterns that were beyond the range of geometrical description. Leonardo, in particular, was fascinated by maelstrom, deluge, and vortex, destructive forces that overwhelm the order imposed by human agency. Ernst Gombrich has suggested that Leonardo's was an intelligence divided between the investigation of geometry and the avoidance of it, finding inspiration for his art in Euclid, but finding it also in a patch of moldering plaster or a cloud.[86] Leonardo's figure compositions arise out of a simulated chaos of countless layers of *pentimenti*. The *Battle of Anghiari* (1503–1505), now only known through drawings and copies, was perhaps the most compelling crystallization of these ghosts teased out of vapors, smudges, and stains. Nothing could be further removed from the way Piero went about his business.

Battle scenes were always concerned with the throng, a violent altercation of bodies. But warfare, ancient or modern, was conducted with machines, equipment, weapons, and armor as well as bodies. In the *Battle*

92 Peter Paul Rubens, copy of Leonardo's Battle of Anghiari, c. 1603, Louvre, Paris.

of Anghiari Leonardo minimized these manufactured items and naturalized their forms. Such items were more evident in the Arezzo cycle, and still more so in three famous battle scenes painted by Paolo Uccello. In these three scenes we find what Piero never provided: the Other Method applied without restraint.

Vasari tells how Uccello was so preoccupied with problems of perspective that he neglected everything, his wife, his livelihood, unfortunately even his art, to pursue it.[87] Uccello was always experimenting, sometimes perverting perspective, as in the *sinopia* for the *Nativity*, where a diagonal vanishing point is suddenly appropriated to foreshorten a frontal structure; sometimes pushing one aspect of perspective to an extreme, as in the *Subsiding of the Flood*, where two slightly crossed vanishing points drag the drowned world in their wake;[88] sometimes adjusting perspective convergence to imply panoramic vision, as in the *Hunt in the Forest*;[89] sometimes adjusting it to reduce marginal distortion, as in the painted *Monument to Sir John Hawkwood*, where a sequence of vanishing points are used to represent one set of parallels; sometimes exacerbating marginal distortion, as in the *Profanation of the Host*, where the centric point is located on the edge of the picture. All these experiments relate to vanishing point constructions. The situation is different in the three paintings commemorating the *Battle of San Romano*.

These were commissioned by Cosimo de' Medici in the mid-1450s to record a near-disastrous engagement with the Sienese in 1432. The three paintings, now in Paris, London, and Florence, originally hung in one

93 Paolo Uccello, Subsiding of the Flood, *Chiostro Verde, Santa Maria Novella, Florence, 1446–1448.*

94 Paolo Uccello, Battle of San Romano: Niccolò da Tolentino at the Head of the Florentines, *1456, National Gallery, London.*

THE PROJECTIVE CAST   **172**

95 Paolo Uccello, Battle of San Romano: The Counterattack of Micheletto da Cotignola, *1456, Louvre, Paris.*

96 Paolo Uccello, Battle of San Romano: The Unhorsing of Bernardino della Ciarda, *1456, Uffizi, Florence.*

97 Paolo Uccello, *perspective study of a chalice, Gabinetto Disegni e Stampe degli Uffizi, Florence.*

98 Paolo Uccello, *perspective study of a* mazzocchio, *Gabinetto Disegni e Stampe degli Uffizi, Florence.*

room.[90] Where are the central vanishing points? The London and Florence paintings give directions that turn out to be unreliable. Uccello lays several broken lances and a dead knight on what one can only call the floor, since it is perfectly flat. All seem to have fallen parallel, pointing more or less to one spot.[91] Apart from this miracle in honor of Alberti, the rest is mayhem.

In two of the three panels, exquisitely foreshortened *mazzocchi* are distributed at various angles. Vasari lists *mazzocchi,* which Uccello "had drawn with their points and surfaces shown from various angles in perspective,"[92] along with "other oddities, on which he wasted all his time." In the Uffizi and the Louvre there are three constructed perspective drawings of *mazzocchi* attributed to Uccello, also a magnificent projection of a chalice that includes three variations of the *mazzocchio* form. There is no indication of vanishing point construction in any of them. The outlines in the *Battle of San Romano* panels, like those in Piero's *Nativity,* are precisely inscribed. Delineations of every facet of the *mazzocchi* are visible throughout. *Mazzocchi* became more or less standard subject matter in demonstrating the Other Method in later treatises.[93] There is a similar perspective sophistication in other repeated items in Uccello's battle paintings, never exactly aligned, twisting and turning in pictorial space as they do when out of pictures, such as the crossbows in Florence, the extravagant but rigid plumes in Paris, and the helmets in all three. These results would be almost impossible to achieve without an understanding of the Other Method.

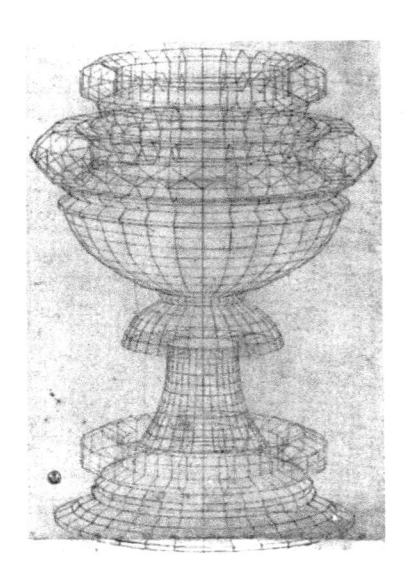

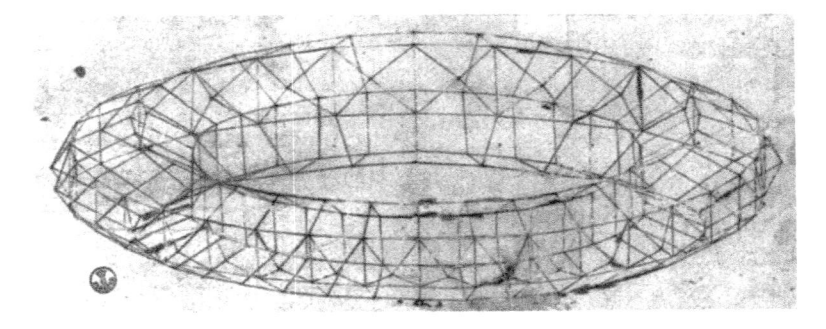

As in Piero's *Flagellation*, supporting evidence is found in a small inconsistency. Wherever they appear in Uccello's *San Romano* panels, the facets on the *mazzocchi* are symmetrically disposed. This is wrong. They should only be so when the vertical axes of the *mazzocchi* coincide with the center of projection. Thus, as they move toward the edge of the picture their asymmetry should increase. The error is easily explained, however. In the textbook *mazzocchi* as well as in the drawings attributed to Uccello, the results are symmetrical because the vertical axis of the ring *does* coincide with the center of projection. Uccello presumably used ready-made drawings of this sort as cartoons. Individually, each is perfectly projected if you look straight at it; together they fall apart and the picture decomposes into several slightly anomalous spaces. Uccello has been called the first cubist. A number of critics have sensed an incoherence in the space of the paintings above and beyond the chaos of the scene. This may be one reason why.[94]

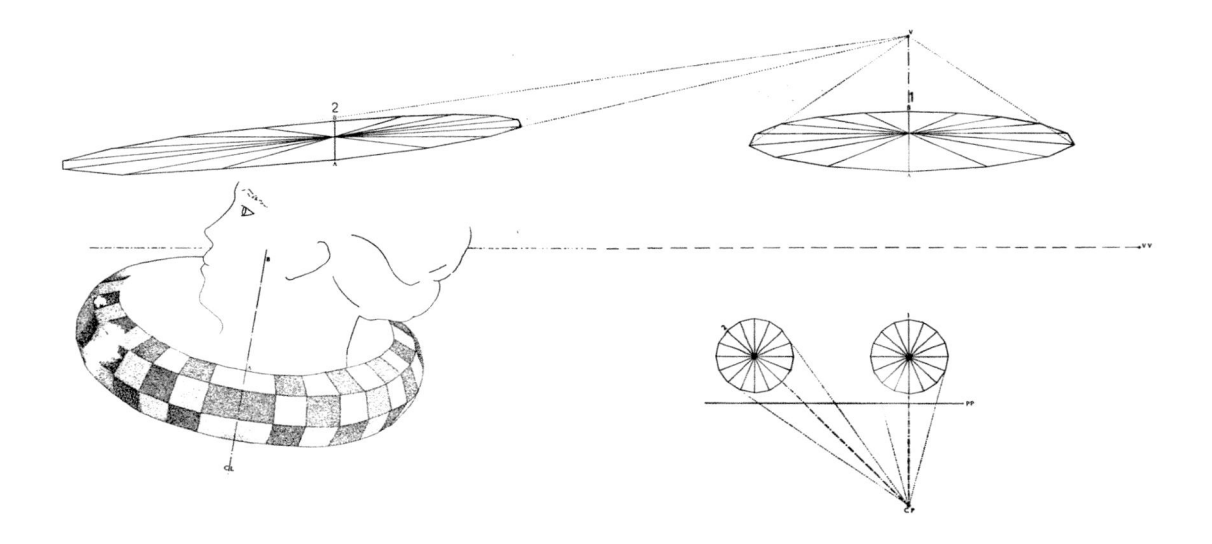

99 Two perspectives of a disk with equal segments: top right with sight line between the center of the disk and the center of projection (CP) perpendicular to the picture plane; top left with the same line oblique (45°) to the picture plane. (The diagram at bottom right shows this arrangement in plan.) The perspective image on the right is symmetrical, that on the left is not and cannot be symmetrical. Typically, the mazzocchio *from Uccello's* Subsiding of the Flood *(bottom left) is perfectly symmetrical, although the image of its vertical axis is nowhere near the central vanishing point (VV). The perspective should only be symmetrical if the point VV lies on the line AB. (Drawn by the author.)*

Exploitation of the Other Method allowed Uccello to portray another kind of chaos, quite different from Leonardo's, where forms neither quite human nor quite geometrical, but in some sort of halfway state, are seemingly let loose from any required orientation. At the same time, the action is crowded right behind the picture plane, quickly receding into its own obscure density, as if the frame into which we look had pushed through a welter of flesh, leather, iron, and wood. Some items, especially several of the bridles, seem to be stuck straight onto the picture rather than seen through it. This, together with the peremptory folding up of the undulating backgrounds so as to deny the horizon and the emphatic underscoring of outlines, led John Pope-Hennessy to conclude that both space and perspective were insignificant in these paintings. "We find," he wrote, "that optical experiment has been subordinated to a rich imagination and naturalism to an all-pervading decorative sense."[95] I, on the other hand, believe that optical experiment was indispensable to depict the local affray of discrete, flying forms flattened into pictures.

Uccello's three episodes from the *Battle of San Romano* were as much concerned with perspective as any of his paintings, but not with the vanishing point. The *Battle of San Romano* is the Other Method writ large, taking over the entire space of the composition. There are thus two ways in which Renaissance perspective geometry organized pictures. Compare Raphael's *School of Athens* with the *Battle of San Romano*.

A Supplementary Note on the Origin of the Other Method

If Piero was not the first to employ the Other Method, was Uccello? Vasari credits him with the invention of a perspective method using plans and profiles,[96] but then Vasari thought that Uccello died in 1432 when he is known to have died in 1475. Richard Krautheimer has argued that it was Brunelleschi's invention.[97] Brunelleschi was an architect, not a painter. He would have been interested in making designs and records of buildings, and he chose architectural subjects for his two perspective demonstrations. He would have been familiar with and predisposed toward profiles and plans, that is, orthographic or proto-orthographic projections. Krautheimer's proposal is at least not contradicted by the recollections of Brunelleschi's procedures for drawing by his biographers, Antonio Manetti and Vasari. The former writes of a peculiar way of recording ancient buildings that involved many ciphers on graphed sheets;[98] the latter states over again, as he had already about Uccello, that Brunelleschi discovered a technique for rendering exact perspective likeness "by tracing it with the ground plan and profile and by using intersecting lines"[99] (probably intersecting in the eye, but it is also possible that they were intersecting on the horizon of the picture).

Since it is almost certain that linear perspective was discovered with the aid of plan and section (still preserved in Alberti's method), they must have played a part in Brunelleschi's demonstrations. The question is, how large a part? Were the profiles simple cuts, acting as stepping stones to the discovery of the vanishing point? Or were they fully developed orthographic projections capable of surveying spaces? If the latter, as Krautheimer maintains, then the vanishing point would be an aftereffect of the discovery, not integral to it. The evidence does not permit conclusive judgment on this matter, but it is certainly possible that the Other Method was first practiced by Brunelleschi the architect, and that the discovery of linear perspective took place sometime prior to the discovery of the vanishing point.

A lot has been made of the fact that Brunelleschi drilled a small hole in the panel on which the baptistery of San Giovanni was painted. You stood in the door of the cathedral (the point of view), held the panel facing away from you, closed one eye, looked through the hole toward the real baptistery, held a mirror at arm's length to reflect the picture and, lo and behold: the magic of complete eidetic correspondence. Those who identify perspective with the central vanishing point dwell on this hole because it must have been drilled through that very spot.[100] Thus targeted, the centric point would be immediately recognized, even if it had escaped notice before. Well, it would be if there were orthogonals urging our eye toward it. The significant thing is that there were not. That is what struck Susanne Lang, who pointed out that even in the frontal, axial view of the octagonal baptistery, there were precious few orthogonals, and those few rather short.[101] In Brunelleschi's next demonstration, an oblique view of the Piazza della Signoria, there was none (afterward, Manetti tells us, Uccello went and made the same picture in the same place). We may therefore conclude that Brunelleschi was not seeking to exhibit the convergence of orthogonals to a point in perspective projection, though he may have understood it. Only later would the centric point become the sign of the technique.

Yet these uncertainties regarding the use of, recognition of, or prominence of the vanishing point in Brunelleschi's perspective demonstrations, and the inference that he may be the inventor of the Other Method, do not suggest any aversion to frontal, axial, orthogonal organization on his part. As an architect, Brunelleschi was concerned to augment these properties, not to mitigate or evade them. Painting is not architecture, but it may still seem strange that a perspective technique that came out of architectural drawing, that was developed by classical architects (even Piero was referred to as an architect as well as a painter by his pupil Luca Pacioli),[102] and that employed parallel projection to obtain perspective results, should be so divergent from classical architecture in tendency. But then why strange? Only because we expect technically related practices to cooperate toward the same ends. This is a prejudice, not a law.

The way architects customarily used, and indeed still use, orthographic projection to make plans, sections, and elevations permits restricted representation by a fairly easy method. So long as what you can get is what you anyway want, there is no problem. The problems will only arise when there is a perceived difference between what you want and what you get. The way in which Renaissance artists used perspective also permits restricted representation by a fairly easy method—Alberti's method, give or take a rule. Alberti understood the requirements of artists in just these terms; they did not want to be bothered with abstruse mathematical calculations or difficult geometrical ideas. The restrictions imposed by easy architect's drawing and easy artist's perspective happen to be much the same and, since the restrictions anyway promote desired properties such as symmetry and orthogonal orientation, there was no problem. Thus the easy methods conspired together to fullfil the need for a certain, definite, recognizable order.

I have described a kind of perspective made with orthographic projection that produced differently constituted pictures within the confines of its more demanding construction. There was another technique of orthographic projection, also more demanding than normal practice, that produced a differently constituted architecture: stereotomy.

Chapter Five   **Drawn Stone**

Stereotomy, which means the cutting of solids, was a seventeenth-century French rubric under which were gathered several existing techniques including stonecutting, which remained the principal concern.[1] The basis of stonecutting was the *trait*. *Traits* were layout drawings used to enable the precise cutting of component masonry blocks for complex architectural forms, especially vaults. Thereby accurate fabrication of parts could be achieved prior to construction. *Traits* are not illustrations and yield little to the casual observer. They are orthographic projections, but they are not like other architectural drawings. They were required only in exceptional circumstances. They seem rarely to have been employed outside France until the nineteenth century. The theory of stonecutting did not play even a minor role in the development of general architectural theories, while in practice it was marginal in every respect.

Stereotomy was at the very edge of architecture. It was also at the edge of mathematical geometry, at the edge of technical drawing, of structural theory, practical masonry, and military engineering. Within architecture, it is impossible to periodize for the same reason. It was on the edge of classicism and every other stylistic category—baroque, rococo, neoclassical, Gothic, and even modern. From within each category the art of stonecutting could easily be dismissed, since it had no central importance to any. It flourished only where definitions blurred, where one thing began to glide off into others; where structural theory met technical drawing, where neoclassical blended with rococo, where mathematical geometry came into contact with architectural composition, and so on. This is what was remarkable about it: being peripheral to each, it was shared by all, like an unrecognized border joining many diverse regions.

It would be impossible, in one essay, to describe the full extent of the territories skirted by this defunct technique. I shall concentrate, therefore, on the way that the geometrical art of stonecutting permeated the divide between classical architecture and Gothic. Although in this sense regional, the study suggests a reevaluation of more general ideas about the evolution of style and about the historical relationship between structure and ornament.

Unfolding

How *traits* were constructed, and what sort of architecture they made possible, may best be indicated by considering one early but outstanding example: the *trompe* built on the garden side of the chateau at Anet by Philibert Delorme between 1549 and 1551, which has always been regarded as a masterpiece of masonry construction. Sixteen years after its completion Delorme gave it pride of place in his *Premier tome de l'architecture*, devoting 24 pages to its explanation. In later books on the art of stonecutting the Anet *trompe* was applauded for its dazzling virtuosity and for its daring structure. These same books frequently reproduced versions of it, together with other worked examples derived from Delorme's *Premier tome*.[2] A backwash of critical comments began toward the end of the seventeenth century. Praise was followed by reminders that though such novelties as *trompes* may demonstrate great skill, they may also offend against architectural propriety. A structure should always look stable as well as be stable.[3] *Trompes*, the most advanced expression of the theory of stonecutting, flouted this rule by appearing to defy gravity.

But the relation between the *trompe* and classical architecture is both more aggressive and more intimate than this criticism suggests. Stonecutting and classicism flourished at the same time, but it took centuries to bring them into some kind of accord. Their gradual harmonization and synthesis can only be observed in France, which was the epicenter of the theory of stonecutting, as J.-M. Pérouse de Montclos has convincingly demonstrated in *L'Architecture à la française*.[4] Because stonecutting is associated with masons rather than architects (Delorme was the son of a master mason), because masons are associated with architecture before the Renaissance rather than after it, and because much of the material in the *Premier tome* was presented by Delorme as already known (he occasionally fobs off his readers by advising them to consult a master mason if they want to know more), the technique recorded by Delorme is regarded as evidence of an earlier, well-established tradition. Pérouse de Montclos sees it as specifically French, originating in the southern part of the country in the romanesque period. Most others have seen it as Gothic rather than classical.[5]

I shall argue that there was no preestablished, indigenous tradition of stonecutting at odds with the newly imported classicism, because the kind of stonecutting associated with *traits,* such as those published by Delorme, was itself of very recent origin and because, in all likelihood, it originated outside France.

A *trompe* is a splaying, conic surface of masonry, reminiscent of a trumpet, upon which projecting towers usually are supported. Delorme said that the *trait* for the Anet *trompe* was his invention. He did not lay direct claim to any other of the *traits,* which is significant because, whatever the history, recent or otherwise, of the rest, the Anet *trompe* arrived together with classical architecture. It had no previous history and cannot therefore be summarily dismissed as a Gothic leftover carried by force of habit into Renaissance architecture, or an inadvertent colloquialism soon to be erased, or an old graft on a new stock.

The *Premier tome* was one of the earliest French books to publish the new Italian style, as well as the first to publish the theory of stonecutting.[6] The amount of space Delorme devoted to each was about the same: 194 sides on *traits,* then 258 sides on classical ornament. Since most of the material on *traits* dealt with vaults, while most of the material on the classical orders dealt with walls, it would seem reasonable to infer, as Geert Bekaert does, that Delorme distinguished two kinds of architecture with distinct rules and different emphases, which could be added one on top of the other.[7] The five orders of Italianate classicism belonged to walls exhibiting the distribution of vertical loads, while the masons' art of *traits* belonged to ceilings. The distinction, emphasized by the organization of Delorme's book, helps nudge opinion toward a Gothic provenance for the theory of stonecutting, since vaults play so dominant a part in Gothic building. Again, while this is true in general of the material in the *Premier tome,* it is not true of the *trompe,* because the *trompe* is a vault pulled down into a wall.

Delorme invented the *trait* for this kind of suspended vault at the age of 25 or 26, just after his return from Italy. His first *trompe* was a small cornerpiece in a house built for a banker called Patouillet in the rue de la Savaterie, Paris.[8] In 1536 there followed the Hôtel Bullioud in the rue Juiverie, Lyons, where Delorme added two different varieties of corner construction as part of an extensive remodeling of a courtyard house. Delorme said that he built countless others afterward. The Patouillet and Bullioud *trompes* were portrayed as ingenious solutions to difficult planning problems on restricted sites. At the same time, Delorme clearly was proud of their beauty and proportion ("belle grâce," "tel grâce et proportion") and in this connection repeatedly made reference to their apparent suspension in midair ("suspendu en l'air"), the characteristic that would be identified by later commentators as incongruous with classicism.

100 Hôtel Bullioud, Lyons, Philibert Delorme, 1536, elevation of courtyard.

101 Philibert Delorme, perspective of the trompe at Anet, 1549–1551, from Le Premier tome de l'architecture.

The Hôtel Bullioud was an early work, so any inconsistencies could be attributed to the architect's youth. Even so, Anthony Blunt, in his monograph on Delorme,[9] was baffled by it. Why, when the *hôtel* already exhibited Delorme's considerable understanding of the classical orders and their application, was it so badly marred by the intrusion of the two bulging *trompes*? To Blunt these seemed to be residues of an entirely unclassical intelligence, scything through and truncating the lower range of pilasters, abrogating the structural logic of the orders which were left dangling, incomplete, and exposed as decorative fictions. The irony is that the *trompe*, which destroyed the illusionistic structure provided by the classical orders, was in fact a real structure of some sophistication.

The conflict between the two kinds of structure is nevertheless undeniable. Nor can it be put down to Delorme's inexperience. The Anet *trompe*, built 13 years later, looks no less vicious a deformation of classical regularity, no less obtrusive a suspension of classical stability.

A recurring justification for the employment of difficult *traits* was that they allowed architects to adapt to circumstances, making it possible to join new building to existing construction without glaring discontinuities, and to get maximum use from a small space, as in the Hôtel Bullioud.[10] At Anet, enormous and ambitious, such problems need not have arisen, yet the *trompe* registers as an extraneous element, added as an afterthought to an already completely conceived palace facade. Delorme tells us that this was, indeed, the case. Anet belonged to Diane de Poitiers, mistress of Henry II. The king's suite was provided with chamber, antechamber, and *garderobe*, but there was no space left, says Delorme, for the very necessary additional item of a *cabinet* where the king could write and meet in secret. So, he then says, "I was put into a state of great perplexity, for I could find no place for the said cabinet without spoiling the lodgings and the rooms that were made on old foundations and walls already constructed. What would become of it?"[11] The answer was that it would be stuck into a nearby corner where a projecting wing joined the main block. The reason for suspending the *cabinet* in midair, and for twisting the *trompe* on which it rested to one side, was to avoid an existing window that lit an adjacent staircase.

Delorme's explanation suggests that the *trompe* was shaped by circumstances. But, having presented the difficult *trait* as the solution to difficulties, he then proudly points out that there is no timber in the construction.[12] It is made entirely of stone, is heavy and therefore much more difficult. All of which leaves the modern reader with the suspicion that the difficulties were as much sought after as found. Delorme, it should be said, used the word *difficult* as a superlative.

Since the range of buildings on the garden side at Anet was demolished two hundred years ago, we have to rely on the illustrations provided by Delorme in the *Premier tome* and an elevation done in 1696 to get an idea of what the *trompe* was like.[13] It is a small incident in the later drawing, and unfortunately the view provided by Delorme, or rather the view as copied by whoever was responsible for the woodcut, is badly drawn (with numerous vanishing points instead of one, and with unforeshortened arcs to represent the outlines of the circular tower). The form of the *trompe* as shown in the view is also inconsistent with the *trait*. The *trait* indicates that the right side of the *trompe*, raised over the portal window, meets the wall in a straight line, not in a curve. Nevertheless, looking at this illustration, anyone would be excused for thinking that the Anet *trompe*—grotesque, frivolous, and amorphous—was not merely an offense against classical principles but an offense against any serious conception of architecture.

The vaulted surface of the Anet *trompe* is dragged down into the wall. Moreover, because the weighty stone drum of the king's *cabinet* is set on top of it, the impression conveyed is of an inverted structure, with the vault (*suspendu en l'air*) holding up the walls, not the walls holding up the vault, as is more usually the case. To make matters worse the aimless waving edge of the *trompe* cruelly bends and crushes the ring of balustrading.[14] As illustrated in the *Premier tome*, it looks like a careless and untidy amputation, but is the plate trustworthy? Delorme expressed some annoyance regarding the printed plates in his own book ("Mais ceux qui taillent mes planches, sur lesquelles sont imprimées les figures, ne les ont si exactement representées").[15]

Luckily there is another drawing of the Anet *trompe* in the *Premier tome:* its *trait*. We can be certain of the superior accuracy of the *trait* for one simple reason; a mistake in the drawing of a *trait* will invariably show up as an internal inconsistency. It can therefore easily be identified as an error, whereas errors in a perspective will just produce peculiar results; improbable but not demonstrably wrong. I have used the information from the *trait* to reconstruct the surface of the *trompe* in another perspective, this time projected accurately, to the best of my ability. I then added a perspective drawing of the *trait* on top of my perspective of the *trompe* in order to show how the *trait* was employed to generate the surface of masonry, since Delorme himself did not really explain this.

The most confusing thing about the *trait* is that there are 15 separate drawings superimposed. Each of these 15 drawings represents a different horizontal or vertical slice through the *trompe;* each is therefore a more

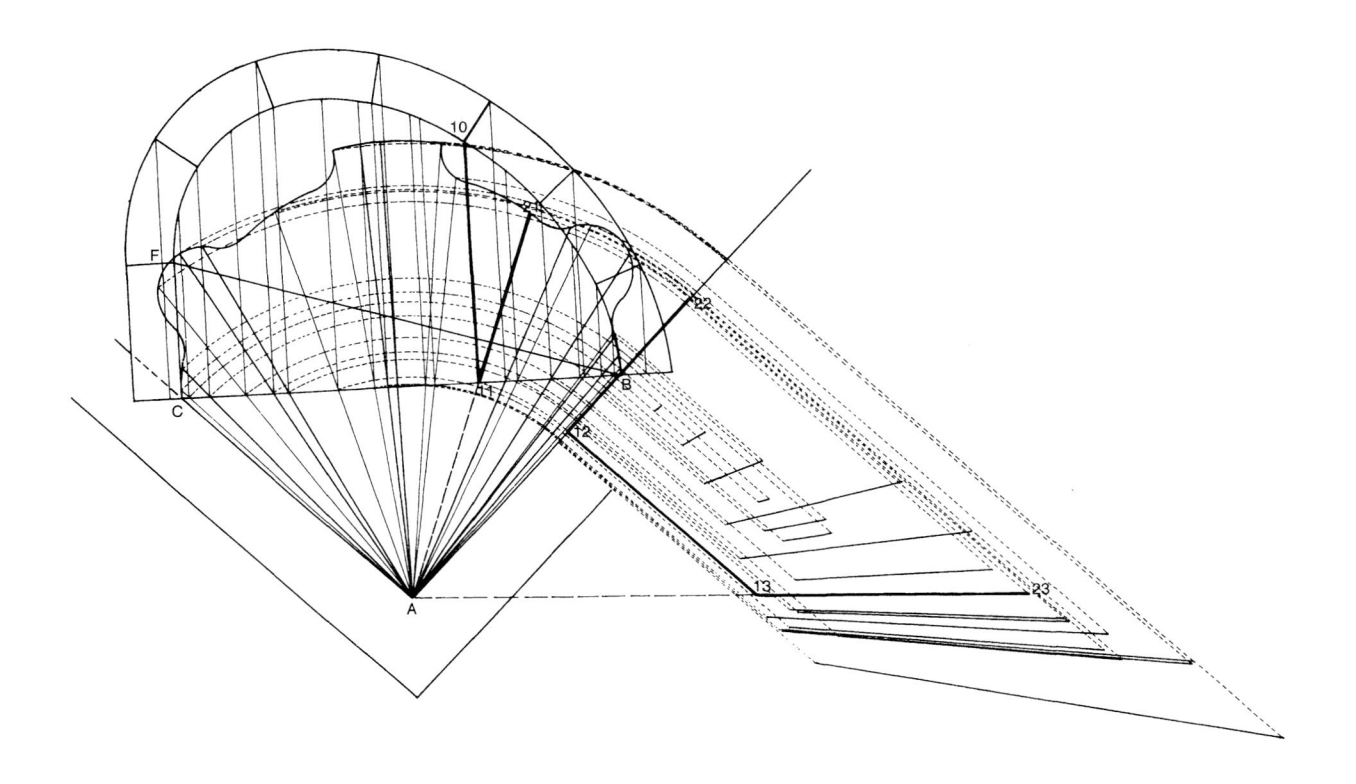

102 Philibert Delorme, trait *for the* trompe *at Anet (redrawn by author from the* Premier tome*). The explanation should be read in conjunction with the following two illustrations. I will concentrate on one line on the conic underside of the* trompe*. A set of such lines makes the surface. The* trait *shows how these lines are generated and then shows how to find their true lengths. Take the joint between the fourth and fifth clockwise stones of the ramped arch (labeled 10). The vertical line dropped from 10 projects into plan to meet the line BC at 11. In plan the generator of the* trompe *surface passes through A-11–21 to meet the outline of the cabinet. In plan all joint lines between stones radiate from A (although Delorme only showed the lines from the diagonal arch to the perimeter of the* trompe*). A-11–21 is then rotated to the right until it hits the wall, as it were, along A-12–22, which then become the baseline for an elevation. The height of the joint in question on the diagonal ramped arch can be read off the front elevation of the arch as 11–10. This height is then transferred to 12–13. Since all generators must pass through the vertex of the cone at A, and all sections through the vertex must be straight lines, the line A-13 produced gives the angle of elevation of the joint. It only remains to find its true length, which can be easily done by projecting up from 22 to obtain the point 23. A-23 is the true length. Sufficient reiteration of this procedure of rotation and* rabattement *along radials through A produces the* trompe *surface.*

DRAWN STONE **185**

or less cryptic "picture" overlaid on the others in such a way as to give the required information. Nearly as confusing is the lack of differentiation between section lines and projectors (projectors being the parallel lines used to transmit the position of salient points from one sectional slice to another).

It is worth bearing in mind that only two sorts of folding are necessary to get from the projected figures in the *trait* to the shape of the *trompe: rabattement* (folding on a horizontal line) and rotation (folding on a vertical line).

13 of the 15 drawings are shown on the right-hand side of the *trait* as the set of dotted, fanning shapes extended from the line AB. That leaves only two others, and they are the most fundamental. The plan (a horizontal slice through the upper part of the *cabinet*, above the level of the *trompe*) can easily be made out. The two walls meet at A. The outline plan of the *cabinet* extends in a complex series of circular arcs beginning at C and ending at B, but it is easy enough to construct. The most significant line in the plan is the straight line CB, passing diagonally across the corner. It is not a tectonic line and has no visible manifestation in the constructed *trompe*, yet it is the principal generator of its warped conical surface. For on CB is erected a conceptual formwork sufficient to define everything else; the "arc droict rampant." This ramped arch with its seven voussoirs represents a section through the masonry of the *trompe* above the line CB, so one has to imagine it folded up out of the surface of the plan until vertical. It can be seen that way in the perspective version.

The arch is cocked sideways with the springing point higher on one side than the other so as to miss the aforementioned staircase window. The shape of the arch was obtained by taking the vertical dimensions of a semicircle on the horizontal baseline BC and sliding them up onto the sloping line BF. The resulting curve, an asymmetrical semiellipse, maintains the advantage of terminating in verticals at both ends.

So we now have a diagonal section of the masonry overlaid on the plan. All the succeeding operations performed in the *trait* require that we continue to think of this as folded upright. The entire construction is made by taking straight lines from a point in the corner (A in plan, and level with the line BC in section), so that they extend through the ramped arch and continue onward until they meet the complex outline of the *cabinet*.

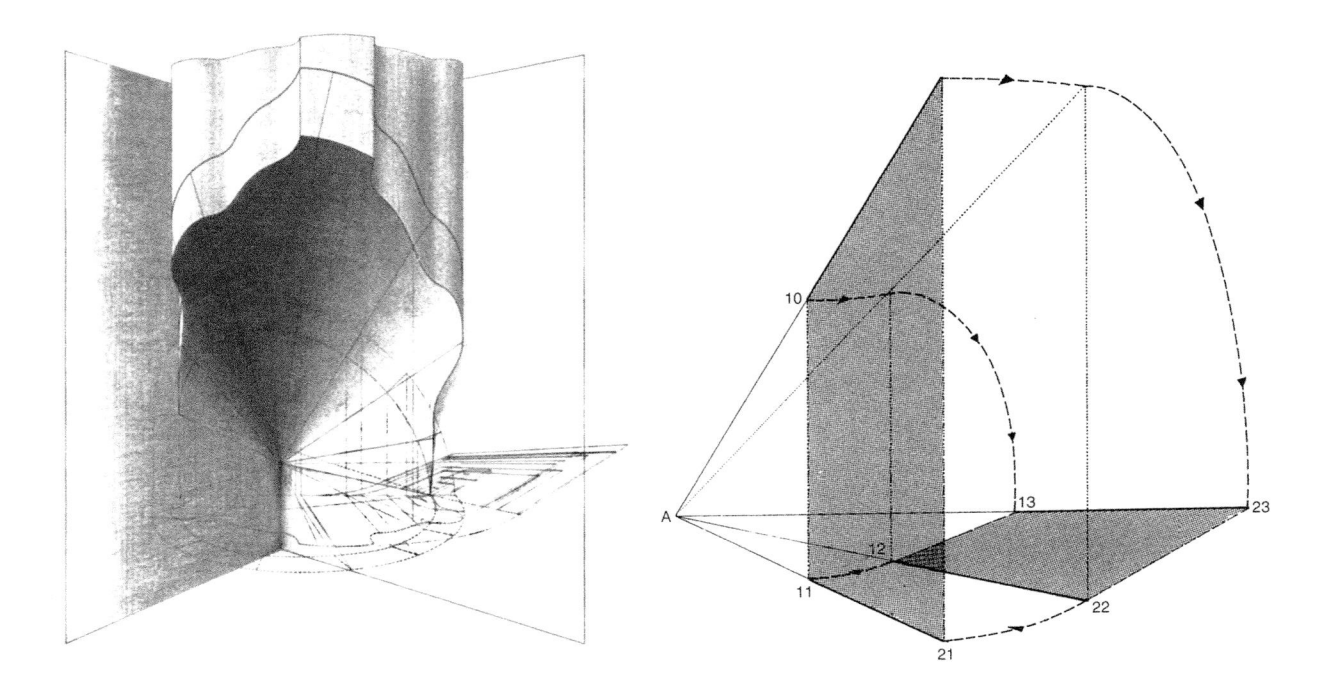

103 Philibert Delorme, the trompe *at Anet, relation between the* trompe *and the* trait *(drawn by author).*

104 Philibert Delorme, the trompe *at Anet, the basic rotation and* rabattement *(drawn by author).*

The *trompe* is composed of 7 long wedges of stone.[16] These have to be cut before the arch can be constructed. For them to be cut accurately, the lengths of all the sides and the angles between all the faces of each block have to be known. The *trait* makes it possible to deduce this, but it only provides the essential outline of measurements, in particular the real lengths and angles of elevation of the long sides of the 7 radiating voussoirs. The 13 vertical sections through the *trompe* radiating from A are all rotated into one vertical plane at AB and then folded down (rabatted) into the horizontal plane of the *trait*. A more detailed account of how this information was obtained, and how the surface of the *trompe* arose from the projection and folding of the principal *trait*, is given in the caption to figure 102.

Five derivative drawings in the *Premier tome* lead from the general definition of the surface to the description of each facet of each component block of stone. They employ an important extra operation, supplementary to the *rabattement* and rotation already observed: development. Development is yet another kind of folding that entails the flattening of a curved or faceted three-dimensional surface into a two-dimensional sheet. For example, one of these drawings shows the underside of the *trompe* with the seven adjacent facets of its voussoirs laid out flat. The joints between some of these stones can be seen overlaid in the same

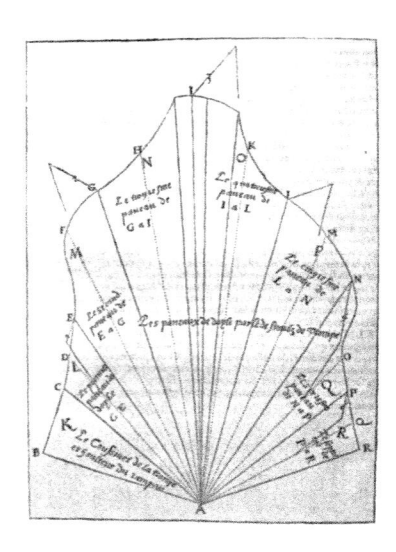

105 Philibert Delorme, the trompe *at Anet, developed surface of the underside, from the* Premier tome.

106 Philibert Delorme, the trompe *at Anet, perspective (drawn by author).*

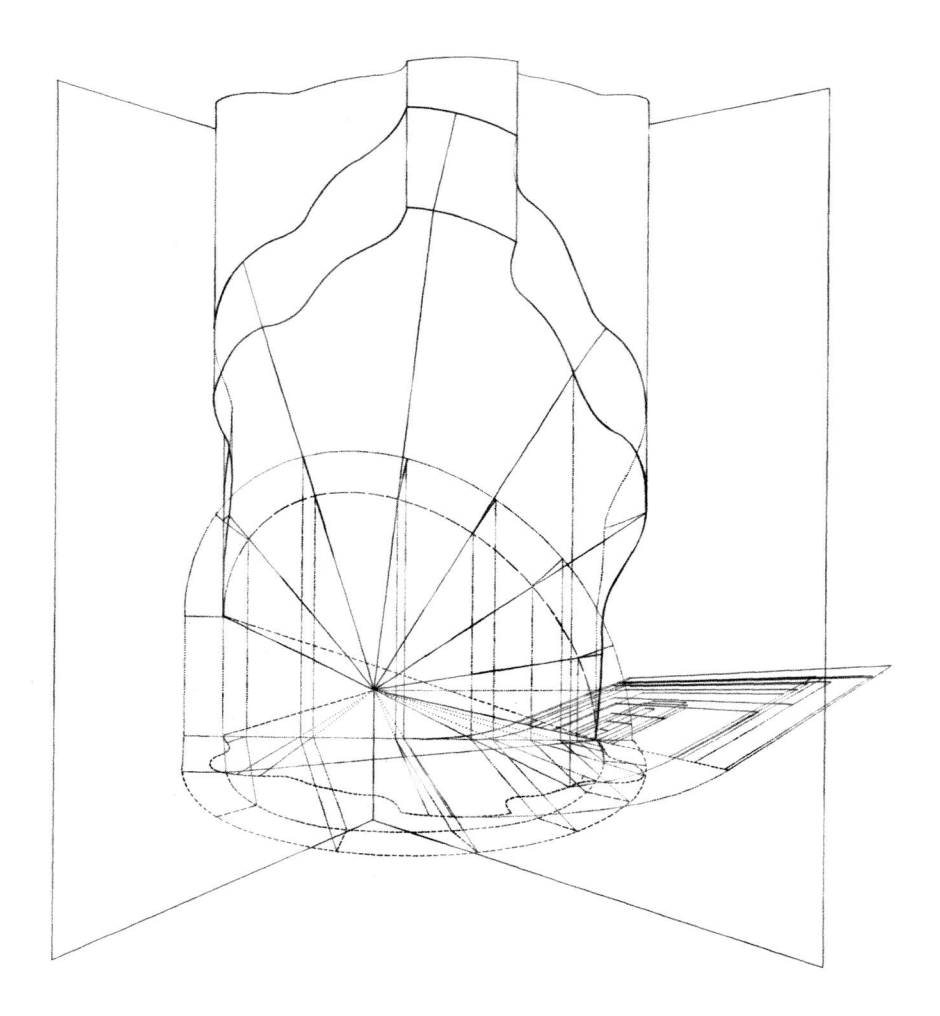

107 Trompe under the organ loft, seminary chapel, La Flèche, Jacques Nadreau, 1636.

plane. Such drawings were to become typical of stereotomy. From here it is a short step to the mason's yard, because each face, when drawn full size, would become a template furnished to the mason as paper, board, or zinc panels.[17]

Each stone of the *trompe* therefore has its individual and unique specification. They are cut and dressed as prefabricated items which, when assembled on site, magically combine into a perfectly unified form. It is not always easy to appreciate the sophistication needed to procure these results. We can glance at the finished work and see, all too easily, how everything fits because we use the completed surfaces to gauge the relations between their own parts. We can immediately perceive any intrinsic coherence and may well blithely assume its accomplishment to have been no more difficult than our comprehension of it.

Obviously, the *trait* could not have been made unless Delorme had a clear mental picture of the three-dimensional configuration to which it referred. It does not follow that he would have known everything about it. In this instance the resulting complex, scything curvature around the edge of the *trompe* is exceptionally difficult to envisage from the *trait*. It is the straightforward consequence of mounting a conic sheath of straight lines onto the upstanding ramped arch, but the intersection with the already elaborate figure of the drum produces an unpredictable distension of three-dimensional curves. It would be hard to judge their quality, prior to construction, without some other, intervening representation. If such there was (there is no evidence of it apart from Delorme's repeated insistence on the value of models),[18] it must necessarily have been derived from the *trait*. The shape of the *trompe* was not merely facilitated by projective drawing, but generated by it.

The antagonism between classical architecture and the *trompe* is in no way mitigated or explained by the retrieval of the *trompe*'s form. All the classical detail has been omitted from my redrawing, so the conflict has simply been avoided. Yet, although the reconstituted surface is recognizably similar to the original view, the general sense of it is much different. What seems a destructive conceit in the original view—clever but ugly—seems in the projected redrawing graceful and, in its own terms, proportionate, just as Delorme said. One might add voluptuous, suggestive even. These qualities are absent from the woodcut view. I did not invent them; they come straight from the *trait*.

Undressing

In painting it is easy enough to free architecture from the dual bonds of the rectangle and gravity, as can be seen in Francesco Salviati's fresco of *Bathsheba Going to King David* at the Palazzo Sacchetti in Rome, which was begun one year after the completion of the Anet *trompe*. In this panel Salviati took the sinuous double curve—a line that had originated in figure drawing—and applied it to architecture. Bathsheba is shown four times in different positions, ascending the stairs and then lying with David in a pavilion at the very top. Salviati's four figures are not all equally curvaceous. On the flat platform at the foot of the stair Bathsheba adopts a languid, elegant stance. Her body follows the characteristic snaking line of the *bella maniera*. As she ascends, her moving body becomes more stiff and taut, losing something of its languor, which seems instead to be imparted to the masonry stair that she climbs. The S-shaped line is now gracing the hard stone, having left the soft folds of her garments and the contours of her torso and limbs.[19]

This is a very painterly effect. Notice that the calligraphic shape of the stair is obtainable only from this one point of view. Properties shift from people to things; bodies freeze into sculptural poses and buildings quicken and twist. It is coyly pornographic, with the serpentine line as a sexual innuendo, exhibiting a promiscuity in the petrified form of the stair that leads to the entwined, copulating pair within the pavilion.[20] But we know, when we look at these scenes where the fleeting play of human desire betrays itself in the countenance of inanimate objects, that they are either accidental or restricted to representation. In this instance the restriction applies not only to the sympathetic registration of a passing mood in permanent fixtures, but to the form of the fixtures expressing it.

The serpentine line was the line of grace. It also represented escape from the confining stricture of geometry.[21] The free-flowing trajectory of Salviati's stair corresponds to no known architecture. Its lazy cantilever, which gives the impression of meandering through space as it unwinds from the belvedere tower, was unrealizable. One advantage of painted architecture is that the laws of statics do not take full toll; they need only appear to apply. Inside the frame of the painting, architecture may divest itself of its rigid, cautious rectangularity. As it is compressed and flattened into pictures, it is made free. A dimension is relinquished and a freedom is gained; a reasonable exchange. I use the word exchange because I want to suggest some kind of bargain struck within an established economy. This economy affects all sorts of transactions between wishes and fulfillments, but the cost increases with the extent of that fulfillment.

108 Francesco Salviati, Bathsheba Going to King David, *Palazzo Sacchetti, Rome, 1552–1554.*

We all recognize this crude precept that permits us to wander freely in the imagination while opting for a confining security in reality.

It is the application of this same formula for localized indulgence that leads to the parceling of architecture into businesslike and gamelike aspects. The business is to make stable structures, and the game is to invoke illusions that make the building appear to contradict its own necessarily stable constitution. Salviati's "capricious and ingenious inventions"[22] are easily assimilated because they are understood to be fictions. When a comparable caprice is realized in stone, its status is not so clear. The game has gone further, perhaps too far. There *is* masonry that defies its own material constitution and the pull of gravity; there is building that escapes the straight line and the circular arc.

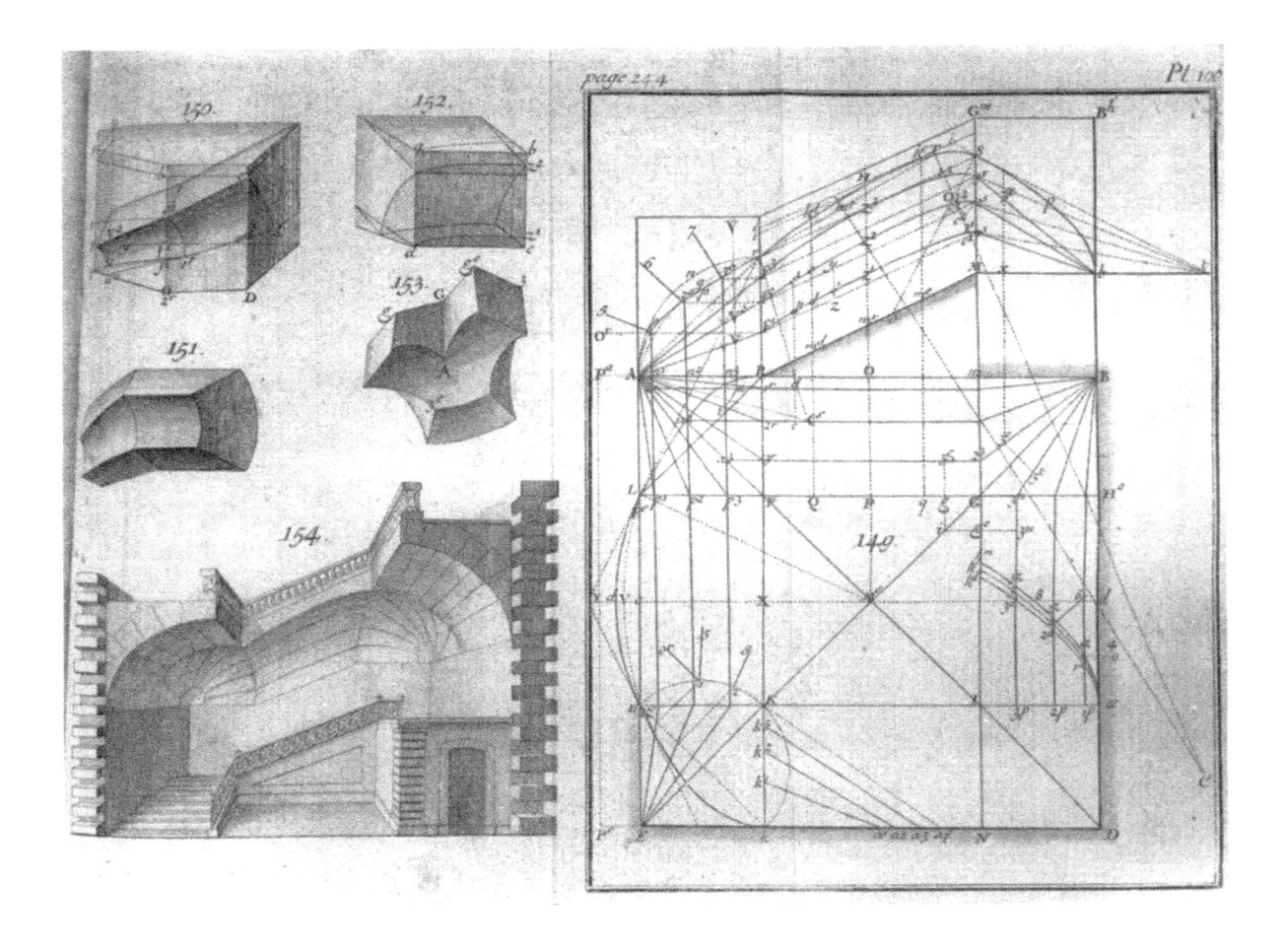

The Anet *trompe* was not an isolated instance of this disturbing effect, nor were *trompes* the only kind of structure to provoke it. In the Benedictine abbey of Notre Dame de la Couture in Le Mans, built between 1720 and 1739 by an unknown architect or mason,[23] there is a stone stair of a type frequently encountered in published *traits* and of which there are numerous built examples:[24] *l'escalier vis de Saint-Gilles suspendu*. In this unusual case, however, a major part of one of the encasing walls was removed to make a vast window that extends over several stories. Consequently, the adjacent flight of stairs has no support on either its inner or outer edge. It could not cantilever from the side wall, nor could it span from landing to landing, so the intrepid architect or mason set it on groins that extend from the only two points of contact with vertical support, in shallow curves that disappear like sinews into the body of sus-

109 A. F. Frézier, escalier vis de Saint-Gilles suspendu, *from* La Théorie et la pratique de la coupe des pierres, *1739.*

110 Notre Dame de la Couture, Le Mans, suspended stone stair, architect unknown, 1720–1739.

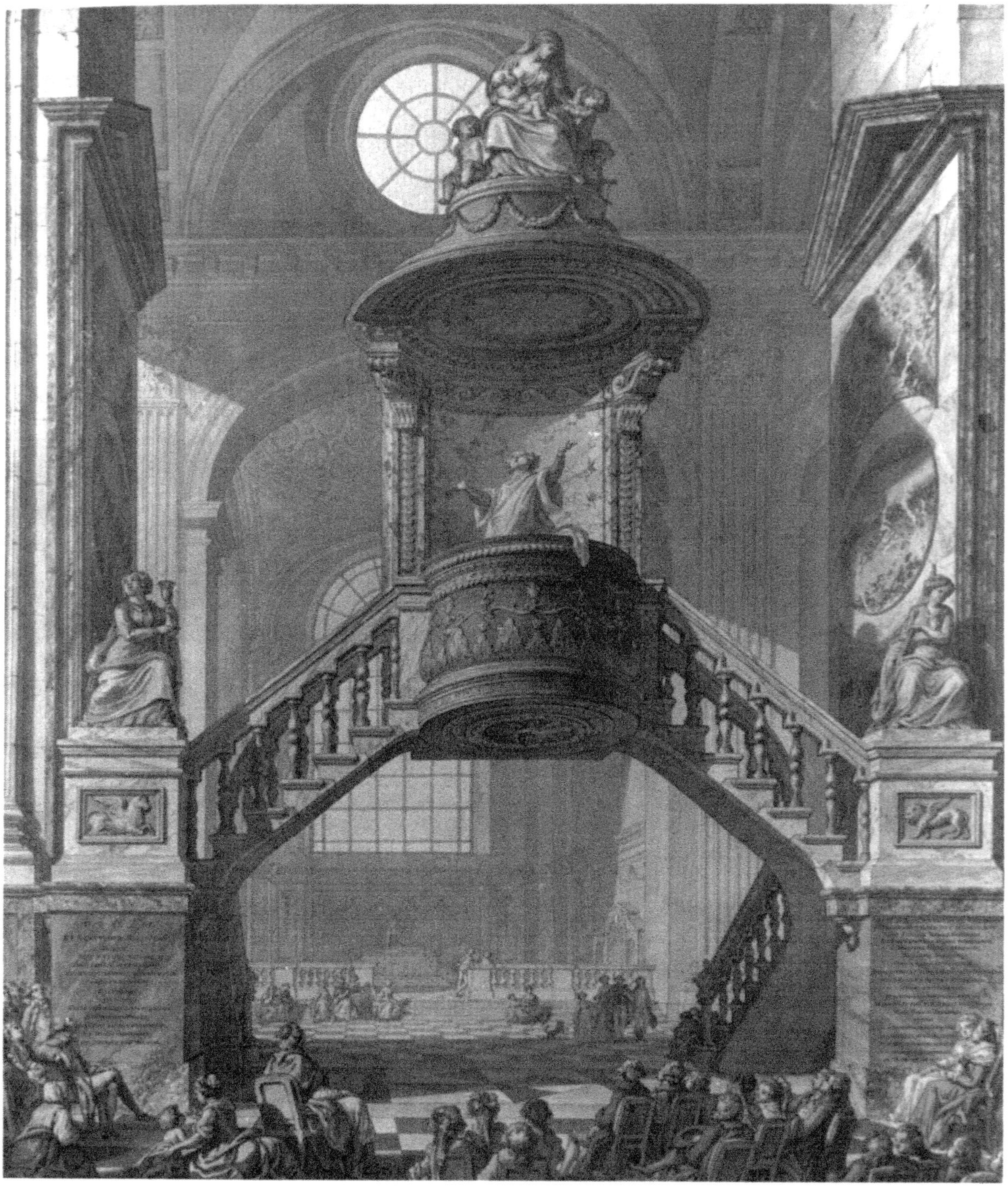

111 Pulpit, Saint-Sulpice, Paris (second project), Charles de Wailly, 1789.

pended masonry, to carry the length and the breadth of the rising stair simultaneously, using the inward thrust of the adjoining flights to stabilize its detached ends. And there it hangs, resting on lines of intersection that have no physical body at all. A project by Charles de Wailly for a pulpit at Saint-Sulpice, Paris, drawn up in 1789,[25] froze the imagined oratorical gesture of a preacher in imitation of the vigorous thrust of the stairs that led to and sustained a large masonry canopy balanced in midair with no vertical support whatever. Proposed in the year of the Revolution, it was never built, but de Wailly, who was a painter as well as an architect, and who had made orthographic drawings of the pulpit[26] as well as the fanciful, dramatized perspective of it in use, had nevertheless bridged the gap between the two arts by synthesizing two quite different aspects of the suspension of disbelief: the gestural (like Salviati) and the structural (like Delorme).

The resemblance between Salviati's fresco and Delorme's *trompe* is unexpected because of the difference between paint and stone. There is another difference that is perhaps less striking but no less considerable. It is not just that these unlikely shapes could be made to exist as structures, but that they were brought into existence with the aid of a box—a rectangular figure just like the one from which we imagine them to have been liberated. Salviati's *linea serpentinata* is the line of free inspiration, unhampered by compass and rule. Much was made of this by the sixteenth-century Italian painter and theorist Giovanni Paolo Lomazzo;[27] it had been authorized by the divine Michelangelo; it was like a flame; no wonder it could catch the ineffable. Vasari recalled that his friend Salviati's compositions were unstudied, quick, even impetuous.[28] How laborious, by contrast, was the geometrical art of stonecutting, all measurement, projection, and instruments. It should produce, according to the painters' prejudice, which tends to be our own, opposite results, but instead turns out to be similar. Stereotomically determined masonry, despite the immanent presence of geometry throughout, shows less obvious trace of geometric regulation than the most whimsical scenarios in mannerist painting. Not only had geometry been at work, not only were the billowing stone surfaces geometrically determined, but they were determined by the very forms that had absconded from them: circles, rectangles, and straight lines. By using more geometry, they appear to have less. This effect is largely dependent on a normative idea about what geometry looks like, but there was, in the more advanced examples of stereotomy, a careful removal of the evidence that would lead the mind back to that normative conception; a kind of undressing. The French word for the procedure for getting a complex stone out of a rectangular block is *dérobement*. It is, says Amédée Frézier, author of the major eighteenth-century treatise on stereotomy, "as if one stripped the proposed figure of the robe with which it [she] is enveloped."[29] *Dérobement* referred only to an optional technique for cutting individual stones, but the entire corpus of *traits* was based on an analogous process of geometrical undressing.

However complex the masonry formation defined by a *trait*, however angular, bent, or strewn the *trait* itself appears, related sections are always perpendicular. Orthographic projection, which pertains between these plane sections, treats the intervening space as a plenum of perpendicular, parallel lines. There may be several orthogonal orientations in a *trait*, but each will make a virtual rectangular box. Each box is a mental construct in space that is then unfolded into the *trait* where it registers as flat.

Let us return to the mannerist painters' prejudice against geometry—now incorporated into common sense—from the opposite point of view. There is a very fine painting in the Metropolitan Museum of Art, New York, by Hendrick Goltzius, called *Job in Distress.* Job was traditionally shown being beaten by the Devil, or with horribly diseased skin, sitting on a dunghill,[30] but in this version there is no very obvious evidence of the man's suffering and humiliation. There is only the doleful expression on his face and the constraint of his body within the edges of the picture. The diagonal turning of the square and the fitting of Job's contracted extremities into it make this one of the most powerful and effective uses of framing in Western art,[31] especially so since it is all a matter of inference. There was nothing in Job's life to suggest torture by constriction, nor are the edges of a perspective picture supposed to represent the edges of a claustrophobic cell. The frame does not normally impose itself on the depth of the picture in quite this way. Usually it is a window directing the observer's view; here it is an instrument of torment closing in relentlessly on the subject. This provides a rare, early illustration of the geometry in painting that we are always inclined to overlook; the prosaic geometry of the flat rectangle on which it is painted. Exaggerated, its subtle control turns into a tyrannical domination.[32]

Goltzius's painting makes the containing square into a prison by cramming Job's body into it. The closer the straight edges press in on the mortified flesh, the more uncomfortable the effect, until it becomes excruciating; remember the story of Procrustes cutting off his victims' limbs so that they would fit into his iron bed, or the Crucifixion.

The process so skillfully implied by Goltzius can be put into reverse, and that is what we witness when we look at the results of stereotomy. In the

112 Hendrick Goltzius, Job in Distress, *c. 1616, Metropolitan Museum of Art, New York.*

trait the body is informed, not by the shape of the geometrical dress but by the procedure of undressing. Whether as tunic (box) or as armature (cross), the rectangular figure must come first, only to disappear.

This accounts for the peculiar way in which Philibert Delorme launches the subject of geometry in the *Premier tome*. After some jaundiced remarks about the relations between clients and architects, he insists that it is necessary for both architects and master masons to know geometry, and then describes the fundamental act, the foundation of any enterprise, as a squaring up. No work can commence

> *without them first laying out a straight line and a perpendicular, or squared* trait *(as it is called by workmen) either on its own, or inscribed in a circle. They can similarly proceed with two parallel lines, providing that, across the middle or at the ends, they throw a perpendicular. One can also draw the perpendicular on the end of a straight line.* . . . [33]

It all comes down to the same thing, and this is not, Delorme continues, only true of architecture. Two lines crossing at right angles, now capitalized as the Cross, lie behind perspective painting, music (the staves and signs of notation), and military engineering. It was venerated by the ancients, as it is by Arabs and Christians. Delorme sketches a brief cosmogony, presided over by the Cross that divided the undifferentiated spherical world at the second instant of creation. The excellence and pervasive presence of the Cross is, he tells us, agreed by philosophers such as Marsilio Ficino. Yet, as we know, Ficino lauded the sphere as the perfect original of geometry and the world, whereas Delorme seems to regard the sphere as merely a potential until it is intersected.

Finally, to sew up the argument for the priority of the rectangular intersection of straight lines, Delorme invokes Christ on the Cross:

> *Perhaps to show that life and health must come to mankind through the death of a single mediator, who was attached to wood in the shape of a Cross, which was the first figure that God the Father gave to the World. But we leave such questions to the Theologians, and return to our lines and geometrical* traits, *which can help the architect so much.*[34]

As well we might, for Delorme had hardly clinched the argument by adding this last example. Instead he had exposed the paradox that vitiated it. But then I would incline to the view that the paradox was more highly prized than the argument. A similarly provocative *non sequitur* was paraded by Delorme in the frontispiece to the *Premier tome.* This was, typically, an ornate engraved frame, upright and classical, adorned with miniature pediments, cornices, and architectural moldings, then bedecked with swags, scrolls, shells, and cornucopias. Inscribed within the rectangular frame was an oval in which went the title, but the focus of attention was the elaborate surround. On it sat five easily identified references to geometry: pairs of dividers rest on the overframe; two *putti,* sliding off the ends of the broken pediments in imitation of Michelangelo's *Day and Night,* hold celestial and terrestrial orbs; lower down, two regular solids, icosahedron and dodecahedron, the two that approach most nearly to the sphere, rest on candlesticks to the left and right. The bedecked frame conveys the familiar idea that the geometry behind architecture is ordained by a higher, universal geometry. It also signifies, by the choice and disposition of emblems, that universal geometry, being spherical, requires considerable Platonic devolution before it gets down to the frame. Now this is exactly what Delorme steadfastly refused to admit in writing.

Nor does he let it rest at that, here in the frontispiece. For, hovering in the interstices between the frames and then again outside, are apparently meaningless hieroglyphics, subject neither to gravity nor iconology. They

113   Gérard David, Christ Nailed to the Cross, *National Gallery, London.*

114   Philibert Delorme, frontispiece to the Premier tome, *1567.*

are, of course, examples of *traits* for an oval dome, spiral stairs, a bowed arch, a helical vault, and a *trompe*. Afterthoughts filling in the spaces left over on the page, their relation to the ornamented frame, with its clear message of devolution, is problematic because they do not belong in the same story. In one story the rectangle is the beginning, in the other the end of the tale. In the classical frame, the rectangle is a derived and visible result: in the *traits* it is an original, soon to become invisible, cause.

The disjunction of these two conceptions of geometry in the frontispiece is comparable to the irrelation between the *trompe* and the facade at Anet. It has been argued that Delorme was versed in a kind of geometry different from that propagated in the Vitruvian treatises. Joseph Rykwert has suggested that there was an oral tradition preceding the printed texts on architecture, based not on Vitruvius but on Euclid. From this oral tradition, centered on the masonic lodges, issued the ingenious geometry of Gothic construction.[35] Rykwert associates Delorme's accomplished use of projection with this Euclidean strain. And certainly Delorme had some sharp things to say, if not about Vitruvius then about his commentators, while he was full of praise for Euclid, commentators and all.[36] The praise for Euclid follows a long passage on the great difficulty and immense mental effort required for the understanding of *traits*, providing further evidence of a schism on the lines suggested by Rykwert.

This evidence is too significant to dismiss, but it is irreconcilable with the geometry of the *trait*, which is no closer to Euclid than to Vitruvius, and which cannot be directly derived from either source. Certainly, there are two geometries vying with each other in Delorme's work, but I would categorize them as Platonic and projective rather than Vitruvian and Euclidean.

Vitruvius contained only occasional and vague statements about geometry,[37] whereas Euclid's *Elements*, the epitome of rigor, contained well nigh all there was to know of it, even in the sixteenth century. The connection between the fabulous Euclid who founded masonry,[38] the drawing of the *traits*, and the authenticated Euclid of the *Elements* would seem a matter of reasonable inference, were it not for one thing: throughout the 13 books of the *Elements* there is not a single reference to, or utilization of, projection, whereas projection is fundamental to the *traits*. No doubt a grounding in Euclid facilitated the understanding and performance of *traits*, but it is hard to see how Euclid could have been their source.[39] So how, then, did the masons get from Euclid to the *traits*?

Let us leave that question and proceed to another, related but no less puzzling. In 1643 there appeared a small, contentious book on stonecutting by Abraham Bosse, *La Pratique du trait*. Bosse was the disciple and ardent promoter of the work of Girard Desargues, whose new method the book described. Desargues and Bosse published several other works on practical aspects of projection such as making sundials, perspective, and geometric drawing (parallel projection). A few plates added to the book on perspective illustrated a theorem—Desargues's theorem, or the theorem of perspective triangles[40]—that was later identified as the first to be presented solely in terms of projective relations. Since this momentous step had been taken in consequence of the study of various sorts of practical projection, and since nearly all of Desargues's output as mathematician, architect, and savant dealt with projection, we might expect wholehearted acceptance of the projective aspect of the *trait* in his work on stonecutting. Yet there was less projection in Desargues's method than in any other. The basic steps of his new method retained none at all. *Rabattement*, rotation, and Delorme's originating perpendicular cross were maintained, but an entirely Euclidean construction of congruent triangles took the place of projection. Desargues's Euclideanizing of the *trait* goes unremarked, although Bosse drew attention more than once in his foreword to the fact that the new method was not "géométrie pratique,"[41] by which he meant that it was not orthographic projection. How is this turn of events to be explained?

The two questions are similar in kind. The second is probably easier to answer than the first, but I am not so much concerned with answers as with what the questions themselves indicate about the status of *traits*. Considered as a branch of mathematical study, the projective geometry inaugurated by Desargues seeks invariant relations between figures that undergo projective transformation. Think of a triangle. If it is projected, the sizes of sides and angles vary from one plane projection to another, but the sides will invariably be straight lines and the projected image of the figure will invariably stay triangular (even when it degenerates into a straight line). Euclidean geometry is principally concerned with magnitudes, but Desargues's projective geometry is not. The *trait* comes somewhere in between. The objective is to find the size of lines and angles; the method is projection, which alters their size.

The *trait* was, no doubt, affected by Euclid; projective geometry certainly received a tremendous fillip from the *trait*, but the *trait* is not reducible to either, nor is it confined within either; it straddles. Desargues recognized that since the objective of stereotomy was to find magnitudes, these could be more expeditiously found by Euclidean means.

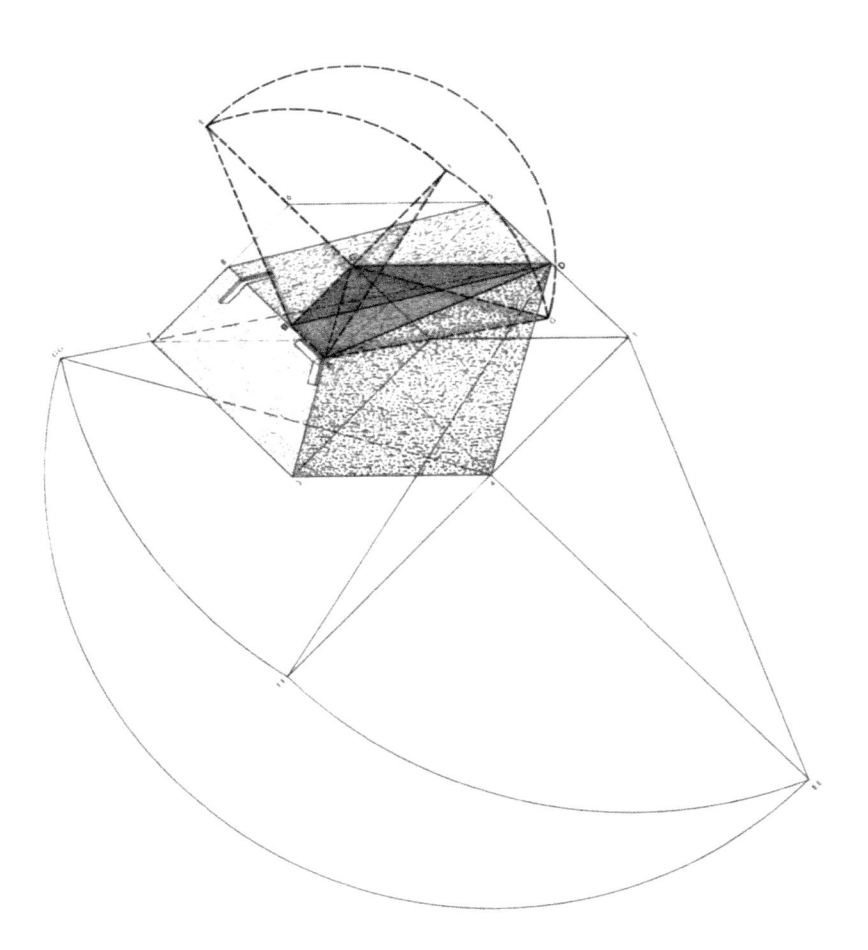

115 Diagram of the first 7 of 11 successive operations in Girard Desargues's universal method for stonecutting, from which projection was eliminated.

The objective is to find the compound angles 3AD and DA8, given the angle 4D5 and the angle of inclination 7-8-5, both the latter easily obtainable from standard plan and elevation. The construction as explained by Desargues/Bosse is shown as 1-D-GG-FF-EE. The rest shows how this cryptic figure relates to the problem.

In brief, this figure represents the faces of the imaginary tetrahedron ABCD inserted between the block of masonry and a circumscribing virtual cube 1-2-3-4-5-6-7-8. The figure ABECFG shows this same construction in its most obvious location, from which it is easily understood to be an unfolding of three sides of the tetrahedron into the same plane as the fourth side ABC. Describing the enlarged, displaced version deepens the obscurity of the method by divorcing the construction from any recognizable visual connection to its object, but it reduces the number of steps required to obtain a result.

Referring to the more accessible figure ABECFG, triangles ABC and BCE can be constructed with the information given. Rotating BE to BF enables construction of the face ABF (rabattement of ABD). Finally, the intersection of CE rotated about C with AF rotated about A at a point G gives ACG, the fourth face of the tetrahedron (rabattement of ACD). The angles required are 3AG and FAB. (Drawn by the author.)

Giorgio Cucci has shown that the division between those for and against Desargues's method for stonecutting followed the lines of a class division between the few academic theorists who approved of it and a larger body of builder-architects, associated with the masonic tradition, who reacted with predictable hostility to Desargues's denigrating their established techniques and their native abilities. It was as if the technique were being taken out of the hands of practitioners and reserved for intellectuals. Certainly, Desargues reduced the visual comprehensibility of the *trait*. All *traits* were abstruse, Desargues's even more so. This was not a problem for Desargues, whose sublime ability to conceive geometrical relations in space should be the subject for another study, but it was for almost everyone else. So the division between antagonists and protagonists of Desargues's method cut several ways: between those who saw and those who thought; between masons and architects; between artisans and professionals; between those who cut stones and those who drew lines.

Paper and Stone

Great works of architecture were made of stone: heavy, solid, obdurate, and enduring. *Traits* were made of paper: light, thin, pliable, and expendable. No working *traits* have survived from the Gothic or classical periods.[42] The only evidence of their existence is from written sources and printed books, lending fuel to the view that *savants* usurped the secret skills nurtured into existence by craftsmen. This rings true of the seventeenth century when Derand, the architect and mathematician from the Jesuit academy at La Flèche, took it upon himself to improve the published work of Mathurin Jousse, the mason-carpenter also from La Flèche; when Bosse claimed that Desargues's method would permit the architect to take control of stonecutting from out of the feckless hands of workmen; when François Blondel, professor at the French *Académie*, expressed identical sentiments; and when another renowned mathematician, and Blondel's successor at the *Académie*, Philippe de la Hire, continued the rationalization of the *trait* embarked upon by Desargues.[43]

It was true of the eighteenth century, when Nicolas Le Camus de Mézières, architect to a veritable showpiece of stereotomy, the Halle au Blé in Paris, confided that a good *appareilleur* was a real treasure ("un vrai trésor");[44] a patronizing but understandable remark, under the circumstances, since he had just informed his readers that the *appareilleur* had sole responsibility for everything to do with masonry, from the execution of plans (*épures*) to the finished quality of the assembled stone surfaces.

116 J. -B. de La Rue, frontispiece to Traité de la coupe des pierres, *first published 1728.*

Just as revealing of the dominance of architecture over masonry was the frontispiece to La Rue's *Traité de la coupe des pierres* (1728), which showed a benevolent and smiling personification, who might signify Architecture, Geometry, or Design, but who was most certainly in command of the *trait,* surrounded by the paraphernalia of construction. Scurrying and flitting around her were nine *putti* attendants doing the work. In this puerile allegory of site labor, three transformations were effected by virtue of a common conceit: the drudgery of hewing and setting was turned into child's play; the exclusively male authority of the architect was turned into a benign matron's kindness; the roughness of the mason's trade was turned into the delicacy of a domestic diversion. Such were the ways in which a palatable iconography stood substitute for distasteful reality.

La Rue's portrayal of free and happy labor under amicable regulation may justifiably be interpreted as pure ideology in the guise of an Olympian fairy tale. By common consent, the masons' task was backbreaking. Little was written about it, but there is indirect evidence enough in support of the idea, such as the forcible impressment of masons in the late medieval period, or hard-labor convicts quarrying and dressing stone during the nineteenth century. It has been suggested that working masonic lodges, until their decline in the sixteenth century, must have been more like nineteenth-century factories than free associations of guildsmen.[45] Stone is heavy, yet in the La Rue frontispiece it is as if the stones borne by children had no more substance than the drawing. The intellectual difficulty of constructing the *trait* had eclipsed the physical difficulty of constructing stone buildings.

There can be no doubt that architects usurped masons. But are we to believe that there had been some prelapsarian state of masonry, prior to Delorme's publication, prior to Jousse's second divulging of the secret,[46] prior to the divorce between theory and practice, prior to the divide between architect-designer and mason-fabricator?[47] Such a condition there may conceivably have been, but how the *trait* could have functioned within it is another question, for surely the *trait* itself encourages the distinction between theory and practice by resolving problems of construction on paper and off site. In the sixteenth and seventeenth centuries we witness a tussle between masons and architects over possession of the *trait*. When the architects wrest it from the masons, we jump to the conclusion that the masons must have been the original owners. It is more likely that the *trait* never really belonged comfortably within either the craft or the profession. It was a thing in itself which, as in the matter of geometry, straddled recognized borders.

The illusions in La Rue's frontispiece extended in another direction too. The trivialization of labor was effected by making light of it; unearthed stones lose weight. In this respect the allegory was accurate, because that was the observable effect of using *traits;* buildings made with their aid possess quite astounding powers of deception. Despite the fact that the stone is exhibited with absolute candor, its plain unvarnished truth is not easily assimilated; it just does not look as if it can be stone. The crepuscular surface texture and the lineaments of precisely joined blocks can be felt and seen, while undulating smoothness, against the nature of the material, suggests something light, bodyless, and pliable.

Stone, with the labor and the weight wrung from it, could look almost like paper. Seemingly, the *trait* had exported some of its own properties to the masonry it defined. Doing so, it demonstrated the victory of intellectual form over material substance. In *La Pratique du trait,* the most

117   A block of stone and its trait, *from Abraham Bosse,* La Pratique du trait, *1643.*

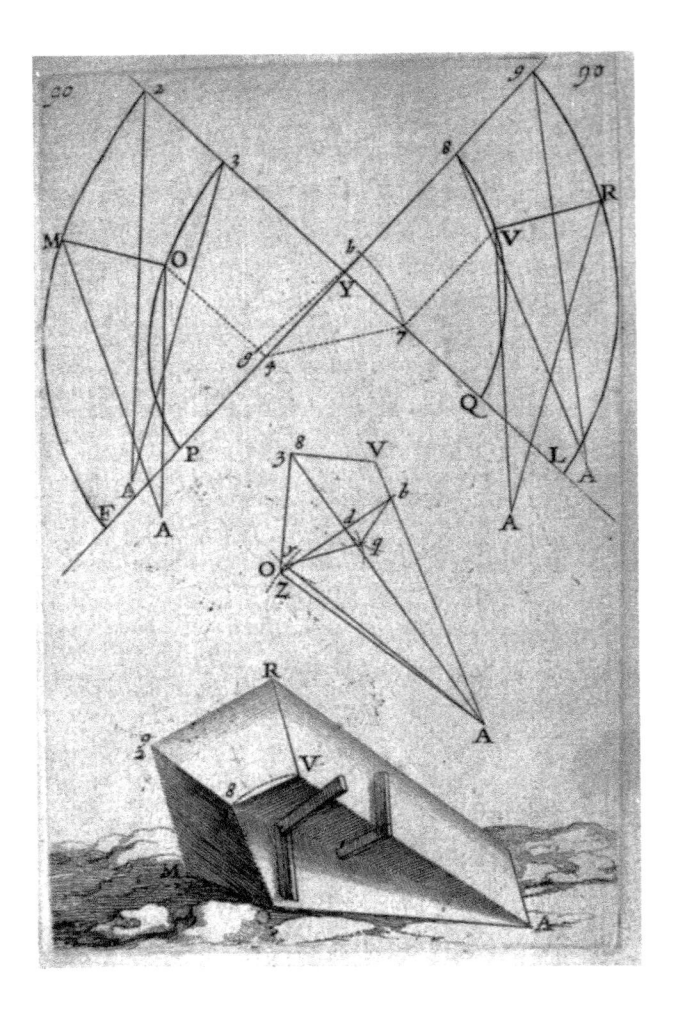

cerebral text on stonecutting, this was summed up in a plate showing a coffinlike lump of masonry with, above it, a halo of lines that contained the principle of its shape: like a soul risen from its body. With the aid of some inference, the *trait* can be pulled onto safe and certain ground. We know where we stand with distinctions like soul and body. But, while acknowledging a strong tendency toward disembodiment in stereotomic drawings, inference could easily lead us in the opposite direction. It depends whether we look at the *trait* or the stone.

Not all the properties of the paper *trait* traveled into buildings. A *trait* is made with nothing but lines. These are of two kinds; the imaginary lines of geometrical construction, most prominent of which are the projectors, and the lines indicating contours of the thing drawn. Delorme made perfunctory and occasional efforts to distinguish the two in his drawings.

118 François Derand, trait *for an oblique arch descending through a cylindrical tower into a dome, from* L'Architecture des voûtes, *1643.*

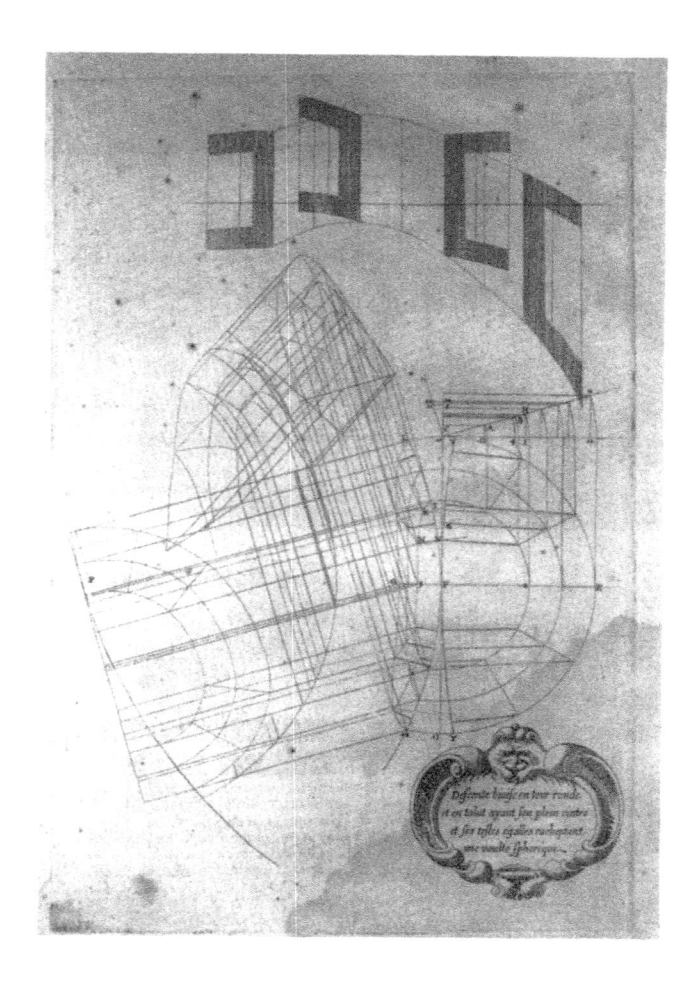

The same was true of Jousse, although he used the evocative term *lignes blanches* to describe projectors in his text.⁴⁸ Derand, who made the distinction clear and consistent for the first time, reversed the emphasis: the imaginary projectors—the impalpable *lignes blanches*—were drawn in solid line while the contours of the stones and vaults were drawn in a fainter broken line. The forms of objects therefore recede behind the manner of their manipulation, since the lines that stand out are the metaphysical cuts and thrusts of sections, projections, and folds.

The tight-jointed masonry made from a *trait* may look like paper with lines drawn on it, but the *trait* can hardly be thought of as a portrait of its masonry. The published drawings on stonecutting had aesthetic properties of their own. This was already apparent in the woodcuts for Delorme and Jousse. In the copper engravings for Derand, Guarini, La

Rue, and Frézier it was surely being consciously wrought, as is indicated not just in the cultivation of the printed *trait* as something distinct from workshop practice, and in the elegant composition of lavish plates, but also in the noticeable decline in the quality of published illustrations during the nineteenth century, by which time stereotomy had ossified into a set of rote instructions, no longer subject to significant internal development, interest having shifted to the allied subject of descriptive geometry. When technical attention drifted from the *trait,* so did aesthetic attention; but this aesthetic attention was never exclusively devoted to the *trait:* it was also directed at the masonry. Both required immense investments of aesthetic intelligence, but there was no necessary connection between the formal perfection of the one and the formal perfection of the other; the one type of weightless grace did not ensue from the other. The theorist-practitioners of *traits*—Delorme, Derand, and Guarini[49] in particular—were able to design with a kind of double vision such that one operation could produce two results in different modes, each of equal finesse. This requires an unusual sensibility that seeks gratification in the drawing but is not seduced by it; that can focus on the deferred consequence as well as on the immediate result.

Having distinguished these two aesthetic locations, consider the one most remote from the practitioner of the *trait*—the stone buildings that were made from the paper drawings. The skeletal ideality of the *trait* is in some contrast to the sensuous character of stereotomic masonry. There is a smooth, skinlike quality, noticed already in regard to the Anet *trompe* but applying to a large proportion of stereotomically determined constructions. Stone ends up like taut skin on firm flesh. I have to admit that I could not cite any credible historical source to validate this anthropomorphism, other than what may be intimated from technical terms such as *dérobement, nu,* and groin, while a vast number could be mustered in support of the anthropomorphism of the classical orders.[50] Does that mean that it is less true? Might it not be more true, though unsaid? We will happily believe (or we will happily believe that others believed) that a Corinthian column is like a young woman and a Doric like a virile man, but that is because these images have been forced onto them by insistent repetition, not because the analogy is close. After all, as Frézier put it, "man is not made to carry a static load."[51] He is not made like a column.

Between Classical and Gothic

Frézier's statement occurs in the *Dissertation sur les ordres d'architecture,* a supplement to his three volumes on stonecutting. He expressed the hope that the preceding volumes, which dealt with real structure and construction, might throw light on the proper use of the classical orders, which he conceived to be ornamental. However, in these same books, he had proposed a Gothic origin for stereotomy. Earlier commentators (Delorme, Jousse, Charles Perrault, François Blondel, La Rue) had said that the *trait* was a modern invention, unknown to the ancients, but Frézier was the first to specify its modernity as Gothic.[52] It is now generally accepted that the French classicists' attitude to Gothic was ambivalent. Condemned as illogical, disproportionate, and fantastical, it was simultaneously applauded for its lightness and luminosity.[53] Frézier too was ambivalent. He advocated a classicism modified by stereotomy, and therefore, by implication, modified by Gothic. French architecture included numerous attempts to synthesize favored aspects of Gothic into the classical style. Frézier's new pedigree for stereotomy accorded with that intention and fits perfectly into the larger picture plotted by historians in our own times.[54] In fact, that picture had already been mapped out by the end of the eighteenth century. In 1792 the architect Delagardette, in his introduction to the mathematician Simonin's *Traité élémentaire de la coupe des pierres*, described modern architecture as a beneficial mixing. According to him, the natural genius and creativity of Gothic architects had been exemplified by their use of masonry. Afterward, the gradual recovery of tasteful ancient models allowed their superimposition on the prejudices born of Gothic fancy until triumphant over it. Then finally the laudable things in Gothic masonry were brought back into ancient architecture, and the resulting *mélange* was neither ancient nor Gothic but modern.[55]

What then was to be borrowed from Gothic construction—its actual stability or its apparent daring? The mystery of building is that many things conspire to stand as one. When Hegel wanted a simile to illustrate the paradoxical combined effect of individual, self-interested human endeavors, he chose architecture:

> *Stones and beams obey the law of gravity—press downward—and so high walls are carried up. . . . Thus the passions of men are gratified; they develop themselves and their aims in accordance with their natural tendencies, and build up the edifice of human society; thus fortifying a position of Right and Order against themselves.*[56]

This helpful paradox was pondered in the article on stonecutting in the *Encyclopédie* (1754): forces pressing downward on lots of small stones made it possible to build great compilations that seemed to defy those very forces.[57] This was as true of classical building as of Gothic. Both could be thought of as victories won against the earthbound weight of stone, yet there was clearly a tremendous difference in the visual effect of the triumph in each case. Stereotomy was associated with Gothic because, from the point of view of any French neoclassicist, it toyed with danger. Classical structure reassures; Gothic structure disconcerts. Classical structure is stable, Gothic vertiginous. The *hardiesse* of Gothic and of stereotomy—an audacity bordering on foolhardiness—appealed, it was said, to naive sensibilities.[58]

The acknowledged real stability of the best Gothic was therefore compromised by a tasteless flaunting of its triumph over gravitation. Practically every French author on architecture said so. Here is how Frézier put it:

> *While there are limits to solidity, there are also limits to the delicacy and lightness of an edifice; for even though it may be solid because of the consistency of the materials, and by the artifice of their arrangement, until the idea we naturally possess of the proportion that must pertain between a support and its load [is satisfied], supports appear to us as too feeble or too wide, we are not able to approve of the construction, our spirit is offended by things that appear hazardous, we want not only a real solidity, but also apparent solidity that does not give the spectator occasion to think that the edifice will tumble. We will admire a man who dances on a tightrope, but in the end one does not approve of his exposing himself to danger, and it becomes painful to watch.*[59]

Startling feats of construction induce worry. The classical orders are applied to restore a sense of security.

However, Frézier made it clear that if the architect could err in one direction, he could also err in the other. There were limits to solidity as well as to lightness. The accusation of unjustified ponderousness was usually directed at the somber and massive baroque monuments of the previous century. Frézier's answer was to reduce the expressive range of the orders by subtracting the effete composite and the primitive Tuscan from the set.[60] With the Tuscan order went all the monstrous caprices of rustication. Rusticated masonry was a conceit, representing the crude roughness of unworked stone in various ways, as if the aspiration to rise above the ground had not quite succeeded. It had always encouraged the grotesque and irregular.[61] It always remained, more or less uncivilized, at the other extremity of classicism. Philibert Delorme himself had published a proposal in which he used every imaginable rusticating device to emphasize the mineral weight of rock as recalcitrant to human will.[62]

119 Philibert Delorme, rusticated temple, from the Premier tome.

We are left with the idea of a median range of classical performances that impose geometry on stones but do not cause alarm by removing the appearance of stability. An anodyne and commonplace statement of an equally anodyne ambition, it is far from the truth. Every building that abides by the rules of classical ornament, every building to which the orders are applied, is in flagrant defiance of it. The reason for this is exhibited with unusual force in Hubert Robert's painting of *L'Arc de Triomphe d'Orange*. Despite its ruinous state, the arch stands like a gigantic, solid menhir. But when we direct our attention to the Corinthian order that gives the impression of having been carved in bas-relief on its blockish sides, then, if we think of the order as if it were a structure, the interior will open up into a gaping fictional void within a trabeated frame. Classical architecture was as much an invitation to imagine the achievement of an exhilarating instability as was Gothic. But classical instability tended to be more a matter of suggestion than demonstration. It relied on the observer entering into a momentary delirium, in the way that a perspective painting may be looked through and not at, as if its space were real. Renaissance classicism was more cautious in execution

120   Hubert Robert, L'Arc de Triomphe d'Orange, *1783, Louvre.*

and less disturbing to the mind than the direct physical apprehension of unsupportable slenderness in Gothic, or the unsupported projections in *trompes* and flying stairs. The mural backing of Renaissance architecture could nevertheless be removed without challenging its rules. That is what the Abbé Laugier proposed: columns and lintels deprived of sustaining walls. By dogmatic assertion of an antediluvian original, a Gothic elan, long lost to classicism, would be restored to it.

The classical orders never did portray complete firmness, yet they portrayed another reassuring property under the same guise. The orders were ornamental pictures of structure, always the same structure; rectangular frames of columns and lintels. Classicism was not a *solidification* of structure, it was a *rectification* of structure. Frézier, who knew very well that static distributions of stress rarely resolved into horizontal and vertical vectors, nevertheless presented the rectangular portal as an archetype.[63] For Laugier and Cordemoy the inherent stability of the rectangular frame was beyond question. Classical structure was less a precaution against the collapse of buildings than a precaution against the collapse of faith in the rectangle as an embodiment of rational order.[64]

Laugier's wavering attitude to stereotomy is explained by his unwavering faith in rectangular form. On the one hand he would rid architecture of arches, severies, ocular lights, *arrière-voussoirs,* niches, and the whole inventory of deformed, showy tricks that had issued from stonecutting. On the other hand he understood that this same technique was able to deliver the otherwise impracticable depressed vaults, "so flattened that they resemble proper ceilings."[65] To this end he recommended the treatise by his fellow Jesuit, Derand, although he did not think much of him as an architect. In Laugier's *Essai,* few buildings were dismissed as peremptorily as Derand's masterpiece, the church of Saint-Paul-Saint-Louis, built in Paris between 1627 and 1641.[66]

So far we have been dealing with the confluence of an apparently Gothic stereotomy and eighteenth-century neoclassicism in largely theoretical terms. A similar reconciliation, just as told by Delagardette, was to be observed in the history of monumental building. Rude intrusions such as Delorme's *trompe* became less frequent during the seventeenth and eighteenth centuries. On the whole, stereotomy was still restricted to particular parts: staircases, ceilings, and the odd, furniturelike appurtenance—parts that, if we wanted to play on words, we might call extraordinary, since they lay outside the orthogonal domain of the orders. At Maisons-Lafitte (François Mansart, 1642–1646), the Hôtel de Ville at Arles (Jules Hardouin Mansart, 1673), the Lonja at Barcelona (Joan Soler i Faneca, 1764–1794), to mention but a few, magnificent demonstrations of stereotomy shelter within classical facades that betray no sign of anything so devious within. The effective territorial divide between walls and vaults, outside and inside, the classical and the stereotomic was never wholly erased, but there were buildings where some sort of fusion between them was achieved. Two such buildings are discussed below, one detested, one admired by Laugier.

First, Derand's Saint-Paul-Saint-Louis, the interior of which was beholden to the works of Martellange and Jacques Lemercier,[67] both of whom had played their part in developing a specifically French type of Jesuit church. Derand sought to consolidate this type.

What Derand took from Lemercier (son of a mason) were the so-called naked (*nu*) vaults, built with astonishing precision. As in Lemercier's Oratoire (1621–1630), there are no structural ribs in the vault, only flat bands of carved masonry between bays, the decorative character of which is pointed up by the interspersing of similarly contrived emblematic cartouches[68] that float on the vault surface in positions that would have been occupied by bosses, were there any ribs. But since there are none, the overall effect is of ornaments embossed on smooth sheets. These naked groin vaults multiply in the aisles, triforia, and transepts, each time stretching over different curves; a tour de force of stonecutting completed in the year Derand retired to compile his treatise on the subject, in which the focus of architectural interest is defined as stereotomy. All great buildings have vaults, he says, and "as soon as one puts one foot inside, one looks upward, toward where one expects to find the greatest satisfaction in the contemplation of forms, and the rare and agreeable diversity that comes from *traits*."[69]

What Derand took from Martellange's Jesuit Chapel at La Flèche (completed 1621)[70] was the triple-tier arched structure framed within a giant order, the arches progressing upward from a semicircle, to something that resembles an ellipse, then to something that resembles a parabola— or a Gothic pointed arch. At La Flèche these curves, compounded of circular arcs, were ungainly and full of visible discontinuities (the *jarrets* or knuckles, warned against by many authors on stonecutting).[71] At Saint-Paul-Saint-Louis Derand removed the knuckles. He also removed the ribs, still to be seen in the quadripartite vaults of La Flèche. Despite their removal, Saint-Paul-Saint-Louis comes across as a thinly disguised Gothic carcass, elongated, soaring, and flooded with light.[72] The orders were much in evidence, but applied like an elaborate overlay of pliable decoration, dependent on the subsumed construction. What Laugier despised was Derand's willingness to adapt the frame to suit the carcass. A similar attitude toward classical ornament, immuring and at the same time cavalier, later informed the stereotomic masonry in the cloister of the Benedictine abbey at Le Mans,[73] where sporadic pilasters extended from walls into ceilings like long, drooping flowers.

There were two ways in which the effort toward a synthesis of Gothic and classical could be biased in favor of good taste as it was understood in eighteenth-century France. The immured orders could be liberated from the wall, as envisaged by Laugier, so that precision-built shell-like vaults would rest lightly on columns in the round; or the more obvious signs of both Gothic and classical might be liquidated in a double abstraction. The first may be represented by Soufflot's Sainte-Geneviève,[74] the second by Le Camus de Mézières's Halle au Blé.

Le Camus's own phrase "pierre de touche du goût et d'intelligence" (touchstone of taste and intelligence) might well be applied to the Halle au Blé, a corn market built between 1762 and 1766 that followed little by way of direct precedent, whether in regard to building type,[75] structure, or construction, classical or Gothic. Its suspended, double-spiral

*121   Saint-Paul-Saint-Louis, Paris,
François Derand, 1627–1641, interior.*

122   Abbaye de Notre Dame de la Couture, Le Mans, architect unknown, 1720–1739, cloister vaults and pilaster.

123   Sainte-Geneviève, Paris, J. G. Soufflot, crypt vaults, 1758–1764.

124 Sainte-Geneviève, Paris, J. G. Soufflot, choir vaults, completed 1779.

125   Halle au Blé, Paris, Nicolas Le Camus de Mézières, 1762–1766 (demolished), section showing suspended double spiral stairs, from Recueil, *1769.*

stone staircases, for instance, were a uniquely daring experiment in structural stereotomy.

Le Camus reduced the classical schema to a compress of framed layers wrapped around the inner and outer surfaces of the annulus of the market building. These flat frames were ironed out until they filled the wall surface and became coextensive with it. Inside the market arcades, at ground level, a double ring of vaults was supported on Tuscan columns.

126   Halle au Blé, cross section, from Recueil, *1769.*

127   Halle au Blé, section showing grain store vault, from Recueil, *1769.*

COUPE de la partie circulaire couverte.

DRAWN STONE **219**

Above, in the grain stores, a single catenary curve abolished the distinction between wall and ceiling. *Arrière-voussoir* and *oeil-de-boeuf* openings glanced into the catenary ring vault with the grace, regularity, and exactitude that was by this time characteristic of French masonry.

Le Camus had pressed and multiplied the classical frame until it turned into a wall. He had also turned the market's continuous, stereotomically defined surfaces into gigantic frames. A revealing painting made by P. A. De Machy in 1765[76] shows the catenary vaults as an open lattice of stone arches while under construction. The lattice was then filled with lighter brick compartments flush with the masonry. In the Halle au Blé the classical frame and the stereotomic vault surface were being propelled in opposite directions toward each other; as classical frames turned into curved surfaces, stereotomic vault surfaces turned into frames. At the same time it would be difficult to point to any feature, with the exception of the Tuscan columns inside the arcade, that could be confidently identified as either specifically Gothic or antique.

The generosity of Laugier's judgment is indicated by his praising the Halle au Blé, although it contravened practically every rule he had laid down in the *Essai*.[77] Like Soufflot's Sainte-Geneviève, it demonstrated the possibility of synthesis. Unlike Sainte-Geneviève, Le Camus's building demonstrated that stereotomy, implicated in that synthesis, could be ungothic and also unclassical.

The Undercutting of Style

The taint of barbarism, persistent throughout the period of stereotomy's blending and synthesis with classical architecture, always points back to a French Gothic origin, as had been surmised by Frézier. Regarded from the vantage point of classicism, the alien character of the *trait*, running parallel to the alien character of Gothic, appeared to merge with it at a sufficient distance. But what would happen were we to approach the *trait* from the other direction; what if we were to look at it from the Gothic end?

There was one architectural historian who did this: Robert Willis. His essay "On the Construction of the Vaults of the Middle Ages," published in 1842, is enough to radically rearrange the accepted picture drawn from the classical point of view.[78] To this may be added an important observation made more recently by Pérouse de Montclos. Mention has been made already of the latter's heterodox opinion on the origin of stereotomy. The reason he removes its birth back into the Romanesque period[79] is because certain Romanesque vaults look far more like the naked, ribless vaults produced from *traits* than do ribbed Gothic vaults. As far as he is concerned, Gothic was an adventitious episode intervening between earlier and later periods of stereotomy. It was not just a matter of appearance; Gothic reversed the basic emphasis of stereotomy.[80] In Gothic, the rib was defined first and the surfaces between filled in afterward, whereas *traits* defined surfaces first. Any groins or ribs then had to follow the determining geometry of the surface. This had been noticed by other authors, Frézier amongst them,[81] but Pérouse de Montclos, arguing that it is fundamental, skipped over ribbed Gothic to find a suitable period for the conception of stereotomy in the preceding, ribless centuries.

Delorme touched on the subject of ribbed vaults in the *Premier tome*. He said that those who knew something of the true architecture were now disinclined to use them, but he confessed to having practiced them himself and did not want to devalue (*despriser*) them.[82] He gave two illustrations. One dealt only with ribs, the other, a star vault on a square plan, demonstrated the existence of a highly sophisticated practice of stonecutting used to negotiate between rib and surface. This second variety, he said, required much more difficult *traits* than other vaults. Why was this so formidably difficult? Because the ribs all had to be made with circular arcs. This in itself was not difficult. It was easy; one only had to ensure that the rib arcs intersected in space where they were shown as crossing on plan. But in order to obtain this result, the consistent geometrical definition of the surfaces *between* the ribs had to be relinquished, and in consequence the overall surface of the vault became a complex of curving facets. Precise cutting of masonry in the cells between the ribs was therefore vastly complicated. Delorme pointed out that while any two adjacent cells could be treated as a continuous arcuated surface, that was the limit. He said that there were some vaults constructed this way in Paris.

Delorme's star vault, which was on a square plan rather than the rectangular plan more typical of Gothic, has been described as transitional in style; flamboyant modified by Renaissance. It may also be described as transitional between two kinds of practical geometry, the result of an earlier, simpler, more limited technique of projection modified by a later, more sophisticated and inclusive technique. The evidence is in the treatment of the ribs. Oddly enough, although the ribs were the first things defined, they were virtual. Delorme took plane sections through the plan lines of the vault as if these were to become framing ribs, easy to build because they were parts of circles, but then proceeded to treat them as groins in an extended surface of jointed masonry. If the vault had been

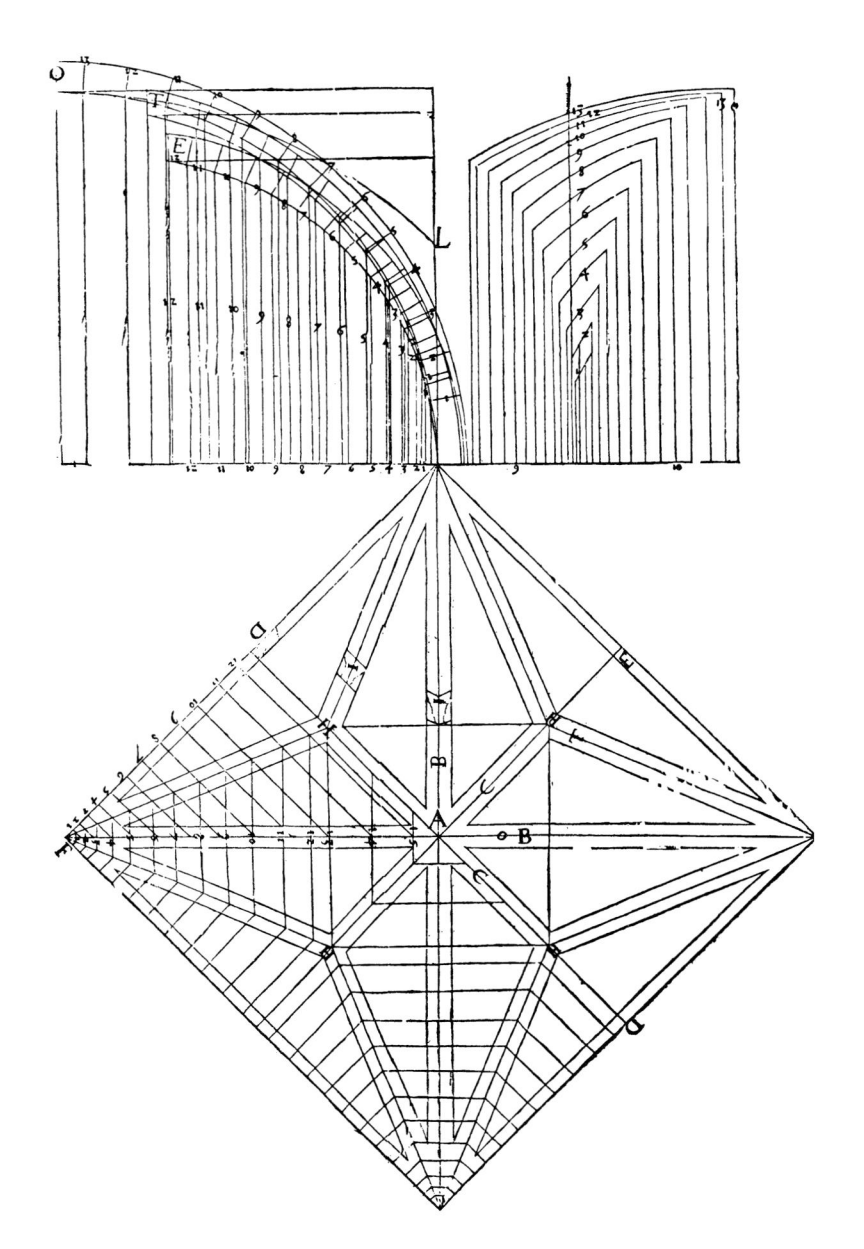

128 Philibert Delorme, trait *for a modern star vault, from the* Premier tome. *The* trait *was constructed as follows: first the plan was drawn using only straight lines to define the net of virtual ribs. The* trait *deals with the one-eighth segment of the vault within ADF. (Delorme also shows an alternative way to divide the masonry of the vault in the lower part of the plan.) Three "ribs" rise from the springing point F: FA, FH, and FD. The widest, FA, is defined as a semicircular arc. The drawing at the top left shows superimposed plane sections of the three "ribs" terminating at O (FA), T (FH), and E (FD). Radii for these three cuts through*

the vault can be chosen at will to give a more domical or more pointed result. The next operation is to divide the longest (semicircular) arc into equal voussoirs (13, plus 2 for the keystone). The positions of the joints on the underside (intrados) are then projected down onto the baseline, and then transferred with dividers to the diagonal "rib" in the plan below, where they are marked with corresponding numbers (there are some misnumberings in Delorme's plate). This done, the parallel lines of the joints are taken from these numbered points in plan across FH to FD. The next stage entails the transmission of information about the joint positions on FH and FD in plan to their sections at top left (using dividers to get them to the baseline and then projecting them up).

With this information in plan and section it is now possible to construct the third drawing (top right), which shows the true profile and angle of elevation of each numbered joint line. Delorme uses the "rib" FH as a median from which measurements are taken left and right. FH is thus represented as the numbered vertical line in this third drawing. Widths of successive joints are taken from the plan, heights from the elevations to the left. Three points on each joint line are now known, and they are sufficient to define a circular arc, which is the true profile of the lower edge of two adjacent voussoirs.

This trait is not a unified orthographic projection. The drawing on the top right is fractured by superimposing three plane sections, and the drawing on the top left gives a twisted picture of the vault cell. Such distortions and disjunctions were the result of the drawings being conceived as full size, at which scale projection is best limited to small fields, since it is easier to carry measurements in dividers than to construct very long parallel lines.

built exactly as shown in the *trait* it would have had no ribs, although ribs were clearly drawn in the plan and the vault was designed as if it had them, so the ribs may be regarded either as decorative extras, or as the fundamental conceptual schema of the construction; a polarized perception that will prove to be particularly significant.

The difficult technique described by Delorme must surely have been known to the most accomplished late Gothic masons. Yet, although he referred to French examples, these are not easy to find. The surviving French virtuoso vaults from this period either have ribs that fly independently of the vault surface, as with the ambulatory vaults at Saint-Pierre, Caen, where a complex of intersecting ribs was simply covered with flat stone lids on top, or they use ad hoc filling to build the cells between precisely cut ribs erected first as a supporting framework.[83] This latter was the kind of vault construction at the heart of Gothic from the twelfth to the fifteenth centuries.

The story of Gothic vaulting is the story of the rib and its multiplication. First diagonal ribs intersect over the middle of a bay, then transverse ribs divide bays, then ridge ribs run laterally and longitudinally, then tiercerons go in between the diagonal ribs, then liernes run across these, interlocking in increasingly dense networks. The progress and ultimate demise of the Gothic style is tied up with this multiplication and diversification of ribs. For every other salient feature of Gothic the theater of development was the Île de France, whereas the multiplication of ribs occurred outside France, in northern and eastern Europe. Its several sources have been traced to England. I do not wish to make this into a counterclaim for national preeminence;[84] developments of a similar sort took place in Germany and Czechoslovakia as well.[85] The point to be made is that they did not occur at the epicenter of Gothic.

Could stereotomy have arisen from the construction of ribbed Gothic vaults, despite their constitutional difference from the vaults of published stereotomy? And, if evidence to help answer this question exists, does it help explain why and when the emphasis moved from rib to cell, from framework to surface? Robert Willis supplied convincing answers to both questions.

Willis, who was Jacksonian Professor of Applied Mechanics at the University of Cambridge, looked at Gothic buildings, on which he collected and published an enormous quantity of information,[86] with the eyes of an antiquarian. He lamented the difference in effect between the perhaps primitive but virile vaulting from the thirteenth century, and the pale, if neat, nineteenth-century imitations. His object in the essay on vault construction was to show that the modern efforts to recall Gothic vigor employed the wrong means to that end, but in order to show this he was obliged to describe the original means. It turned out that what had been so effective was a particular sort of ignorance.

Willis's method of inquiry was unusually rigorous, made so by his assuming the absence of what he proposed to find. Since he had noticed from recent examples, such as William Halfpenny's cuni-conoidal Gothic,[87] the limp and wan character of Gothic vaults defined projectively, he approached the subject already suspecting that projection had been destructive of Gothic style. Lacking documentary proof, but unsatisfied with vague accusations, he sought to discover, from the evidence of the buildings themselves, vaulting that could not have been made without the aid of projection, to see if this confirmed his suspicions, which it did. His conclusion was that from the beginning of the fourteenth century projection left its mark on Gothic building. Further, by a formidable combination of scrupulous archaeological scholarship and forensic reasoning he was able to present a sequence of instances showing the interaction between the formal attributes of vaults and the methods of their geometric determination. So convincing was Willis's account that its substance and conclusions were quickly adopted by almost everyone with an interest in the subject. A year after publication in England they were incorporated into Abel Blouet's *supplément* to Rondelet's *Traité de l'art de bâtir*, and they became an important ingredient in Viollet-le-Duc's *Dictionnaire*. Willis's ideas remain diffused to this day in established opinion on the development of vaults, though Willis himself is either buried in the footnotes or raised from the dead for other reasons.[88]

I cannot do proper justice here to Willis's brilliant achievement, but some examples taken from his essay will help us toward our own conclusions. John Willis Clark, commenting on his mentor and namesake's studies of Gothic, said "he treated a building as he treated a machine: he took it to pieces."[89] And he did, or rather, waiting on others to do the deed, he would hasten to rummage through the broken stones. He himself admitted reluctantly that "the best instructor of all . . . is a building which is being pulled down."[90] Thus did the demolition of Lanfranc's Tower at Canterbury Cathedral and St. Saviour's in Southwark provide him with crucial evidence of their builders' working methods. He noticed remains of profile outlines inscribed on the joint surfaces of rib voussoirs; proof that they had been precisely carved to shape before erection, not after. He then narrowed his sights on the *tas-de-charge*, that is, on the set of stones just above the springing of the vault, where several ribs coalesce into one horizontal slice of stone. Traces of molding outlines and center lines on their joint surfaces indicated that they too were carved prior to erection. The masons must therefore have been able to predict the direction and curvature of the diverging ribs in the upper slices of the *tas-de-charge*, requiring geometry, requiring drawing, but not necessarily requiring projection as such.[91]

As time passed, as ribs multiplied, as the geometry of the vault became more complicated and ambiguous, the role of geometric drawing became more clearly evident: the plan was the generator, and plane sections through ribs were the means of its elucidation, exactly as in Delorme's simple Gothic example. Willis invented an instrument for measuring the curvature of vault ribs *in situ*.[92] Taking his cue from Delorme, Derand, and Frézier, he wanted to test whether all Gothic ribs were, indeed, circular arcs. he found that they were, although in later vaults they were sometimes compounded of two or even three curves of different radii between springing and ridge. One good reason for this was that in a lierne vault, where many ribs intersect, the sweep of successive ribs would need to be adjusted so as to meet others already defined. Most telling of all was the treatment of precut boss stones, set at points of intersection.

It was a matter of considerable interest to Willis that aligning marks for boss stones were carved on horizontal flat surfaces facing up into the roof. Invisible and with no structural function, these "surfaces of operation," as he called them, could only have been made to facilitate the marking up and cutting of the boss. The geometry of lines on a surface of operation is, of course, identical to that of a plan drawing, and Willis realized that the complex reticulation of fourteenth- and fifteenth-century lierne vaults was the product of simple, linear plan patterns, achieving this insight without knowledge of the substantial body of corroborative Gothic drawings collected since.[93] He was able to point out peculiarities that resulted from this reliance on plan, divining that a certain sequence of ribs zigzagging this way and that across the severies of the choir vault at Gloucester Cathedral arose from a straight line in plan. He showed how ribs were always built perpendicular to the floor, whatever their location on the vault, which was the same as being built perpendicular to the plan, observing where this resulted in uncomfortably oblique angles between rib and vault surface.

At Wells Cathedral, Willis also noticed where simple plan shapes like squares were anamorphically distorted on the vault surface. Projection may have been used to obtain these transformations, but he was disinclined to regard them as conclusive evidence of a fully developed projec-

129   Gloucester Cathedral, choir vault, William Ramsey (?), 1337–1367.

130   Gloucester Cathedral, detail of severy in choir.

tive technique, because the use of plans, plus plane sections through ribs, without other kinds of drawing, could produce such results accidentally. It was, however, for these very same choir vaults at Wells and Gloucester that Willis finally conceded fully fledged projection to have been necessary, because they could no longer be considered as networks of independent ribs intersecting in space; they appeared to be latticed surfaces. All the ribs seemed to be attached to a giant cambered sheet, a segment of a cylinder that stretched from springing to ridge down the entire length of the choir.

Thus Willis identified the distinguishing feature of the Wells and Gloucester lierne vaults, as also those at Bristol and Winchester cathedrals and Tewkesbury Abbey. Although the ribs did not comprise a cylindrical surface (in which case some would have to be elliptical), they were so close that the difference is imperceptible from the floor. Only when it is possible to inspect the exposed upper surface of the ribs from under the roof, as in the Lady Chapel at Gloucester (built much later, between 1457 and 1483, but modeled closely on the choir vault), is it possible to discern that there are, in fact, aberrations, the whole being made up of slightly buckled segments between tiercerons.[94]

All things considered, it seems likely that the Gloucester and Wells vaults were designed with more elaborate versions of the procedure already outlined by Willis. If such was the case then they were not *derived* from a consistent surface, but they had *arrived* at a close approximation of one, and this, as far as we can tell, by means (the defining of ribs as circular arcs) that had to be tampered and fiddled with to achieve apparent superficial consistency, and which rendered its actual achievement impossible.

So the new effect of complex nets on consistent surfaces, though a geometrical property, must have been obtained in the face of the masons' geometrical technique, not hand in hand with it. A strange state of affairs, this, but not unique. In architecture, technique and effect do not stand in a simple causal relation. Neither is determined exclusively by the other. Sometimes a technique will produce unexpected but desirable effects; sometimes, as probably with the west country lierne vaults, the desire for a particular effect will impose on technique, demanding of it more than it can easily deliver.

The effect in question was not isolated. Net vaults and the English perpendicular style developed together. The name perpendicular applies to wall and window tracery. The choir walls at Gloucester and Wells were panel surfaces with embossed vertical mullions. The once emphatic distinction between column and wall had been erased. An attenuated grid of tracery was now distributed over an extended surface in both walls and vaults, producing an appearance of surface continuity, noticed by another nineteenth-century historian, E. A. Freeman, as characteristic of late Gothic.[95] But the structural and technical implications were very different; columns can disappear into the surface of a stone wall far more easily than ribs can disappear into the surface of a stone vault.

The technique of projection finally caught up with the effect of superficial consistency in association with fan vaults. Fans were certainly not invented with the aid of projection, for fans already existed under tomb canopies as toy vaults, sometimes carved from a single, solid stone.[96] In such circumstances they were treated as sculpture. There are even small architectural versions of monolith fans in the Dean's Chantry, St. Mary's, Warwick.[97] But the origin of the fan vault did not determine its destination.

One of two radical, transforming steps was taken when the fans were enlarged beyond the limit of single stone construction, which happened in the Despenser Chantry, Tewkesbury Abbey (1375), and the cloisters at Gloucester (begun 1378), where, for the very first time in Gothic, the vault was made as a jointed masonry shell instead of a frame of ribs with infill.[98] It is often said that this was the moment when Gothic lost its vigor. The rib, supporting nothing, multiplied into a fretwork of purposeless decorations. Logic had departed. The complaint, worn thin through repetition, is anyway very one-sided. If the rib was not the structure, what was? If the logic no longer resided in the ribs, are we to assume that it had disappeared altogether? Could it not have moved somewhere else? Rather than sigh once more over the sad fate of the virile rib, emasculated by a regrettable change of taste, let us try to answer these questions.

Over a century after the completion of the fan vaults at Gloucester, the second of the two radical, transforming steps was taken: the fans intersected. The impulse to press complex nets of ribs into simple, consistent surfaces has been observed at Wells and Gloucester, and also the difficulty of obtaining that effect with the means at hand. The difficulty was resolved by using the fan—an arc rotated about a coplanar axis. In plan it is always a circle and any axial section will always be the same curve. Line up fans in rows and a quatrefoil aperture is left in between, which at Gloucester was closed with a single flat stone.[99] The method of sectional drawing, which, following Willis, we may presume to have been well

131 Gloucester Cathedral cloister, fan vaulting, begun 1378.

established by this time, would still suffice for the Gloucester fans because there were so many conformities of shape to help the imagination grasp what was going on between the section lines. Were the fans to intersect, it would no longer be adequate.

In the much larger fan vaults of the early sixteenth century, at King's College Cambridge, Bath Abbey, and the Peterborough retrochoir, the fans intersect so as to reduce or eliminate the quatrefoil gaps in between, which otherwise had to be bridged by different means. The lines of intersection must have been predetermined accurately for the component stones to be cut accurately, and inspection of any of these vaults shows that they were cut with tremendous precision.[100] The ability to cope with this hitherto imponderable geometrical difficulty is put beyond a shadow of a doubt by the brilliant handling of the odd, U-shaped plan for the retrochoir that was fitted round the east end of the Norman cathedral at Peterborough. Twenty fans of jointed masonry, built in the early sixteenth century, probably by John Wastell, simply spread until they meet. There is not a straight line or a flat surface to be seen. Even the central ridge courses along in shallow waves, slicing through the splayed lips of the pouting fans.

Now, and only now, have we reached the point where geometrical projection must surely have attained a level of accomplishment comparable to that displayed by Delorme a few decades later in France, since, as in the major part of Delorme's treatise, drawings had to be made to describe the intersections of *surfaces,* not just the intersections of lines. The intersection of two similar fans parallel to one another produces a plane undulation that cannot be constructed with compasses or other mason's instruments without the aid of projection—a fully comprehensive three-dimensional projection, not the simpler variety restricted to two-dimensional slices.

The newly available, expanded technique of geometrical description was exploited to the full in the Henry VII Chapel at Wesminster Abbey, built between 1500 and 1520 by one or several of the King's Masons, most likely Robert and William Vertue.[101] In the nave vaults of the chapel, the most prominent fans were not firmly attached to the wall but hung, pendant in space, turning a complete revolution in their lower portion (Sitwell calls them parachutes).[102] They then rise to intersect with their neighbors. A row of smaller pendant fans hang from the central ridge. Willis's justly celebrated drawing makes this abundantly clear. Looking up into the vault itself is more likely to induce vertigo than clarification, except looking straight up, when the vivacious, coruscating inverted landscape transforms back to the lifeless, flat doily pattern of its plan.

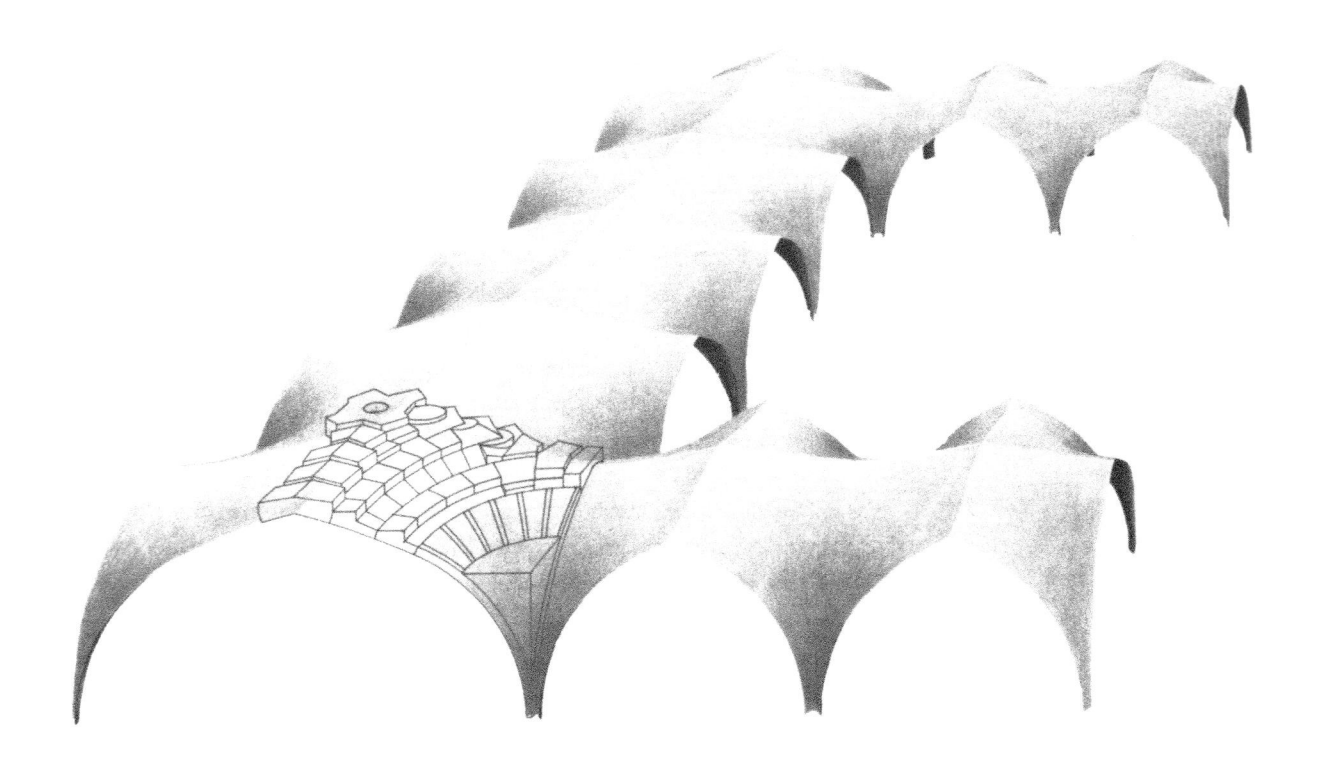

132   Peterborough Cathedral, retrochoir, John Wastell (?), early sixteenth century, intrados of twenty intersecting fans (drawn by author; masonry diagram after Willis).

133   Peterborough Cathedral, retrochoir.

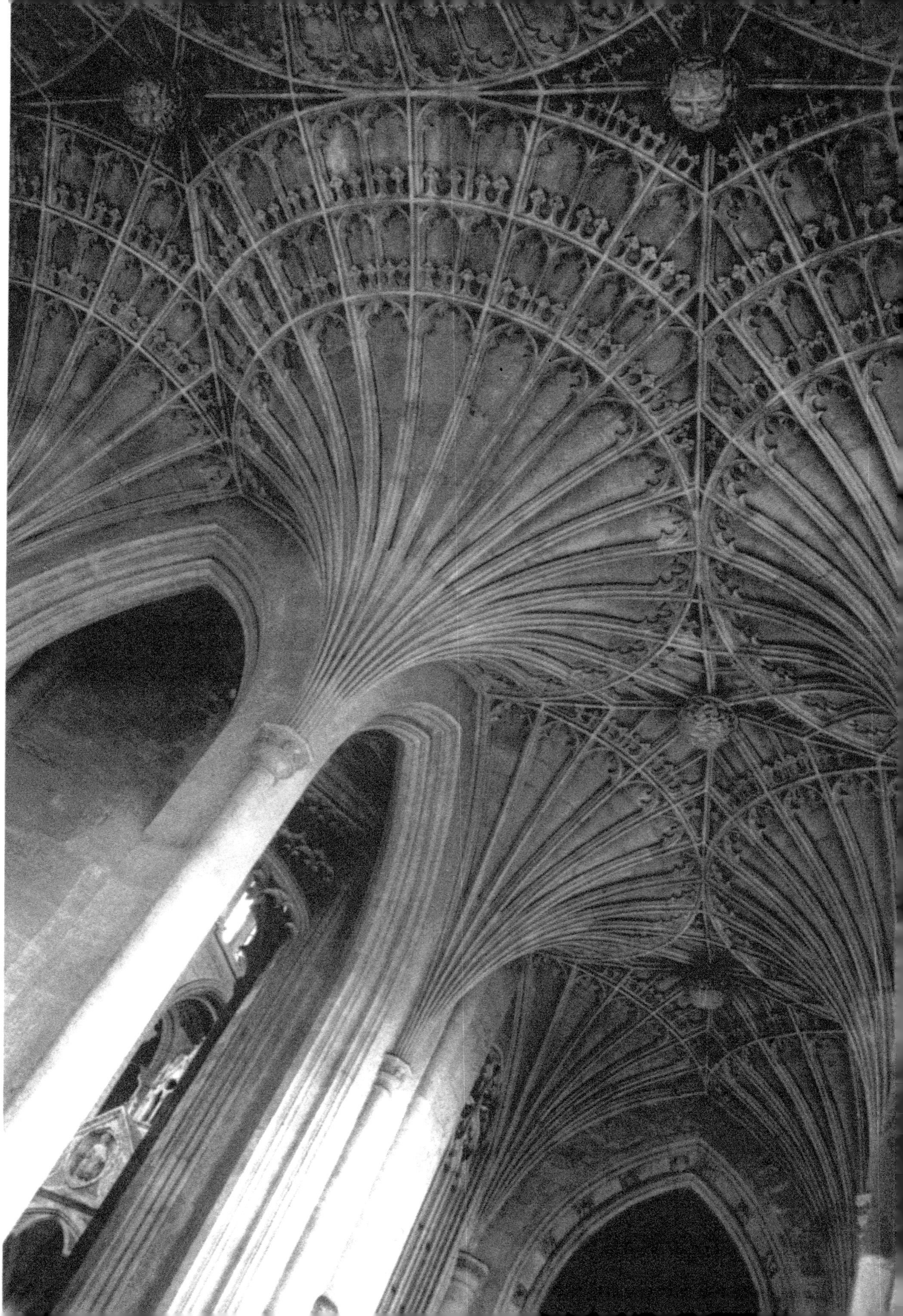

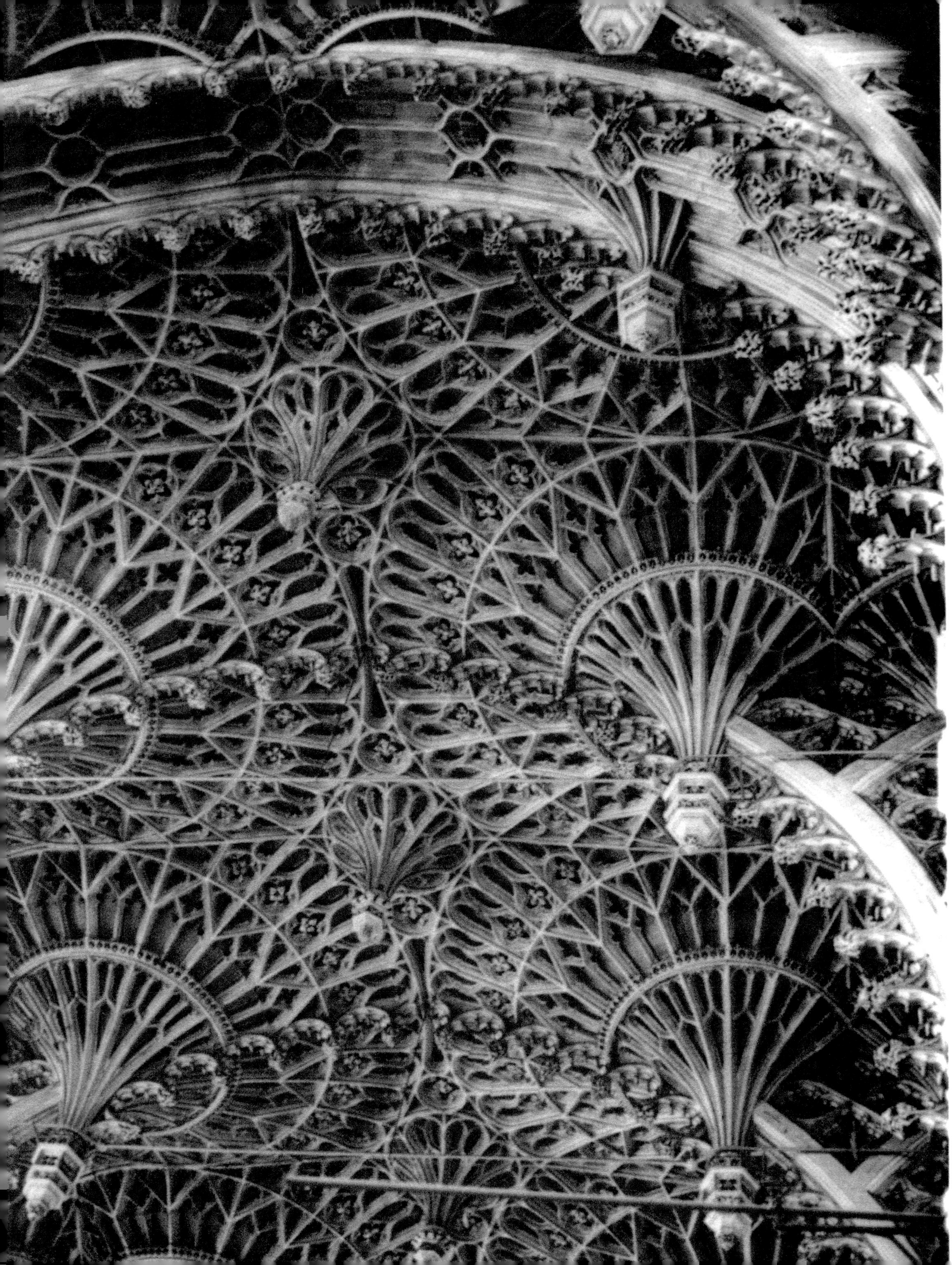

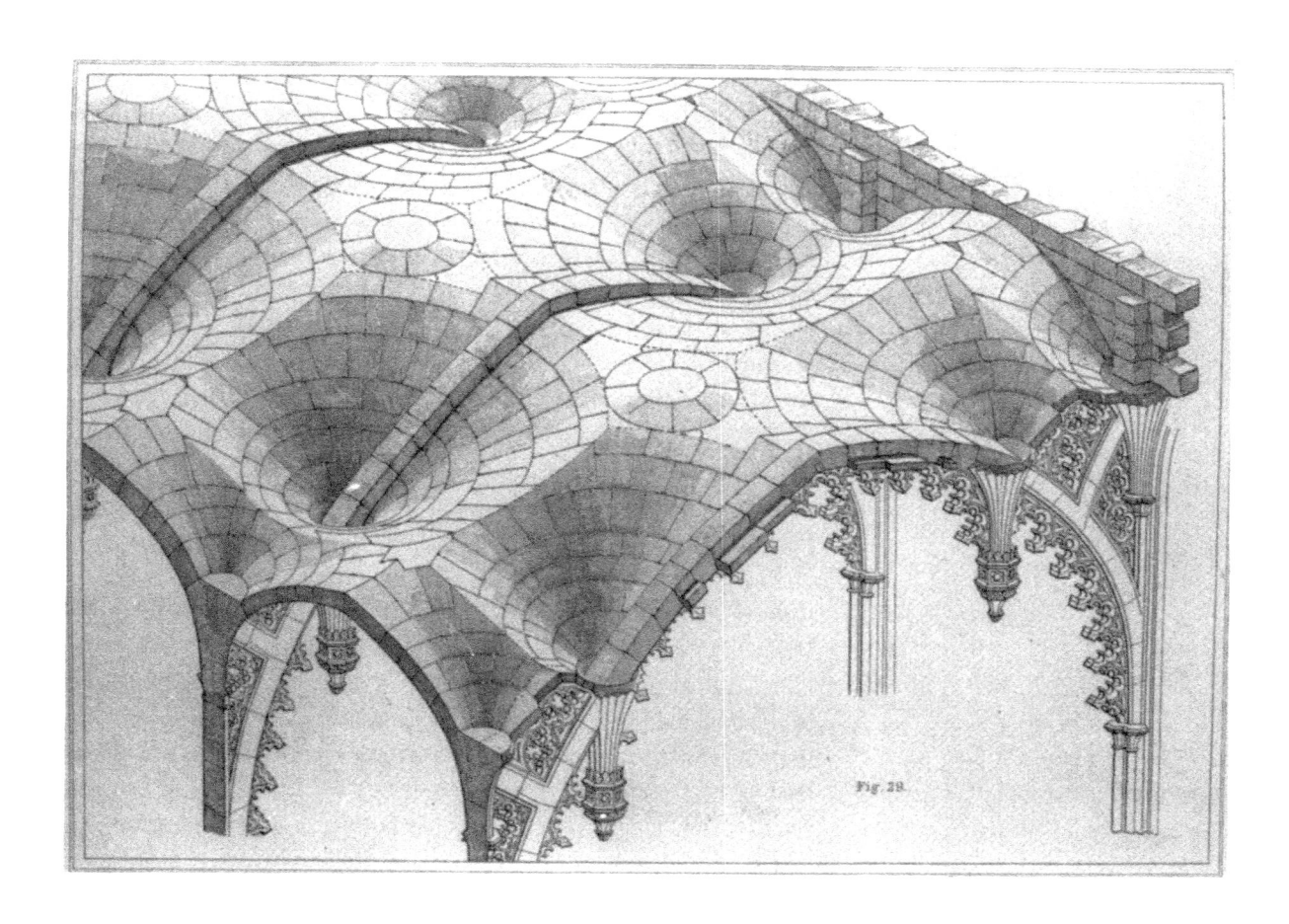

134 Henry VII Chapel, Westminster Abbey, Robert and William Vertue, 1500–1520, pendant fan vaults.

135 Extrados of Henry VII Chapel vaults, drawn by Robert Willis, 1842.

At the beginning of this chapter I said that the geometrical art of stonecutting was always marginal to architecture, and applied to buildings that were themselves marginal in some way. The Henry VII Chapel is certainly on the edge of Gothic, but the word marginal is inadequate to define the position ascribed to it; extreme would do better. It was at an extremity, not on a boundary. It did not develop into anything. It was not on the verge of something else. "It leaves the impression," writes Peter Kidson, "that there was very little left for anyone else to do afterwards."[103] It was, to use the words of previous commentators, the last word, the apotheosis, the end of the line, the culminating triumph, the terminus, at once brilliant and decadent, of an epoch, the most consummate achievement of the masonry of the Middle Ages.[104]

There are few creatures as pathetic as the swan that sang while dying, and there are few buildings that solicit so much pathos as the Henry VII Chapel playing itself out. It is the key English example of an efflorescent *Sondergotik,* spread throughout Europe, moribund, already victim to a debilitating skin disease (the cancerous proliferation of ornament), and soon to be dispatched by the Renaissance. Particularly revealing is the attitude of John Harvey, who has done so much to alter the common perception that English perpendicular architecture of the fourteenth and fifteenth centuries was a degenerating aftereffect of Gothic proper. A more equitable status is conferred on perpendicular within the bounds of Gothic, but the line has to be drawn somewhere and he draws it just on the far side of the Henry VII Chapel, which is still Gothic but not perpendicular;[105] belonging to an as yet nameless classification of very late Gothic that displays "the impasse which had been reached."[106]

Why, though, was this point judged to be impassable? The answer always hinges on the relation between structure and decoration. In 1849, John Henry Parker agreed with Ruskin that, in early Tudor architecture, ornament was "used to such an excess as completely to overpower the features of a building." The example he chose was the Henry VII Chapel.[107] For Harvey, also, the profusion of decorative ribs and tracery there "was carried to an extreme where it obscured the structural sense of the building."[108] The thing to note is that both of these writers, and many others in between, recognized that the Henry VII Chapel, considered as a structure, was remarkable. It was a work of genius; no other word will do. Therefore the problem is not that it is a bad structure, but that the structure is no longer visible, not anyway visible in the way we are accustomed to see it—as hierarchically ordered members. The French classicists' objection to Gothic was that it did not exhibit solidity; the Gothic scholars' objection to fan vaulting was not much different. The relation between support and supported was no longer made manifest as it had been in traditional ribbed construction. The impasse resulted from this dissociation. Decoration, divorced from structure, turned into vapid lacework.

Is this really what had happened? When Lethaby, to quote yet another opinion of the same stamp, wrote that, "notwithstanding the redundancy of ornament with which the building is overlaid, both without and within, the essential structure itself is of a very high order of constructive imagination,"[109] he was making a distinction that is ingrained in all our thinking. The great majority of commentators have accepted, with Lethaby, that ornament must always be an overlay, added to essential building and obscuring it; Willis thought so too. But it is precisely the physical integrity of decoration and structure, particularly in the vaults, that is so striking at Westminster. I do not wish integrity to be taken in some vague metaphorical sense; I mean it quite literally. For one thing, the filigree of liernes and ring ribs is nothing more or less than the generator of the fans and the loci of their rotation repeated over and over. The pattern is a diagram of the way in which the surface was construed. For another, it is usually supposed, again following Willis, that the masons at Wesminster only cut the tracery in rough before the erection of the fans, so that continuity could be ensured between the moldings on adjacent stones by carving the detail *in situ*.[110] This too is a reasonable inference, and yet Peter Foster, the Surveyor of the Fabric, able to view the vaults at close quarters from scaffolding, noticed that while there seems to have been precious little movement in the vault, the tracery is not always exactly aligned from stone to stone, and that in some places one side of the moldings had been shaved to remove discontinuities, raising the extraordinary possibility that the vaults were precut down to the last detail.[111] Ornament of this kind was hardly an afterthought.

George Herbert West, a historian well acquainted with the French rationalist conception of Gothic, was able to give an unusual twist to the usual judgment on the chapel. The vaults, he said, "would be just as sound if all the ribs were pared off and a smooth surface left."[112] Taken at face value this is questionable. There is more depth in the ribs than in the panel surface and it is by no means clear that the shell is structure and the ribs decoration. The two probably work in concert.[113] The removal of all excess stone from the uniquely refined extrados indicates that the masons were aiming for extreme lightness. Take away the ribbing and the whole thing would very probably collapse. But West appreciated that the "apparently excessive elaboration" had led to a paradox rather than an impasse, bringing Gothic back toward bare Roman vaulting. The multiplication of the rib had *caused* the structural surface.[114] West said it; Willis drew it. The chapel vault has two sides. The extrados as drawn by Willis is no exaggeration. The masonry is as smooth, as precise, and as continuous as he showed it. More revealing than the dark side of the moon, one is inclined to move over it with some trepidation, aware that the stone panels are only 3 to 4 inches thick with a clear span of 35 feet.[115]

Delorme's *trompe* and the Vertue brothers' pendant fan vaults, worlds apart in their use of ornament, are two sides of the same coin. The intrados of the Henry VII Chapel vault may seem very elaborate in contrast with any of the later French examples already discussed, yet its smooth extrados certainly belongs to the same family. One need only see it from floor and roof to appreciate the similarity behind the difference, and that is what West was getting at. The source of the affinity was the *trait*, which throughout its existence provided the means to make structures that acted over surfaces, structures that therefore needed no framing and so failed to show the proper signs of strength misleadingly established by architectural precedent. Breakdown of the clear, hierarchical logic of structural members gave the impetus for the development of stereotomically defined surfaces, but, as Willis showed, the breakdown was itself prompted by a strong presentiment of surface consistency.

Unfortunately there is next to nothing to suggest what or how the Vertue brothers or their English contemporaries drew in order to build. In the first years of the nineteenth century, rumors were circulating that working drawings for the vault of the Henry VII Chapel had been found. They have since been lost. Their authenticity was in any case called into question at the time.[116] A drawing of a monumental tomb for Henry VI in the Cottonian collection has been attributed to Robert Vertue, although the connection between the mason and the drawing is by the slenderest suppositional thread.[117] It is, anyway, not much help; a schematic box was laid out in axonometric projection, but there was no projective coherence in the rest of the drawing,[118] nor in any other English architectural drawing of this date. I do not wish to gloss over this problem. It takes some explaining, but not, I would say, anywhere near so much explaining as the existence of the Henry VII Chapel vaults without sophisticated projective drawing.[119] Besides, the drawn evidence has not entirely disappeared; some traces remain in the little lapidary plans discovered by Willis on the upturned "surfaces of operation." The most indubitable evidence however, is provided by the thin, jigsawed eggshells of the intersecting fans themselves.

If developments were as outlined here, then the decorative rib did not work as the mainstream of Gothic commentators have supposed it to work. It was not an extra burden borne by, and utterly dependent on, true construction; quite the contrary. Proliferation of the decorative rib gave rise to new jointed stone building of unprecedented precision, led to the design of immensely sophisticated compressive shell structures, and instigated the geometrical art of stonecutting with which this chapter has been concerned. Take away the decorative rib and you remove the original reason for the existence of these as well. Only afterward did the plain surface become its own justification, but its epidermal smoothness was intimated by virtue of its supposed opposite—dense ornamentation. Thus did stereotomy arrive: between scene changes, one might say. It was no more categorically Gothic than Renaissance.

Excavations in Asia Minor have led some archaeologists to revise the long-held view that metal was first mined and smelted to make tools and weapons. The whole arduous business seems instead to have been undertaken to make necklaces and bracelets. Stereotomy provides an instance of the same sort, but closer at hand. Invariably presented by its advocates and practitioners as eminently useful for the resolution of pressing practical problems, it was nearly useless. The only problems it helped solve were those solicited by its availability as a technique. Other, less troublesome methods could always have been found. A technique without any ulterior motive that could plausibly be described as a purpose, the reason for its development and perpetuation have to be sought elsewhere: in the sheer virtuosity of its performance and the sybaritic, sheer smoothness of the resultant masonry. Even when, during the nineteenth century, stereotomy was incorporated into the more purposeful, matter-of-fact discipline of descriptive geometry, these qualities did not disappear from its products.

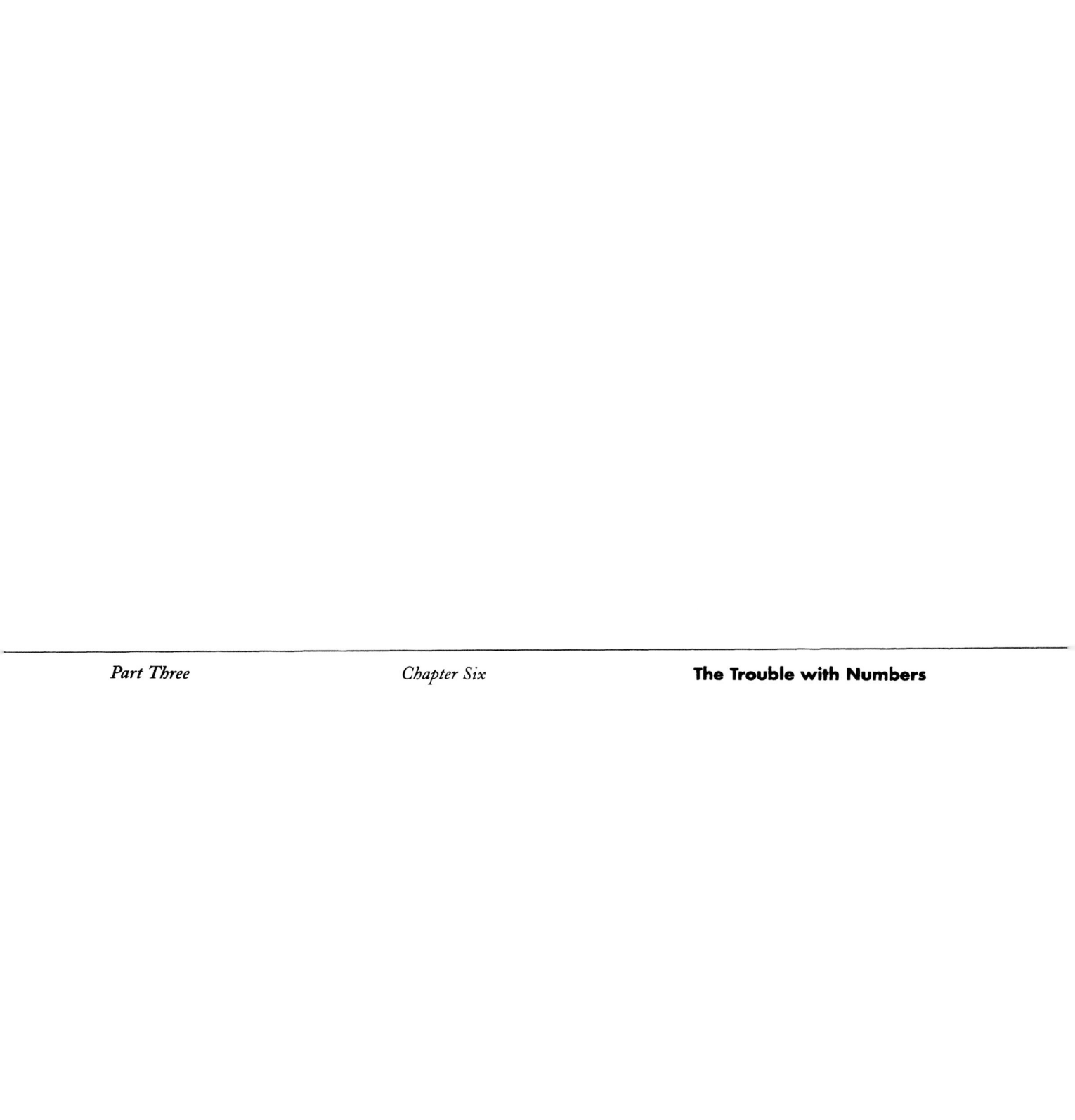

Part Three *Chapter Six* **The Trouble with Numbers**

O f all the ways in which architecture can be compared with music, the most familiar and the most persistent has been the comparison between musical harmony and architectural proportion. This once was held to be more than a mere comparison. In architectural theory, from the fifteenth through to the eighteenth century, the harmonies of the musical scale served as evidence that exact ratios underlay our perceptions of beauty in all things. Since measure was as important to architecture as it was to music (although for quite different reasons), similar harmonies, as exact as those found within the musical scale, were sought to account for the harmonious disposition of parts in buildings. Ratios from the octave scale were invoked to help justify systems of architectural proportion. Modest, exemplary designs were sometimes published, such as the Attic column bases by François Blondel and Bernardo Vittone, the dimensions of which translated into a five-note chord and a snatch of melody, respectively.[1] In his little book *Architecture harmonique,* published in 1679, the royal organist René Ouvrard insisted that architecture should be attuned to music: "There is an analogy between the proportions in music and in architecture, such that what offends the ear in the one, wounds the eye in the other, and such that a building will not be perfect unless it be on the same rules as those of the composition or grouping of harmonies in music."[2]

136 François Blondel, harmonically proportioned Attic base with its corresponding chord, from Cours d'Architecture, *1683.*

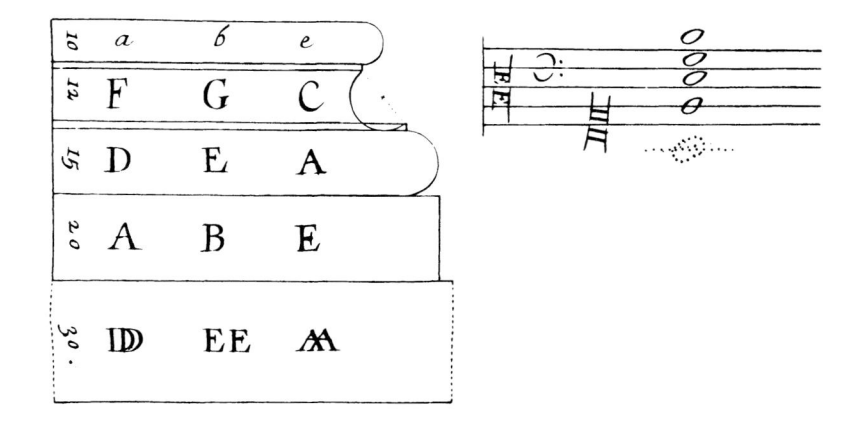

Anyone who has read Wittkower's *Architectural Principles in the Age of Humanism,* in which the Renaissance texts expounding the theory of architectural harmony were first made widely known, will realize how pervasive this notion then was, and anyone who has read Le Corbusier's *Modulor* will see that the analogy may still be brought into play by an architect seeking to establish a measure for beauty. But for most of us the notion of universal harmony now seems sentimental and vague. Wittkower devoted a chapter of his book to its dissolution during the eighteenth century. What put paid to it was, he thought, the increasingly subjective strain in aesthetics. Those of a relativist turn of mind found it difficult to believe in so objective a standard. As if to confirm this, the opening passages of the *Modulor* refer to the efficient division of sound into conventional pitches at artificial intervals, not to the universal truth of natural harmony.[3]

Wittkower's views on the demise of universal harmony fell in line with the views expressed by some of the eighteenth-century writers whom he had cited, which was curious because they had to serve as both perpetrators and castigators of the revisions. They felt that something that once had allowed a complete and unified picture of the world had been lost, perhaps irretrievably, perhaps not, but it had gone and they missed it. There can be no doubt that we too have constructed for ourselves an idea like this about the period prior to the rise of science.[4] The question is to what extent this idea of lost unity is a retrojection by succeeding generations, each augmenting the planted evidence of those who have gone before. As such it would not constitute a proof of something's existence and extinction, so much as a predisposition to believe that it had come and gone. Renaissance proportion may be regarded therefore as a test case. This is how I propose to treat it. If it was unified within and across the borders of different subjects, then that would tend to confirm received opinion. If, on the other hand, unity was conspicuously absent, it would suggest that our portrait of the Renaissance has been heavily overpainted.

The various attempts to align architecture with music and music with cosmology during the Renaissance period indicate that many writers did assume a universal correspondence. Proportion was the thread used to stitch everything together. It was therefore proportion that would provide the bonding for a coherent universe, as well as the link between music and architecture. Demonstrate the presence of one species of proportion in all three and the job is done; and since the greater part of the job was done by Wittkower, there should be little more to add. But there is one quite effective way to change our understanding of the relation between Renaissance music and architecture; instead of comparing architecture to musical theory, compare it to musical practice.

There are two distinct kinds of proportion in music: mensuration and harmony. Mensural proportion is about the lengths of notes and about tempo. Harmony is about the pitches of notes. In the Renaissance period, mensural proportions were derived from the theory of harmonic proportions. This might be presented as Renaissance unity triumphant, or as the inescapable sameness imposed by the Renaissance scheme of knowledge, but it is in any case what we have come to expect.

The Architecture in Music

On 25 March 1436, as Brunelleschi's great cupola was nearing completion, a ceremony of dedication was performed by Pope Eugenius IV at the cathedral in Florence. The event was made the more momentous by the presence of the pope, and also by the performance of a piece of music composed specially for the occasion by the pope's most famous chorister, Guillaume Dufay. This was a motet called *Nuper rosarum flores* (Recently Rose Blossoms), alluding to the name of the cathedral, Santa Maria del Fiore, and to the laying of a golden rose on the high altar by the Pope during the dedication. The exact circumstances of the motet's performance are not known, although it was said that a great number of singers and instrumentalists were present, and we can be fairly certain that the choir stood directly beneath the dome.[5]

A musicologist, Charles W. Warren, claims to have found significant correspondences between the cathedral and the motet, showing that the borrowing of proportions could work both ways, from architecture to music as well as from music to architecture. According to Warren it was the form of the dome that affected the motet.[6] He points out that Brunelleschi was said to have been the first to transfer the harmonic proportions of ancient music into his buildings, so setting up a circle of reciprocity between the two subjects. Warren quotes Wittkower's article on "Brunelleschi and Proportion in Perspective,"[7] which showed a possible connection between perspective diminution, architectural layout, the harmonic series (i.e., $1, 1/2, 1/3, 1/4, \ldots$), and the closely related superparticular ratios (i.e., $1/2, 2/3, 3/4, 4/5, \ldots$) that contain the Pythagorean musical consonances of the octave ($1/2$), the major fifth ($2/3$), and the major fourth ($3/4$). This is already an impressive array of correspondences. Warren provides more. He believes that while architecture borrowed rational harmony from music, music borrowed its proportioned structure back from architecture.

Nuper rosarum flores is what is known as an isorhythmic motet. It is in the isorhythm and in the allied mensural techniques that Warren finds analogies to the cathedral. There are four sections to the motet, each comprising a preliminary duet followed by a rich polyphonic sequence. The cantus firmus (a kind of musical spine), sung by the tenors in the four polyphonic sequences, is the same every time, only it is sung faster or slower. The ratio between the note lengths of the four tenor settings ends up as 6:4:2:3. In fact these different tempos are nowhere near as audible as one might expect.[8] The more prominent countertenor parts keep a similar pace throughout, while a sense of rhythmic acceleration is imparted by the lower voices speeding up the reiterated theme. The effect is surprisingly subtle, even recondite, and, together with other proportioning devices, has been said to be less a musical invention than a notational one that could only be developed in written pieces.[9] The proportional schemes of music were the more prominently displayed in fifteenth-century notation because different ratios were sometimes indicated with different colored notes in the score: black, red, white, and blue. Often they were, as one contemporary writer put it, more easily seen than heard. Willi Apel, author of the most comprehensive study on the subject of musical notation, points a disapproving finger at the "unparalleled complication and intricacy" of late fourteenth- and early fifteenth-century notation. "It is in this period," he writes, "that musical notation far exceeds its natural limitations as a servant to music, but rather becomes its master." He goes on to suggest that the marked visual character of certain scores indicates "the hold upon the imagination of the composer that the purely manual business of writing exercised in those days."[10]

Johannes Tinctoris, in his treatise on mensuration, *Proportions in Music*, written sometime before 1475, refers rhythmic proportion back to Pythagorean origins. So when discussing the doubling of tempo, he recalls the story of Pythagoras finding the octave in the sound of two hammers, one weighing 12 pounds and the other 6 pounds, although this has nothing to do with mensural ratios.[11] The pitches of notes are tied to the "natural" harmonies—move them away and you get more and more aggravated dissonance—whereas the tempo of music is guided by no such natural sanctions or limits. What we witness in these motets is the duplication of music's hallowed harmonic proportions in its tempo. It is music that mimics one aspect of itself in another.[12]

With this in mind, let us return to Warren's treatment of the architectural connection. The ratios 6:4:3:2 are, he says, approximately (very approximately) those of the series of nesting squares that provide the geometrical armature for the plan of the dome.[13] There are 8 ribs, 7 braccia thick, and so there are 8 × 7 breves in each of the four sections of the motet. The prominence of the number 7 in the motet structure relates to the symbolism of the seven pillars of the church, also referred to in the dedication ceremony. The relation between the duets and the four-part polyphony was made similar to the relation between nave and dome. The quadripartite form of the motet contains an analogue in musical meter of the dimensions of the four parts of the cathedral: nave, transept, apse, and dome.[14]

Warren's interpretation has been challenged.[15] The music is organized numerically;[16] what is disputed is the particular relation to the cathedral building. The least plausible aspects of Warren's hypothesis are those dependent on his plan and section of Santa Maria del Fiore, subdivided with a modular grid that leads a life of its own inside the building like a giant climbing frame, while the best evidence that Dufay borrowed from Brunelleschi's architecture comes not from the clear visible articulation of the architecture itself, but from an official document describing totally invisible properties of the dome, such as the relative thicknesses of its two shells.[17] We have already seen that isorhythm was a compositional device, the order of which was more easily read in musical scores than heard in musical performance. Perhaps Dufay incorporated what he read about the dome rather than what he saw of it, but this is still conjecture. While some of the relationships claimed by Warren may exist, they do not tie the motet to the building in the satisfying way he envisaged.

There are threads that link architecture to music in this period, such as Dufay, together with another famous musician, Gilles Binchois, advising on the design of a chapel at Mons in 1449;[18] or Tinctoris saying he named his book *Proportions in Music* in honor of his patron's gloriously well-proportioned chapel;[19] or the simile used somewhat later by Girolamo Mei, and repeated by Bardi and Galilei, in which modern polyphony was likened to several men trying to topple a stone column by pulling ropes strenuously in several contrary directions, merely to cancel out each other's efforts.[20] But the links are tenuous.

A correspondence between Brunelleschi's dome and Dufay's motet would have been satisfying because it would have corroborated quite well-established ideas about the workings of Renaissance art and thought. These ideas are supported by documentation and research; but one must select from among the available material. The material that is not considered may be excluded because it is irrelevant, or it may be excluded because it is disconcerting. We need to know which.

Proportional isorhythm is associated as much with the fourteenth-century *ars nova* as with the Renaissance. It was easily audible in this earlier music. Dufay and Dunstable, harbingers of Renaissance tonality, were among the last to make use of it. John Dunstable was an English composer active between 1415 and 1435; his isorhythm is even less audible than Dufay's. The effect of motets such as *Preco preheminencie* and *Veni Sancte Spiritus* is of an undulating drift of overlapping voices with no immediately discernible punctuation or structure at all.[21] Little is known of Dunstable. Various epitaphs described him as a "new Ptolemy" or "the confederate of the stars," with the ability to unfold the secrets of the heavens.[22] He is said to have been an astronomer as well as a musician. Rhetoric makes the case for unity; these things belong together. Vitruvius advised architects to consort with astronomers and musicians because they too were concerned with the subject of harmony.[23] The proportional arrangement of Dunstable's isorhythmic pieces, tabulated by Margaret Bent and Brian Trowell,[24] is mostly 6:4:3 or 3:2:1. The Pythagorean harmonies that guided the music of the celestial spheres, and that could be transferred to architecture, also reverberated in the time signature of earthly music. While the circulation of similar figures is what we would expect to find in a unified system based on correspondence, it is not easily heard; the numerical agency that gave corporeal things their beauty was not manifest in them, but hidden. Should it therefore be dismissed as an intellectual conceit—no more than a mannerism arising from the preoccupation with writing music rather than performing it, as Willi Apel maintained?

Warren calls the architectural structure of *Nuper rosarum flores* "extramusical." This is why he thought it likely to derive from Brunelleschi's building (other authors refer to the "plans" of these motets, and to the way in which they are "built").[25] But according to him it was not just an added layer of cleverness, propping up the music with adventitious justifications. Because it was extramusical, it had the power to alter the course of musical development. If some aspects of polyphony unfolded quite naturally from within music as it progressed, others were prodded into existence from outside. It is, says Warren, "the extramusical aspect of the motet that appears to anticipate the style of the tenor mass." In other words, the added ingredients in *Nuper rosarum flores* that had no justification except as recondite symbols were in the ensuing decades integrated and digested within the musical sense of Dufay's mature works—the great cyclic masses such as *L'homme armé* and *Ave Regina Coelorum*. In the process they were not simply neutralized, but helped bring into being audible qualities that gave the first intimations of a truly Renaissance music. An identical point is made by Howard Mayer Brown in *Music in the Renaissance*.[26] In mensural polyphony there are subliminal structures, such as isorhythm, that are recognized to have potentially powerful indirect effects. And yet, although they are in the work, under the surface, they are also in some way foreign to it.

Inscrutability

I now ask a question that is both naive-sounding and difficult.[27] If the pure forms and whole-number ratios were thought to be the source of true beauty in created things, then why were they not more obviously present in art? Why were these shapes and proportions not featured more prominently in compositions; why were they always underlying structures or partially expressed forms? Whatever artists might have believed about their intrinsic and fundamental beauty, they were not inclined to display them as such. The situation was similar in painting, architecture, and music. The beauty in the motets and masses of Dunstable, Dufay, and Josquin Desprez,[28] in the facades of Alberti, in the villas of Palladio, and in Renaissance painting generally, could not possibly be attributed to the more or less indiscernible presence of privileged forms and numbers; not unless we understand their indiscernibility or partial expression to be more effective than their full presence. Everything hinges on this little qualification. If it is acknowledged, it alters the argument drastically, for then we would have to recognize that ideal forms are not, in themselves, ideally beautiful. But this is the only reasonable view, otherwise a square sheet of paper or a tennis ball would surely put the works of Michelangelo in the shade.[29]

It could be argued that, in painting, an overriding concern with the human figure precluded direct and obvious presentation of ideal shapes and ratios. Such was not the case in architecture, yet opinion on the explicit or implicit character of proportion in Renaissance buildings is currently divided and uncertain. We can find critics like Colin Rowe, who portrays Palladio's use of proportion as relatively clear and uncomplicated in contrast to Le Corbusier's.[30] There are those on the other hand who complain that even when the rational idea is as clear as a bell, it cannot be perceived in three dimensions. This is what Roger Scruton says of Brunelleschi's Old Sacristy at San Lorenzo: look at the plan and section, and the cube and hemisphere are easily understood; stand in the building and they are not. And so he concludes, "As soon as we place the experience of proportion first, giving it precedence over its mathematical 'explanation', we find, not only that the mathematical laws are inadequate to account for what we see, but also that the concept of 'proportion' is plunged once again into the darkness from which the Pythagorean theory seemed to raise it."[31] If we accept the need for explicit exhibition, Scruton's skepticism is justified. Even in two dimensions the point would hold. There are, for instance, two equally convincing analyses of Alberti's facade for Santa Maria Novella, one published by Wölfflin in 1889, based on the repetition of similar figures rotated 90 degrees, the other published by Wittkower in 1949, based on superimposed squares.[32] External evidence might lead us to think the latter a more likely original, but what, otherwise, is there to choose between them? They both rely on auxiliary lines (imaginary lines) to reveal privileged relations and virtual figures that could not easily be inferred from direct inspection.

137 Proportional analyses of Santa Maria Novella, Florence: left, *by Wölfflin, 1889;* right, *by Wittkower, 1949. (Redrawn by the author.)*

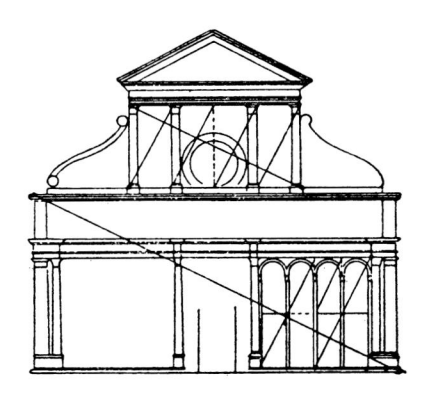

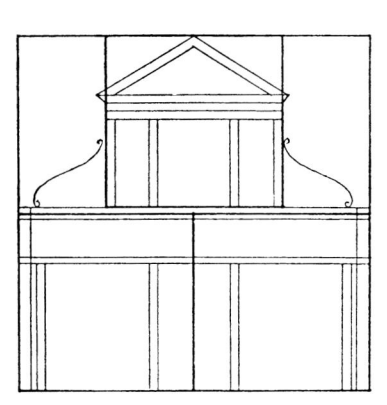

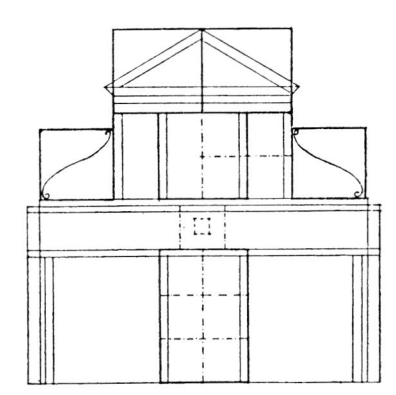

There are also indications in Alberti's own writing on proportion that it was not something to be superficially exhibited. He distinguished between an aspect of beauty that belongs to the overall order of a building and is a matter for philosophy, and a beauty that belongs to the various parts and is a matter for experience. Experience may grasp relations directly, but where the effort is too great and the order too large, philosophy takes over.[33] He refers to the proportion that informs beauty as a "hidden cause," a "secret argument," as something "inherent and implanted,"[34] the effects of which are vivid, although the causes that give rise to them are not understood by the ignorant. The profusion of books and articles over the last century trying to uncover architecture's secret skeleton[35] surely also indicate that regulating lines and dimensions have, in buildings of any formal complexity, tended to be recessive, not dominant. But although proportion is often inscrutable, it is not necessarily ineffective because of that.

In practice, Renaissance artists were not trying to show proportion in things but trying to bury it in things in imitation of nature, where it was as hard to see. Just as the proportion in isorhythm was "extramusical," so the proportion in architecture was extra-architectural, brought to bear on it from outside. And we may judge from what Alberti wrote and built, or from what Cesariano, Serlio, and Delorme drew,[36] that its purpose was not to exhibit its own format but to engender beauty of another sort that was directly available to the senses. The ideality of pure form is thus compromised because it is more important that we sense the presence of an underlying order than it is to see it clearly. In fact, if it were seen without effort, without the impedence of proliferated form, the exercise of showing it would be futile. Its presence has to be sensed like that of a ghost. The imagination has to conjure it up, and for centuries artists, critics, and historians have obliged, approaching the work of art with an inventory of imaginary elements with which to construct its fugitive truth. What are these imaginary elements? In the visual arts they are the traditional elements of geometry, no more, no less.

Take the "line of fate" that Leo Steinberg finds in Michelangelo's *Last Judgment* fresco. Skewering through a melee of bodies, it aligns five things that string together to make a story. Its role in organizing the composition of the fresco is debatable. Steinberg sees this and says to his doubting colleagues:

> *I propose the following operation: apply an inch rule to our diagonal on a good sized reproduction. . . . You discover—I am still awed at the sight of it—that the midpoint of the face in the skin (the flayed skin of*

Michelangelo himself) marks the centrepoint of the bend, exactly halfway between the highest and lowest points of the traversing diagonal. Again, such precise ratios do not materialize accidentally, and no one who has ever constructed a picture will doubt that these metric relations were planned. In fact the centrality of the shed skin on that two-way track must have been envisioned by Michelangelo before the painting began.[37]

Art needs to be imaginatively undressed in order to be appreciated, and so does the world: "Please observe, gentlemen, how facts which at first seem improbable will, even on scant explanation, drop the cloak which has hidden them and stand forth in naked and simple beauty."[38] This is no critic talking about painting, it is Galileo talking about statics. We search for inherent structure to find original causes. No one is certain that our inserted diagrams represent inherent structure, nor is anyone certain that inherent structure is the ensign of an intelligent plan, but we are inclined to think so because we can make artifacts that way and see afterward, with some difficulty, that they were made like that. Reality may then be presented as neither opaque nor transparent; it is well dressed, like art, like us, disclosing only so much of its inner essence.[39]

The full force of the oxymoronic phrase "inherent and implanted" can now be appreciated. It was coined by Giacomo Leoni, Alberti's eighteenth-century translator into English, opting for a reading of "infusa de la natura" in Bartoli's Italian translation[40] that accentuates its ambiguity, but which is not so far from Alberti's original "aut impressum aut insuffusum sentimus." Alberti was discussing the curious fact that three young girls, one stout, one slender, and one of middling build, can all be truly beautiful (bringing to mind the similar case of the three orders of columns based, according to Vitruvius, on a man, a matron, and a maiden). Their loveliness is gauged not according to some single measure, but by the degree of internal metric correlation between the parts of the body. It is this internal correlation of parts that is referred to as inherent and implanted, a combination of words that highlights the puzzling uncertainty as to whether the patterns behind appearance are discovered by us or are forced into place by us, who know what we are looking for and find it because we put it there by inadvertent mental projection. Were it not for the prejudiced attention we give to things, whereby we construe them as if they were made from ideas, would we ever see lines and figures in them? Vision may be achieved through the intromission of light rays, but certain kinds of visual understanding are achieved through intellectual extramission. This may explain why we still imagine vision as a projection from the eyes, although for centuries we have known it is the other way round.

Perhaps, then, classical art, maybe most historical art, deserves some redefinition: it does not expose the ideal, it plays hide-and-seek with it. The game, however, is as serious in its implications as a game can be. Beyond all the antagonisms between art and science, there is an abiding complicity; a complicity with early modern science, all natural philosophy, and theology too. Classical art, it is often said, does not merely imitate nature's superficial appearance, it imitates nature's constitution as well. In synthesizing these two kinds of imitation, in aiming to show that what we recognize as real derives from what we cannot properly see, it provides reassurance that the world at large is similarly underpinned. By communicating appearances that do not allow complete comprehension of their own insinuated structures, classical art solicits the belief that all things in their fullness and variety are subsumed, organized, and animated by abstract relations. So, in the highly cultivated obscurities of art, amongst other things, on the threshold of perception lies the planted evidence, the treasured thought: clarity can be found under everything, if our vision is sufficiently informed and penetrating.

Perspective and Proportion

Brunelleschi probably invented linear perspective, and in retrospect this seems a more momentous achievement than his building of the Florence cathedral dome. Linear perspective revolutionized painting and made a considerable impact on architecture too. How, though, did it relate to the theory of proportion? Clearly, the perspective transformations of actual dimensions into projected dimensions are coherent and systematic. Equally clearly, they are subject to continuous enlargement or diminution depending on the distance of the item from the eye, and the angle and position of the screen onto which the image is projected. During the fifteenth and sixteenth centuries there was no mathematical way to account for such dynamic morphology outside of the system itself, which was, after all, a species of geometry. An article by Wittkower, published in 1953,[41] proposed to show that "binding laws of proportion" were operative in perspective, and that these binding laws were understood by fifteenth-century practitioners such as Piero della Francesca, Leonardo da Vinci, and Brunelleschi. He shows that Piero and Leonardo recognized how, under particular conditions, equal measures in real space were transformed into a harmonic series of measures in perspective.

A double bonus can be got from this. First, it links perspective to music, as the ratios between pure musical tones are the supplements of the harmonic series that is allegedly present in perspective (the harmonic series begins $1, 1/2, 1/3, 1/4, \ldots$; the pure tones are $1:2:3:4\ldots$). Thus Leonardo could explain in 1492, "I give the degrees of the objects seen by the eye

138 Proportion in perspective, showing the generation of harmonic ratios when the viewing point O is one bay width distant from the picture plane. The top drawing shows a line of equal bays mapped into perspective under these conditions. To the right is a horizontal section through the centric ray. The resolution into four congruent triangles (heavier line) indicates that the image of the fourth bay width G–4 ends up as 1/4 its true length. Likewise the image of the third bay is 1/3 its true length, etc. Thus, in the perspective below, the line D4-C4 is (D1-C1)/4. From this it can be seen that successive perspective verticals and horizontals form a harmonic series 1, 1/2, 1/3, 1/4, etc. But, measured in the same way, the ratios between the successive orthogonals traveling toward the vanishing point, such as D1-D2:D2-D3:D3-D4, are not harmonic. The sequence here is 1/2, 1/6, 1/12, 1/20, 1/30, etc., which is the difference between successive harmonic ratios (the sequence could thus be regarded as a line divided harmonically that had its root in the vanishing point). The diagonals, D1-C2, D2-C3, etc., have no direct relation to a harmonic series. (Drawn by author.)

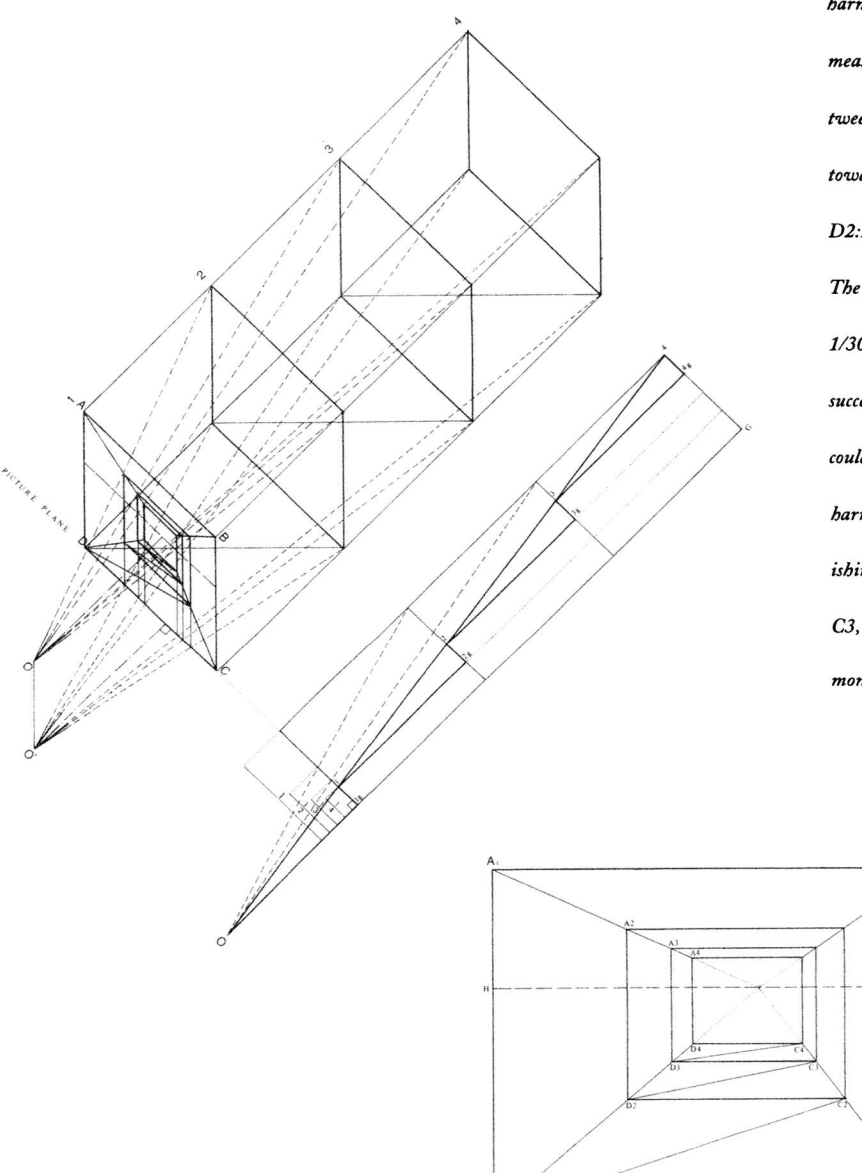

139 Santo Spirito, Florence, Filippo Brunelleschi, started 1436, the nave.

as the musician does the notes heard by the ear."[42] Second, it links both of these to architecture. Wittkower maintained that axial views down the naves of Brunelleschi's longitudinal churches, Santo Spirito and San Lorenzo, reproduce these fundamental ratios directly on the eye; an observer standing in line with one pair of flanking columns would see the rest shrink by successive harmonic intervals. He concluded that in these buildings, "everything was done to make the perception of a harmonic diminishing series in space a vividly felt experience."[43] He was right to portray this link between mystical number, cosmic structure, music, architecture, and perspective (and therefore painting) as important. It was the only one of its kind.

Even if we accept the unlikely idea that proportions registered in vision are the same as those registered in a perspective picture plane (which Wittkower assumes to be the case), harmonic diminutions will only be experienced fleetingly in Brunelleschi's naves. Alter the position of the viewer and whole-number ratios disappear. You still get harmonic ratios in the picture plane, but they are not the musical ones; most of the time they will be incommensurables. Alter the position or orientation of the screen and the harmonic series vanishes altogether. But looked at from the front, with screen and eye lined up with the columns, and measured on an interposing orthogonal, flat surface, the relation holds. Wittkower's observations make sense because the vast majority of paintings from the early and high Renaissance insist on frontal alignment while many of the buildings encourage it.

We must allow that frontal, axial architecture, not in itself new, may have been bolstered by the momentary numerical correspondence that it fixed in place. But the reader of Wittkower's article is left with the impression that harmonic proportions permeated perspective. They did not. Even in the privileged frontal view that arose as a matter of course out of Alberti's central perspective construction, the transformation only worked on lines parallel to the picture plane. Nor do Renaissance artists seem to have been too struck by this property. Very occasional reference was made to it, and it was probably understood by only a few.

However, the most striking aspect of perspective, and the one that architects and even painters continued to find perplexing, was its disruption of measured proportions. The treatise writers, following Alberti, recommended caution in making perspective drawings of buildings.[44] Moral judgments were already insinuated through terms such as degradate, invariably used to describe foreshortening, as opposed to *perfetto*, used to describe an orthographic elevation. Albrecht Dürer, in his study of human proportion, spends half the work describing a range of comely bodies with the aid of a system founded on the harmonic series, and the other half producing caricatures and monsters by subjecting the former to peculiar perspectival transformations. He called some of the constructions used to perform this task "falsifiers" (*Falscher*).[45]

But the full force of opprobrium was reserved for the act of seeing. Brunelleschi's linear perspective construction had miraculously wrested this from the subjective impressions of the beholder and rendered it reproducible. Nothing could be more strange than this constant railing at the eye—the deceitful eye—yet it draws attention to the peculiar status of Brunelleschi's invention.[46] Renaissance perspective techniques were dependent on simple Euclidean constructions, but, despite this, perspectival and Euclidean geometries were at variance, as we have already seen. Perspective could thus have cast a shadow over its parent geometry, but that is not what happened. Instead, a clear distinction was made between a geometry that belonged to the essential structure of the world—Euclidean geometry—and a geometry that recorded appearances—perspective geometry. Acknowledged for its brilliance and its novelty, recognized as, in all likelihood, an advance on ancient practices, perspective was nonetheless kept subordinate to Euclid. This was exactly how Lomazzo described it: "Perspective, being subordinate to geometry and as it were the daughter thereof, is a science of visible lines."[47] Hence its account of reality was regarded as a deformation, a trick of the light.

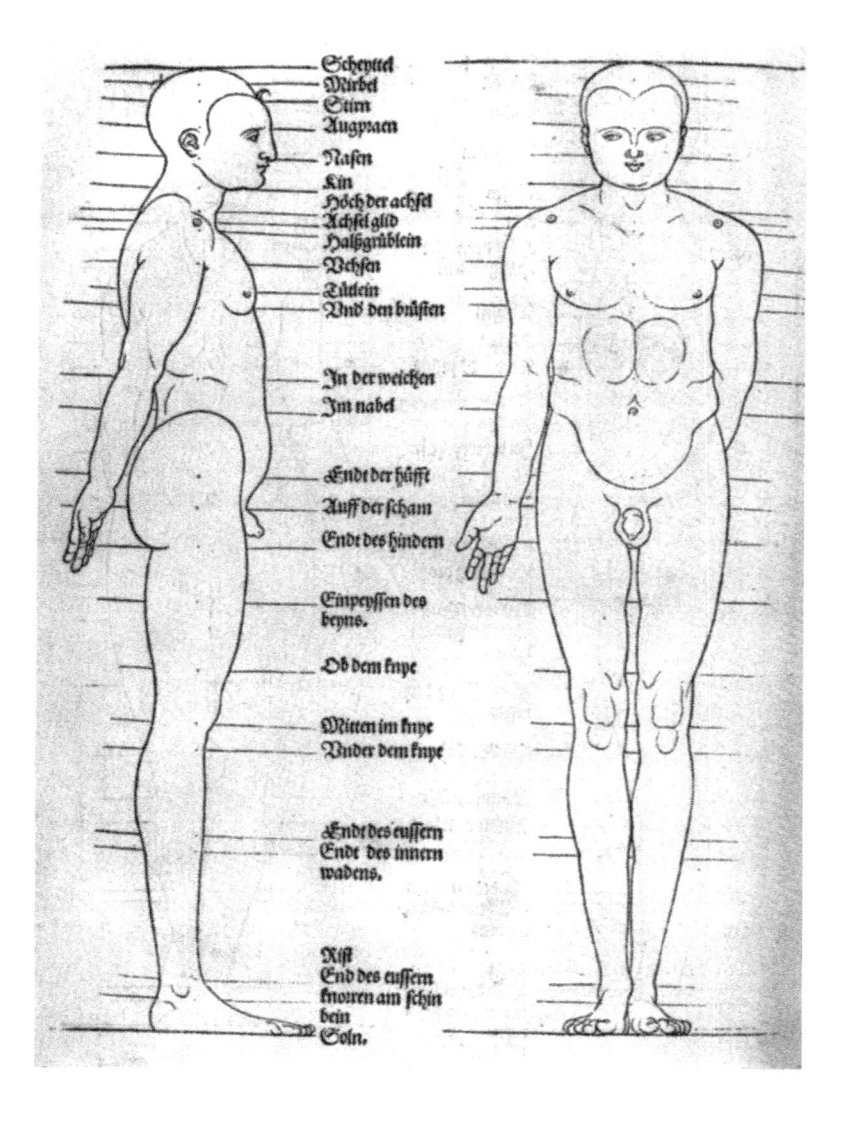

140 Albrecht Dürer, figure from Vier

Bücher von menschlicher Proportion,

1538.

THE TROUBLE WITH NUMBERS   255

What set perspective apart from other systems of representation, or indeed systems of any kind with the sole exception of Euclidean geometry, was its internal consistency. Because this consistency did not appear to relate to any other kind of consistency, because its most obvious characteristic was its distortion of measured relations, it was seen as a record of the way in which vision perverts the ultimate truth in the constitution of things. Nowadays, the same point is made by certain advocates of axonometric projection, which, they claim, shows things not as they appear but "as they really are."[48] I allude to this only to show the persistence of the idea that perspective is a systematization of untruth, and that the real truth lies in the direct measurement of objects. It is a questionable doctrine, but it shows that architects and painters have been loath to give up their reliance on direct measure, which is the most straightforward kind and the one over which we feel we have most control. There are obvious advantages in defining architectural proportion by direct measure, but in the face of perspective it also gives rise to difficulties in reconciling the real with the apparent, the measured with the seen. Perspective is proportional, but not classically so. Its coherence is of a different sort.[49]

There is still some confusion about this. Perspective is said to be the product of a rational, calculating, mercantile mentality. Michael Baxandall has it that fifteenth-century Italian viewers of art, immersed in trade and accounting, had "become adept through daily practice in reducing the most diverse sort of information to a form of geometric proportion: A stands to B as C stands to D,"[50] and therefore understood better what they were looking at. Unfortunately, such a viewer would not be predisposed to make sense of perspective since perspective obliterates ratios of this kind. Alberti himself recognized as much when he condemned the practice of some painters who drew each successive row of tiles in a pavement $2/3$ the width of the previous row. It is logical, said Alberti, but it does not correspond to the way in which things are seen.[51]

Yet if perspective destroys classical proportions, why were so many classical architects involved in its advancement (think only of Brunelleschi, Alberti, Serlio, and Vignola)? Perhaps it was because architects were as deeply preoccupied with apparent size as they were with true measure and so sought means of reconciliation.

Over two hundred years after Brunelleschi's demonstration, another architect, Girard Desargues, neglected in the history of architecture[52] but celebrated in the history of geometry, finally devised a law of proportion that was always true within perspective configurations. Desargues achieved this by concentrating on relations within the picture plane. His

141 Vittore Carpaccio, Birth of the Virgin, *1504–1508, Accademia Carrara, Bergamo.*

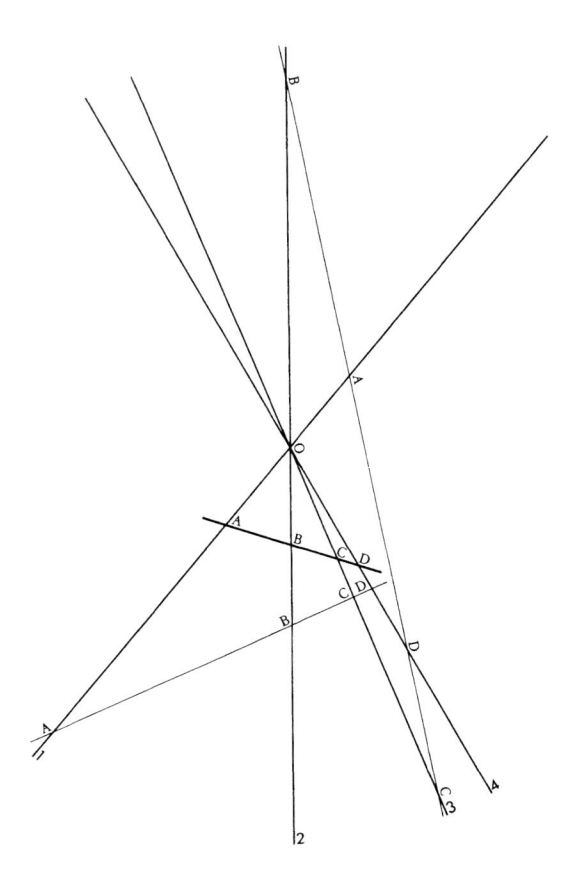

142 Girard Desargues's theorem of four-point involution: for any four fixed lines intersecting at a point O (e.g., O-1, O-2, O-3, O-4), any other line crossing the array of four lines at A, B, C, and D will be divided such that AC/BC : AD/BD is a constant (drawn by author).

theory of the four-point involution determined the ratio between the parts of any straight line cutting four other straight lines that converge on a point (and that may therefore represent parallels in the Euclidean space behind the picture plane). Thus, for any line intersecting at A, B, C, and D with four fixed lines all passing through a point O, AC/BC : AD/BD is constant. It is called the ratio of cross ratios. With Desargues's discovery of this invariant, the proportion within perspective was at last being treated on its own terms rather than being forced into accord with the more familiar arithmetical and harmonic ratios.[53]

What we must never forget—what is nevertheless so easy to forget while reading about perspective—is that the proportionally regulated character of perspective pictures is obvious. Anyone can see it at a glance in the receding walls, coffered vaults, ceilings, and tiled floors of fifteenth-century painting. In place of a commensurable system of numbers, perspective provided a coherent system of lines: visible lines. But while it took two hundred years to develop a mathematical account of perspective's proportionality, that same proportionality was manifest in the drawings of the more accomplished artists from the very beginning, constructed but not calculated.

Perspective as exemplified in the relatively dumb practices of fifteenth-century artists contained a judicious mixture of visually coherent distortion and conserved shape. Frontal alignment guaranteed a scale relation to real lengths and angles, while telescopic foreshortenings played with their organized geometric collapse, easy to see, not too difficult to construct, but impossible to reduce to numbers. And so a practice that for all the world was geometry challenged geometry's most venerable companion: rational proportion.[54]

The theory of proportion extended only so far into practice, and this for reasons exactly opposite to those provided at the time by theorists. It was not the recalcitrance of material in the hands of the artist that impeded the full expression of the idea, it was the quickened practices that defied full incorporation into theory, even of something as abstract and regulated as perspective. This is surely a significant reversal.

Dürer was the only Renaissance artist bold enough to attempt the incorporation of perspective diminution into a general theory of proportion. The most conspicuous message from his book on the human figure, *Vier Bücher van menschlicher Proportion* (1528), was that the numerical divisions necessary to describe corporeal beauty could be complemented with perspectivally distorted measures in order to enable the description of corporeal ugliness. More careful observation reveals that Dürer was willing to introduce into the description of normal figures the very perspective transformations of equal measure that he himself had labeled as falsifying. In his first example, an ideal/typical male, he took harmonic fractions from 1 (fundamental) to $1/40$. He found that these applied well to the extremities of a reasonably proportioned body, but different fundamentals had to be used to describe different parts. For example, he could not use the same series for both the torso and the lower leg. So he introduced an oblique line across four lines (radiating from a point), which transformed three equal measures into three perspectivally diminished measures. These were then used as the fundamental lengths for three separate harmonic series.[55]

In his architectural designs, Dürer applied other perspective transformations to staircases and columns.[56] The metric adjuster that he used was of the type that would be illustrated in numerous architectural treatises to compensate for the diminution of statuary, inscriptions, and such like

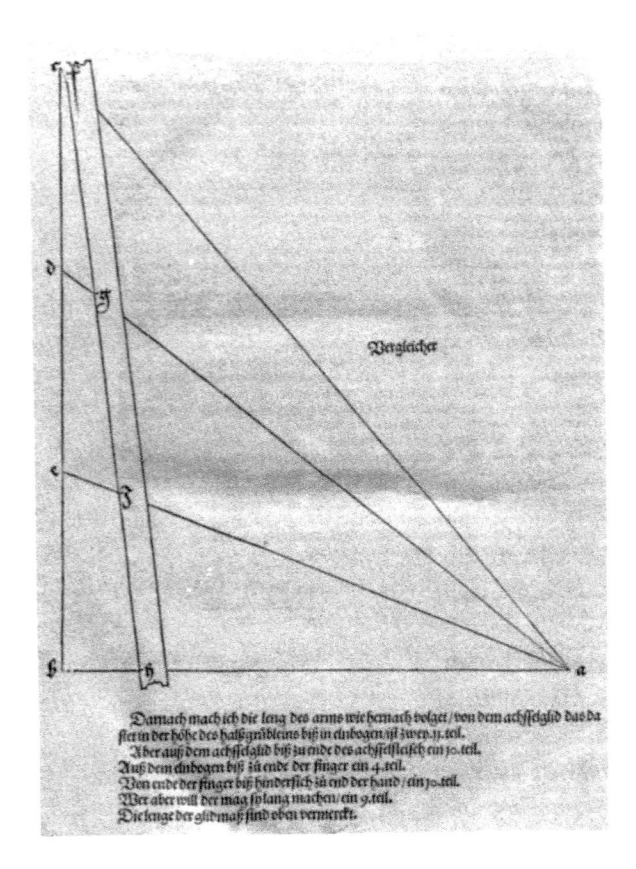

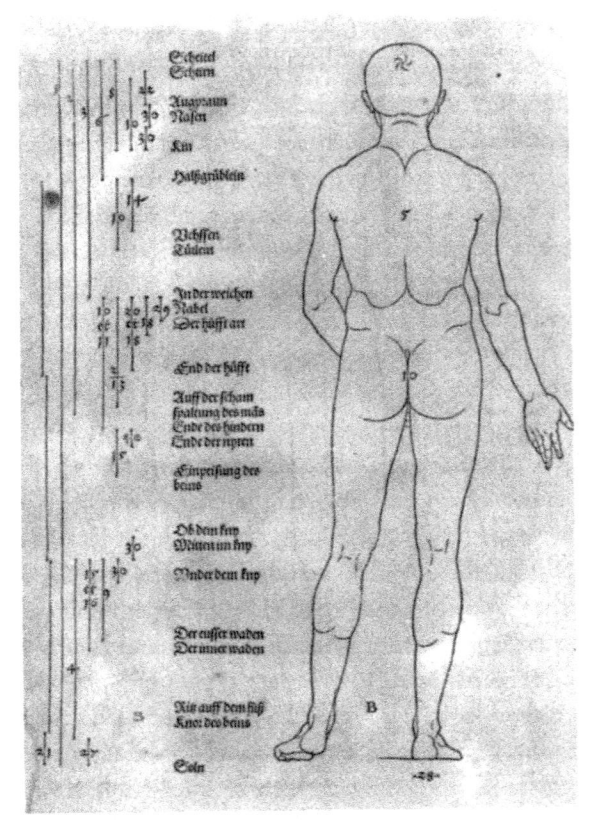

143 Albrecht Dürer, perspective adaptation of equal measures, from Vier Bücher von menschlicher Proportion, *1538.*

144 Albrecht Dürer, figure proportioned with the perspective adaptation shown in the previous figure (unlabeled sequence at far left), from Vier Bücher von menschlicher Proportion, *1538.*

in the upper reaches of buildings. Dürer himself sometimes used it this way too, but in his design for a corkscrew stair, where he employs it to increase the step height, it has nothing to do with counteracting the eye's deceit, because, as he tells us, the stair is encased and cannot be seen all at once. He had anyway supplied an alternative plan that used perspectivally adjusted measure radially, to accelerate the spiral toward the middle. So Dürer was not using this device to put rational proportion back into the experience of seeing by reversing the effect of distance on the eye, as everyone else recommended, but redeploying it to pull proportion one step away from numbers.

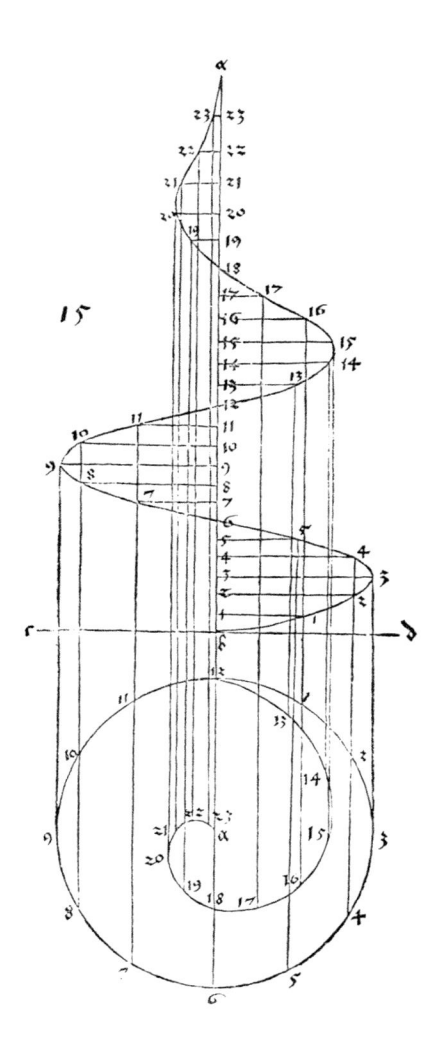

145 Albrecht Dürer, diminishing spiral stair with risers proportioned by perspective adaptation, from Underweysung der Messung, *1527.*

Tuning and the Demise of Rationality

In painting, proportion was slipping out from under the aegis of number. In music, on the other hand, the presence of number was redoubled by portraying integer ratios derived from musical harmony in musical rhythm. But that was not all. The questions that were raised about isorhythm were questions about its audibility. If isorhythm was seen but not heard, we should now deal with what was heard but not sanctioned in the same music.

While Pythagorean harmonic ratios were being recycled into musical rhythm, and while they were being exported to architecture as the basis of classical proportion, they were mortally threatened at their source. It can hardly be an accident that the composers who made most effective use of isorhythm were amongst those who made much wider use of harmonies outside the Pythagorean scheme, thus upsetting the theoretical basis on which all music was supposed to rest. It was, again, Dunstable and Dufay who figured prominently. What they did was populate their compositions with three kinds of note groupings that had, until then, been used sparingly since they were regarded as dissonant or imperfect. These were vertical thirds (C and E played together, for example), vertical sixths (C and A played together, for example) and triads (C, E, and G played together, for example).[57] The new sound, dominated by these unsanctioned harmonies, seems to have originated in England, for when it spread to Burgundy and Flanders, then the centers of European music, it was called the *contenance anglois*. Dufay began to make use of this more lively tonality in the 1430s, and *Nuper rosarum flores* was one of the earliest of his compositions incorporating it. Warren does not mention the thirds and sixths in the Dufay motet. They cut across the carefully constructed picture of unified correspondence that he is at pains to conserve.

It is a pity that so little of the sense of this music can be conveyed in print. I would urge any interested reader to listen to the unfamiliar, wavelike modulation of Dunstable's motets, to Dufay's majestic and mellifluous cyclic masses, or to the driving complexity in Johannes Ockeghem's work. But the *contenance anglois*—always considered to be vivacious—did not liven things up much. Indeed Renaissance polyphony was less mobile, less sanguine, less obviously rhythmic, and more unearthly than what came before or after (compare the just-named examples with Machaut on the one hand and Monteverdi on the other), which may be why music plays so small a part in our conception of the Renaissance. Considering that *Nuper rosarum flores* was written for such a happy occasion with such "frisky" harmonies, it is a rather melancholy piece. As the pulse of the music disappears into an inaudible undertow of mensuration, the thirds and sixths pile in only to give a vague relieving brightness to these diffuse accumulations of sound. What this achieves is an effective blending of dissimilar qualities, very beautiful, quite unlike anything else, perhaps as distinctive and as compelling to the ear as perspective was to the eye.

The reasons why thirds and sixths were not admissible as perfect concords are as follows. The Greeks had established a logic within the scale of notes, recognizing that the sweetest sounds were produced by the strict numerical subdivision of a plucked string under constant tension. This discovery, traditionally ascribed to Pythagoras, is the root of all ensuing ideas of natural harmony. In music, the world's secret numerical armature lies exposed. It can be illustrated and mimicked elsewhere with fruitful results, but it can be experienced most directly in the scale.

However, the numerical subdivision of the stretched string only operates up to a point. The system appears so wonderfully consistent at first sight because the ratios are always related to the fundamental note (1:$^1/_2$, 1:$^2/_3$, 1:$^5/_6$, etc.). But the ratios between notes within the system get distressingly complicated and cannot be arranged in such a way that all whole tones hold the same relation to one another. For instance, the ratio between a pure fifth ($^2/_3$) and a pure sixth ($^3/_5$), one whole tone apart, will be $^9/_{10}$, while the whole tone between a pure fourth ($^3/_4$) and a pure fifth will be $^8/_9$, a difference that is audible and unhappy. In antiquity, a host of partial solutions to this insoluble difficulty were devised.[58] The Pythagorean scale, the one that preserved the most whole-number rationality, was widely accepted in medieval music.[59] It just so happens that in this scale the thirds and sixths absorb all the inequalities and sound distinctly sharp. In monody this posed no insuperable problems, but in polyphony, when three, four, or five voices sing different notes simultaneously, it did. Hence they were used sparingly as passing dissonances on the way toward harmonious resolutions.

No one knows quite how musicians dealt with thirds and sixths in the early fifteenth century, although it is usually assumed that in practice they tempered, that is, adjusted, the Pythagorean scale to suit the music.[60] We do know that throughout the sixteenth and seventeenth centuries such adjustments became the subject of extensive investigation and heated debate. Opinion ranged from those who held that polyphony was a corruption of music's original purity because it forced harmony out of its proper numerical shape, to those who, acknowledging as a matter of course the indisputable virtue of pure harmony, nevertheless sought ways to adapt the scale to contemporary practice, which meant departing from pure harmony. Sometimes this departure could be quite radical. Vincenzo Galilei, who on the one hand advocated a return to ancient practice, on the other described an expedient to obtain approximately equal intervals in the fretting of stringed instruments (he was a lutenist) that retained no pure notes at all.[61] The situation remained confused and fraught for over two hundred years, with different tunings for different instruments, and more and more complex ways developed to combine modern harmony with pure tonality, such as the keyboards with 19, 31, and 53 notes to the octave.[62] The tuning with which we are familiar, called equal temperament, was not generally accepted until the eighteenth century. It is a more precise way of getting what Vincenzo Galilei was after. Each semitone has exactly the same ratio to adjacent notes throughout the system. That ratio is $\sqrt[12]{2}:1$. Only the octave is preserved; everything else is irrational.

146 Comparison of musical scales visualized as the subdivisions of a monochord (so the differences are equivalent to differences in position of the frets on a guitar fingerboard). E shows equal temperament tuning, as in modern instruments. P is Pythagorean tuning and Z is Zarlino's attempted revision of the Pythagorean scheme. The triangles indicate the repetition of similar ratios within the system, $1/\sqrt[12]{2}$ in equal temperatment and 8/9 in the Pythagorean scale. It is easily seen that the greatest variations occur in thirds and sixths. (Drawn by author.)

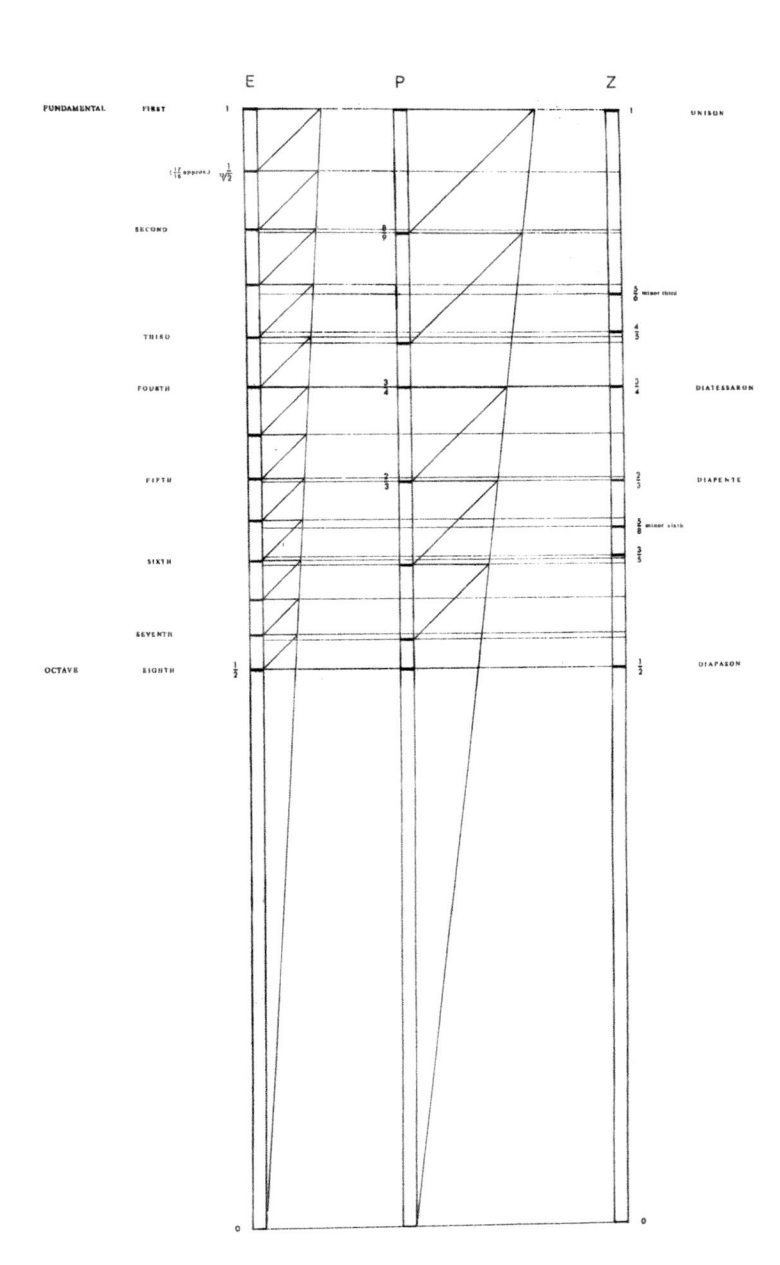

The history of Western tuning is the history of the extinction of whole-number ratios. It may seem peculiar, then, that the theory of proportion in the visual arts, especially architecture, should have rested on this shifting foundation; even more peculiar that the period during which the Pythagorean musical analogy held sway in the visual arts should correspond so exactly to the period during which the Pythagorean harmonies were being modified in music itself (the early fifteenth to the early eighteenth centuries). But most extraordinary of all was the relocation of numbers in Dufay's motet. As the Pythagorean ratios were invading the rhythmic structure of *Nuper rosarum flores*, they were being challenged within its harmonic structure. As they were being represented, they were being dissolved.

We might therefore surmise that the expansion of the Pythagorean system was due less to a profound faith in its timeless validity, than to a fear that if it stayed put it would be liquidated. And so it went underground or abroad. We can either believe that the propagators of whole-number proportion, including those who expounded it as the guide to architectural beauty, from Alberti to Ouvrard, Blondel, and Lobkowitz, were all unaware of what was going on in music and were simply handing down the same conventional formulas, which is hardly credible,[63] or we must search for some reason to account for the evacuation of whole-number ratios from musical harmony and their resettlement in architecture. Since the combination of isorhythm with thirds and sixths in fifteenth-century polyphony provides a vivid instance of rational proportion's displacement under threat, it is not unlikely that harmonic theory passed into architecture as an evacuee rather than as a colonizer.

Double Measure

What had happened in music and in the visual arts was similar. Elements of both were susceptible to mathematical treatment, and yet the two most conspicuous such elements—perspective and the musical scale—were proving recalcitrant to rational calculation. The striking thing is that, in architecture and painting as in polyphony, a countervailing rationality was established at the same time, and by the same people: Brunelleschi and Alberti introduced "harmonic" architectural proportion as well as the linear perspective that distorted it,[64] just as Dunstable and Dufay used proportional isorhythm as well as the thirds and sixths that destroyed numerical proportionality. There is thus in fifteenth-century art a strong compensatory aspect. What some take to be its essence was already departing from it, and was then ghosted to such an extent that the specter of the usurped was more clearly recorded than the presence of the usurper.

However, no point would be served in claiming that the true basis of music was the new tempered scales, or that the true basis of painting was perspective construction, with the incalculable or irrational divisions in these as the foundations of art,[65] while relegating the harmonic proportions to the status of a sentimental recollection. Leaving aside the question as to whether anything is fundamental to art, it would be difficult to imagine either musical temperament or perspective developing the way in which they did without the Pythagorean scale and Euclidean geometry to develop from. In both cases a point was reached when knowledge derived from these originals took over from them.[66] The resolution of equal temperament as $\sqrt[12]{2}:1$ between each consecutive semitone, which was worked out by the Dutch mathematician Simon Stevin in the first years of the seventeenth century,[67] might be compared to Desargues's identification of the ratio of cross ratios in perspective some thirty years later. There was one crucial difference between the two events, however: Stevin's formula for equal temperament, although neglected at the time, had some impact on musical practice, whilst Desargues's theory of the four-point involution was irrelevant to the practice of painting. I would argue that this was because perspective had already achieved its reformulated measure in the hands of fifteenth-century draftsmen whose graphic constructions outstripped calculation.

We must avoid the pitfall of associating these more advanced mathematical accounts with more advanced art.[68] At the same time, we must reject the assumption that ideas expressed in Renaissance texts provided full and authentic interpretations of Renaissance art. That would mean that all the distortions, rejiggings, adjustments, tempering, and modifications that were so much part and parcel of artistic practice could only be understood as lamentable adaptations to defects in the way we see, or to the recalcitrance of matter, when these were not the problems at all. The problem was that there existed a nameless area within art, neither properly rational nor properly intuitive, defined neither by theoretical fundamentals nor by the free play of artistic imagination. Let us call it the area of putative proportionality.

Something without a name is difficult to define. It had always been possible to slide between the authority of numbers and the judgment of the senses, to argue the case of the one against the other, to play them off against each other, or simply to regard them as different ways to the same end.[69] Donatello and Michelangelo trusted the eye rather than the abacus or compass; Aristoxenus and Mei trusted the ear rather than the numerically subdivided string. At the time, the perplexing aspects of perspective and temperament were to a considerable extent dispersed into a broader scheme of ideas whereby intuition took over from reason as things became more complex and incalculable. It is only the privilege of hindsight that allows us to identify a different species of order in them; to see an essential difference where it had once been hard to see anything but an increased complication, confusion, or degradation of form.

Another similarity may already have been noticed. Rational proportion in music, architecture, and painting was inscrutable, yet the increased prominence of thirds and sixths altered the sound of music so much that even the untutored would have been aware of the difference. Likewise, perspective was there for all to see in painting. The new practices were acting directly on what was visible and audible, altering the perceived constitution of the artwork. Meanwhile, the countervailing rationality was buried with full honors inside it.

I have described the Pythagorean ghost as more prominently recorded than these vivid effects. How so? The answer is that it was more prominently recorded in writing. It would have been almost impossible to write of temperament in the fifteenth or early sixteenth centuries except as an unhappy practical necessity, however important it may have been for the development of polyphony.[70] And while there were many books on the subject of perspective, they were mostly filled with drawings and descriptions of how to construct drawings, and very little else. The points at which perspective became legible rather than visible were the points at which it could be joined tentatively to established ideas about the role of geometry, number, and proportion in art, or where it could be treated as an extension of medieval optics,[71] or where it could be tied to ideas of composition borrowed from rhetoric.[72] Piero della Francesca chose not to write much about any of these, which is perhaps why the humanist Daniele Barbaro dismissed Piero's perspective treatise as written for idiots.[73]

Returning to the more general question, if we ask whether the practice of proportion was unified in Renaissance art, the answer must be no. We may dwell on the similarities between perspective and temperament, but these only corroborate the skill with which painters and musicians could synthesize two inimical kinds of measure in their work, one dominant, one recessive. It does look, therefore, as if our portrait of the Renaissance has, indeed, been heavily overpainted, or perhaps it would be better to say overwritten, since writing has been the medium organizing our historical perception of this matter more effectively than any other. There was no single species of proportion, no unified idea of it, no general relation of theory to practice, no seamless weave of numerical correspondence and similitude.

But the absence of unity in the treatment of proportion should not be treated as a fault. Let us regard it instead as a characteristic. If studied neither as the vessel of progress nor the container of lost truth, Renaissance art shows itself riven with anomalies. These anomalies are not mere accidents arising from incompetence, thoughtlessness, or unconcern; they are integral. We might even say that they are the subject of the works in question. The unity in a Dufay motet, an Alberti building, or a Leonardo painting is an artifact. It does not derive, as a matter of course, from an underlying unity; it is constructed by the artist, often with immense effort, as a surface quality. It is the final effect, not the first cause. It is the most expensive, the most unlikely of results. The unity of Renaissance artworks, individually and collectively, is real enough, but its value and its fascination derive from its being the last thing we should reasonably expect to find. We may therefore accept that the most accomplished works were unified presentations, without accepting that they were the product of unified ideas. All the more impressive, then, that the underlying incompatibilities were overcome so completely that it has never been necessary to recognize their existence. Maybe this was among the greatest achievements of Renaissance art: to begin the assassination, interment, and memorialization of its own announced regulative principle so silently and yet so effectively that no one noticed.

The most explicit indication of the changed circumstances came not from artists but from an astronomer, Johannes Kepler, who revised the ancient notion of the music of the celestial spheres after discovering that planetary orbits were elliptical. Accelerating and decelerating on eccentric paths, the planets could no longer emit pure tones. They would undulate (like sirens wailing, says D. P. Walker). Their six-part combination would produce a polyphonic, sliding dissonance that, once in an aeon, might pass through an instant of perfect harmony.[74]

Perrault's Critique

We are often reminded that Claude Perrault's critique of the accepted theory of architectural proportion has to be put in the context of the celebrated dispute between Ancients and Moderns.[75] Perrault was a progressive, a Modern. No one in the architectural community had so little patience with the excessive veneration paid to bits of ruins, nor heaped so much scorn on arguments from authority. Yet his commentators also notice some equivocation, a drawing back from the progressive cause, a hesitation in extending the rational argument beyond the critical point. Perrault himself was the first to acknowledge the paradoxical character of his own conclusions.[76] The central paradox was that while he assaulted all the traditional arguments and explanations of classical architecture, he did not assault classical architecture, he tried to conserve it. His ambition was to remove a well-established edifice onto new footings. Perrault did not attack proportion; he attacked ideas about it. Out went the idea that proportion was fundamental. Insofar as it was numerical, it was the least essential aspect of architecture. It had nothing to do with underlying truths of nature; it was a man-made convention overlaid on things to enable us to describe them. This, it turned out, was not a reason for abandoning it. Fully aware of its arbitrariness, Perrault realized that only our insistence on its perpetuation kept it alive. He concluded that since nature provided no distinct rules for architecture, we must provide them instead.

Perrault's arguments tend to be euhemeristic, explaining things away by showing them to originate from trivial, not great causes. There is always a chance that the subject will disappear as its legitimations are erased. I do not think, however, that Perrault's euhemerism is directed to that end. His undermining of classicism was precautionary rather than destructive in intent. I have demonstrated that some momentous practices in Renaissance music and painting were, from the start, unassimilable to other key practices within those same subjects and at variance with the precepts of theory. Critical awareness of this was muted, and awareness was virtually nonexistent within the arts. But this too may have been more of an advantage than a disability, for it allowed the coexistence and interaction between the two kinds of practice to develop without let or hindrance. Such tacit ignorance was prudent indeed if, as I am suggesting, the accomplishments of Renaissance art included the ability to concoct unity out of disparity. But as time went on, it would become more and more vulnerable to exposure. Could it be that Perrault was performing radical surgery, cutting classicism off from its own refutable basis in order to save it? The question is at least worth raising, even though the evidence is circumstantial.

We cannot be certain, because Perrault did not present the case that way. He did not point out the difference between whole-number rationality and other kinds of proportion and coherence, nor did he make any reference to the problems besetting whole-number proportion in modern harmony. He poured scorn on René Ouvrard's opinions, but in his book on the five orders he went so far as to accept the Pythagorean idea that numbers make beautiful sounds, only to deny that it had any relevance to the perception of beauty in architecture, because seeing is not analogous to hearing.[77] Perrault nevertheless knew what had been happening. His manuscript essay "De la musique des anciens" portrayed the ancients as ignorant of the great beauties that were later discovered by multiplying the parts in musical compositions and by treating thirds as harmonies rather than dissonances.[78]

He severed architecture from music, but the implications of their divorce are by no means obvious. It might have either one of two very different results. If architecture was cut off from the Pythagorean theory of harmony then, sure enough, it would lose the justification for whole-number rationality. If on the other hand it was cut off from the confusion of sixteenth- and seventeenth-century tunings, it would be dissociated from the collapse of whole-number rationality. Perrault wrote of the former circumstance in the latter situation; he wrote like a radical and acted like a conservative. That, it seems to me, explains why he not only retained but increased the scope of simple, whole-number proportions in his version of the classical orders.

His handling of perspective is equally ambidextrous. In fact he dealt not with perspective construction but with the related issue of visual adjustment. He said that all the treatise writers, with but few exceptions, agree that the eye deceives. Because of this they advise the architect to compensate for the defects of vision by making more distant things larger than ones close by. Since the same writers are prone to superstitious reverence for numbers, they first accept Pythagorean ideas unhesitatingly and then find reasons for altering them in practice, which is self-contradictory. He noted that, fortunately, not many of them were foolish enough to take their own advice seriously.

Here Perrault's wit was more incisive than decisive; a dilemma was exposed but not resolved. If numbers were not the architect's first guide, what was? Perrault introduced a psychological argument to make some sense of this question. Vision does not distort things, it relates them to one another. As infants we may confuse the size of objects with the size of the visual impression they make, but, with experience, we learn to discriminate with considerable precision their actual size and shape independent of distance and obliqueness, which is why a carriage driver can tell at fifty paces whether he can drive his coach through a gap, even though there may be only a couple of inches to spare. Far from being easy to deceive, sight is hard to cheat. Judgment is implicit in vision, but judgment of comparabilities, equalities, and symmetries, not of numbers hidden in measurements.[79] Perrault used this reasoning to draw the logic of proportion to the surface of the building. It was neither behind the building as a "hidden cause" nor in the cone of space intervening between the building and the observer's eye.

There are those who complain that, in treating proportion as a convenience rather than the source of true beauty, Perrault took the mystery out of architecture. I do not think so; if anything, the unexampled elegance of his own version of the five orders is made more mysterious. He took the mystery out of architecture's explanation, not out of architecture. In doing so he moved attention from invisible causes to the visible constitution of building. That was a risky enterprise for two reasons. First, by driving a critical wedge between what was said and what was done, it cast doubt on the effective illusions of rationality that help architects do what they do. Second, through rational argument it excluded rational procedure. Architecture could not be justified from cause to effect, since judgment would rest on the finished appearance of buildings, not on their adherence to preordained rules. Rules, recognized to be necessary, could no longer provide any guarantee of quality.

Chapter Seven **Comic Lines**

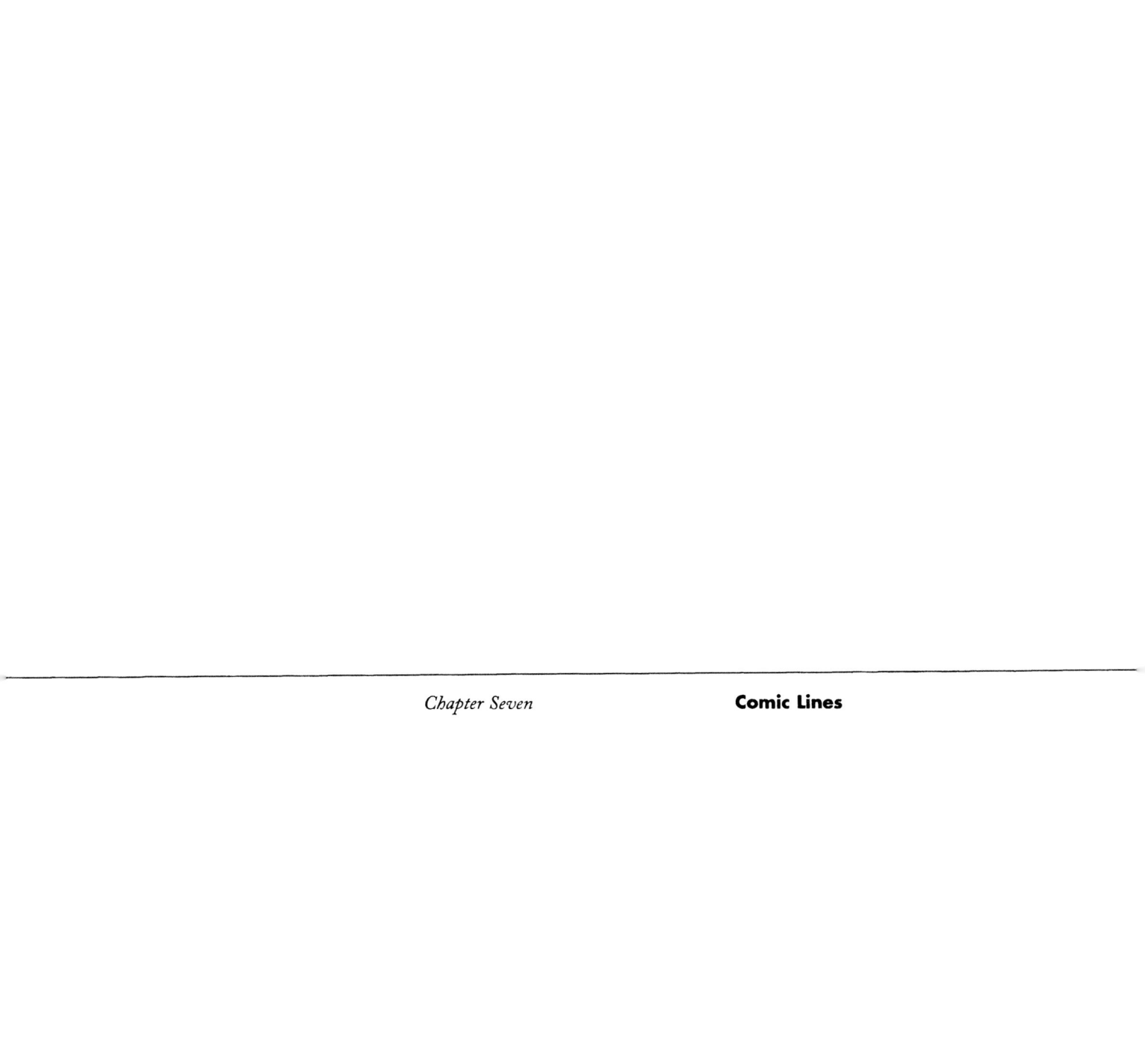

By pulling his guideropes harder, he was able to bend the skiff appreciably; this was possible because his tiller did not simply control a flat rudder aft, but bent the long keel, from the fore-end, to right to left, upward and downward, according to his directional requirements.

Alfred Jarry, Exploits and Opinions of Dr. Faustroll, Pataphysicist[1]

Le Corbusier developed the Modulor system of proportion between 1943 and 1955, designed and built the chapel at Ronchamp between 1950 and 1955, said the Modulor was in the chapel, and then challenged anyone to find it there. In his book on Ronchamp he wrote that his building was "informed by all embracing mathematics," and declared: "The Modulor everywhere. I defy the visitor to give, off hand, the dimensions of the different parts of the building."[2] In his book on the Modulor he wrote of Ronchamp: "It was a pleasure, here, to allow free play to the resources of the Modulor, keeping a corner of one's eye on the game to avoid blunders. For blunders lie in wait for you, beckon you on, tug at your sleeve, drag you down into the abyss."[3] Le Corbusier was very fond of quoting Albert Einstein's generous epigram that the Modulor was a way of proportioning things that "makes the bad difficult and the good easy,"[4] yet he himself seems to have been in at least two minds about it. Only ten days or so after distributing Modulor tape measures to his assistants, he forbade their use.[5]

But the Modulor was given free play at Ronchamp. The word "free" occurs often in this building's history. Canon Ledeur tempted Le Corbusier to accept the chapel commission with an assurance of total creative freedom. Le Corbusier described what he did there as "totally free architecture,"[6] and discussed it in a chapter entitled "Free Art" in *Modulor II.*

Of all his work, Ronchamp is the building that seems to have escaped architecture's limited inventory of forms most completely. The apparently successful escape from the restraint of drawing board geometry is an important ingredient of its perceived freedom. It is also the reason why Modulor proportions are hard to find, for the Modulor belongs to a long line of theories on the subject of architectural proportion that presuppose comparisons between flat rectangular shapes, and, since these are few and far between at Ronchamp, we are led to believe that the Modulor must have some deeper, hidden presence in the architecture.

147   Notre-Dame du Haut, Ronchamp, Le Corbusier, 1950–1955, view from south.

The Modulor began as an ambitious but prosaic attempt to solve the problem of coordinated dimensioning throughout industry and ended up as an equivocal statement of faith in an order that lies beyond cognition. Meanwhile Le Corbusier, helped by many others, developed a system of proportioning, less elastic than most because it was tied to a specific set of measurements. A string of dimensions in a constant ratio was extrapolated from 2.26 meters, finally agreed upon as the height of a man with arm upraised. The ratio was the golden section, vaunted since the mid-nineteenth century as the mathematical fount of physical beauty. The golden section is irrational,

$$1 : \frac{1+\sqrt{5}}{2},$$

so the dimensions had to be approximations. The Fibonacci series, an additive series in which the ratios between successive terms converge toward the golden section, made the calculations much easier. (Start with any two numbers, add them together to get a third number, then add the second and third to get a fourth, and so on—for example, 1, 1, 2, 3, 5, 8, 13, etc.). Because inconveniently large gaps were left on the ladder of dimensions, a second sequence from 1.13 meters was intertwined with the first. The two syncopated sequences, one twice the other, were called the red and the blue series.[7]

The most notable features of the Modulor were the attempt to make practical, anthropometric dimensions ("measures . . . related to the stature of man") identical with aesthetic proportions. and the doubling up of the series—a simple but elegant way of producing further proportional resonances while reducing the distance between consecutive Modulor measures.

All this can be gleaned from the two rambling volumes of the *Modulor*, published in 1950 and 1955. They give the impression of being hastily edited, personal notes that sometimes read like confessions where one would expect the impersonal language of the log book. In one respect, though, these journals, although full of self-congratulation, are generous, even selfless, since they expose to public view thoughts that are normally kept out of sight. Candid exposure of the ego can turn heroes into clowns. Le Corbusier, acknowledged for decades to be a great architect, was dabbling in mathematics, about which he understood little. He tells us so. He makes mistakes. He records them when others point them out, but persists, and ends up convincing himself that the original mistakes were somehow more true than the corrections.[8]

What makes study of the *Modulor* worthwhile, despite its amateur mathematics and strident proclamations, are the insights it offers into the dilemmas of architectural design, especially the difficulty of being both certain and free. Any rule carries with it the eventual prospect of reduced liberty, but new rules can be surprisingly unruly, clearing away customs and habits that have stood in the way for ages. Thus, for a time, perhaps quite a long time, new rules can offer a way round the obvious. In the first place, the Modulor would "change everything, opening the doors of the mind to the free flow of the imagination."[9] Then, perhaps, in the longer run, it would provide an instrument on which to play inexhaustible themes. Both possibilities were contemplated by Le Corbusier. The instrument that he had in mind was not such as a violin, which has no

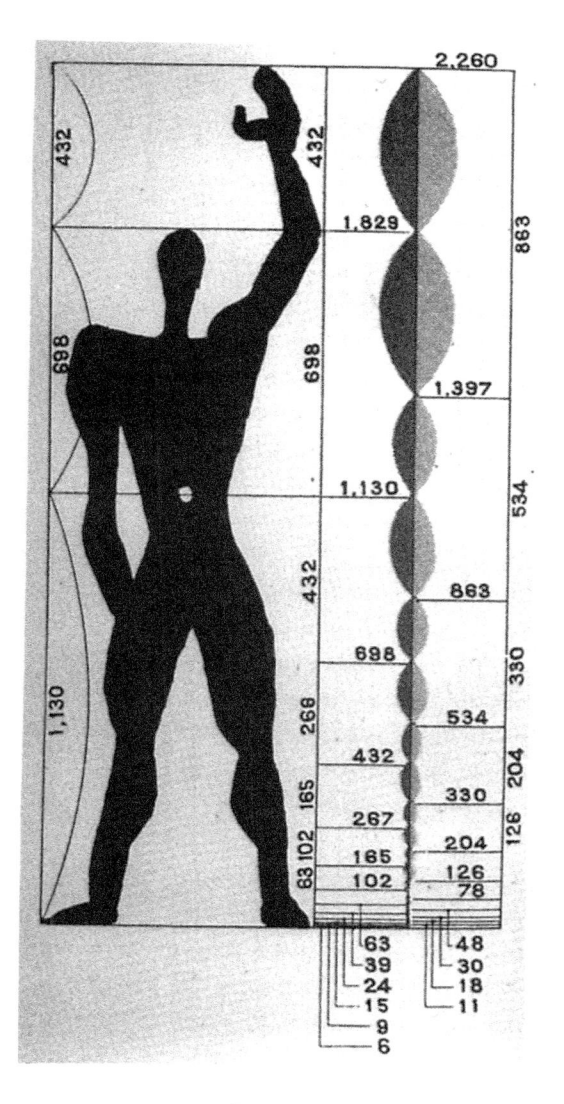

148 Le Corbusier, The Modulor, *red and blue series.*

inbuilt restriction of pitch, but a keyboard, which is constructed in accordance with the rigid intervals of the musical scale. For him this was the important thing: the tempered scale had expanded the horizons of music by defining the intervals between notes more exactly. So when Le Corbusier's assistants produced miserable results with the aid of the Modulor (which they did), he would say that he had given them a tuned instrument that they were incapable of playing properly. At first he liked to think that the Modulor might sweep away the imperial and metric scales of linear measurement, just as equal temperament had swept aside less adaptable musical scales.

These relatively sophisticated, though extravagant, arguments show that the issue of freedom—the artist's freedom—was made prominent by Le Corbusier in the Modulor as well as at Ronchamp, although we are still no nearer to understanding how the rectilinear system of proportion was applied to the curving chapel.

The world according to Le Corbusier was divided. As it got more so, intimations of joining the broken halves became more frequent. Le Corbusier had always held out prospects of resolution, and there was always a strong hint that he, personally, was to be the agent of reconciliation. Latterly his identity shifted from the administrator who makes the world work properly to the artificer who puts the image of the world back together. The recent critical attention devoted to Le Corbusier's powers of synthesis in his later work is quite justified, yet we might contemplate the possibility that Le Corbusier exaggerated divisions so that his portrayal of himself as potential unifier should hold good. This method of flattering one's own vanity is not unheard of. It produced Le Corbusier's version of Solomonic wisdom; a cut must be made before a wound can be sewn; the deeper the cut, the greater the healer's achievement. The situation then gets confused as it becomes difficult to distinguish what was anyway divided from what was forced apart to further the fantasy of redeeming synthesis, just so that it could be said: "The world is dying for want of integrators ready to save us before it is too late."[10]

Some of Le Corbusier's postwar interests also suggest that he dreamt up conflicts to fuel his desire to be healer of the wound; such were his allusions to alchemical symbolism, emphasizing the difficult, never accomplished union of opposites,[11] and such his proud recollection of his heretical Cathar forebears who had believed in a world divided unbridgeably between good and evil.[12] While these archaic dualisms play no direct part in contemporary thought, they do dramatize a tendency that still prevails in human affairs and is made almost inescapable through our use of language: the opposite of a man is a woman. Language enables us to deny this, but makes affirmation easier.

Just look at the way in which our understanding of Le Corbusier's career is facilitated by pitting opposites against one another: functionalism versus symbolism, the international versus the regional, the autocrat versus the populist, the technocrat versus the poet, rationality versus irrationality, progress versus origin, science versus myth, the machine versus the hand, calculation versus intuition, surface versus mass, the straight line versus the curved, geometric versus free form. It is a useful list, the more easily applied because most of these distinctions can be found in Le Corbusier's writings. Le Corbusier can be presented as two consecutive sides of an argument, passing from one proposition to another in a phase change that extended from 1928 to 1950. He was thus twice able to create architecture anew, as Jencks put it,[13] first in accordance with the left-hand terms, then with the right-hand terms of these antonyms. Many judicious commentators point out that this picture of the two Corbusiers is oversimplified, but even those who reject it as a general explanation of his work often resort to it in explanation of Ronchamp, which is rightly regarded as an extreme case. "Ronchamp expresses a liberation of Corbusier from Corbusier," proclaimed the Princeton professor Jean Labatut in 1955.[14]

In this same picture the Modulor registers as a last, faltering attempt to discover a rational basis for architecture, an attempt that was superseded before it was complete. Le Corbusier's decade of investigations into proportion formed an indefinite, turbulent boundary between the two phases. So, if the Modulor is present in the chapel, we can interpret it as a residue that provides a historical link to the prewar Le Corbusier concerned with calculation and precision, but having no power to define or modify the strange architectural forms of Ronchamp; a credible but uninteresting conclusion that identifies two things juxtaposed rather than synthesized.

Genesis

As soon as it was complete, Ronchamp was classified as irrational. James Stirling and Nikolaus Pevsner condemned it as such.[15] The only notable change after forty years is that contemporary critics tend to regard its irrationality as a positive attribute. Danièle Pauly, who has published the most extensive researches on Ronchamp, calls it a symptom of the revolt against reason: "a rejection of the overly rigorous forms and the orthogonal system that had prevailed for decades."[16]

Closely allied to its alleged irrationality is its subjectivity, typically understood as a transmission of the architect's personality into the building. His entire past is brought to bear on the act of pure imaginative fantasy that gives birth to its form.[17] In order to preserve the irrationality and subjectivity of the work, it is necessary to maintain a strict division between the event of conception and the means of realization. Le Corbusier always presented Ronchamp as a sequence of investigations that passed from sketches to models to projective drawings, in that order.[18] Each stage brought its own technical problems because Le Corbusier would insist on the maintenance of the original idea. The orthographic drawings had to replicate forms found in the earliest sketches.

Not much is said about these later technical problems. The story of Ronchamp's conception, on the other hand, has been made into a romantic epic. Pauly spends a long time unraveling the development of the sketches, which she treats as the locus of creation. Again it is Le Corbusier who called the tune, delighting in recollections of that charged moment when, after a period of subconscious gestation, with charcoal and crayons to hand, he was able to "give birth onto the page." The later history of the project was only a conserving of this miraculous object born complete. Ronchamp is the most celebrated and best documented example of design by sheer force of imagination. Creation is described as exclusively cerebral, although the created building turned out to be very corporeal. Even the assistant in charge, André Maisonnier, whose difficult task it was to transform the sketches into a building, wrote that Le Corbusier designed it "d'un seul jet."[19]

There is no denying that the earliest sketches are remarkably like the scheme as built. Yet, although Pauly is convinced that Le Corbusier's sketches were exact transcriptions of mental images, she demonstrates the extent to which successive alterations modified the shape, if not the organization, of the building between June 1950 and September 1953. This raises a question: are we to regard the shape, which altered, as less significant than the general layout, which did not?[20] The layout is unusual; the shape, incredible. It became more incredible as it gained definition; more outlandish as the walls curled, warped, and inclined inward, and as the roof began to twist further up to one corner and belly further down into one end of the interior.

Le Corbusier, retreating from the grandiose ambitions announced in the early part of *Modulor I*, was later to explain that the Modulor was not a way to create things but a way to refine them. A conception is born without any idea of measure attached to it. From this soft, fetal state it can

149 Le Corbusier, charcoal sketch plan of Ronchamp, June 1950 (FLC 7470).

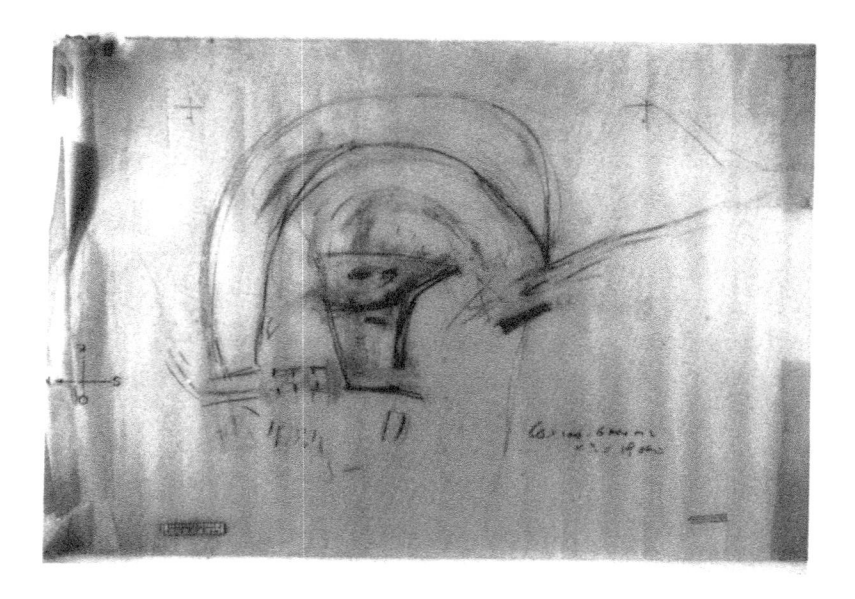

then be stretched onto a Modulor rack.[21] At Ronchamp the effort was spent on the conservation of the fetus. The later stage of rationalization seems never to have occurred, which leads to the same baffled conclusion, that it is difficult to see how the Modulor could be "everywhere."

But while it is difficult to see how the Modulor and Ronchamp are connected, it is easier to see how they are alike. For instance, both were brought into being the same way: hermaphroditically; each was the product of self-insemination. The Modulor was conceived on the day Le Corbusier announced to his young collaborator Hanning his "dream" of a new measure. Hanning was then given some cryptic instructions that Le Corbusier would always insist were fundamental to the Modulor system, against massive evidence that they were not:

> *Take a man-with-arm-upraised, 2.20 m. in height; put him inside two squares, 1.10 by 1.10 metres each, superimposed on each other; put a third square astride these first two squares. This third square should give you a solution. The place of the right angle should help you decide where to put this third square.*[22]

Neither recognized at the time that this was impossible. Hanning tried his best to comply but only managed to produce a diagram that was geometrically incorrect.[23] From then on, successive collaborators were pressed by Le Corbusier to return to this original which had less and less to do with the system he ended up with, until eight years later two more

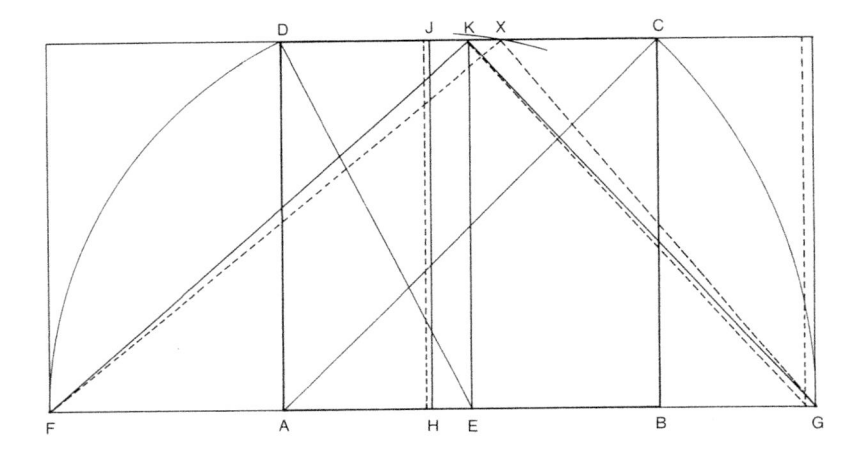

150 Hanning's diagram of 1943 and its errors. Hanning's construction is shown in solid line. A true double square is shown in broken line. The angle FKG is not a right angle. The "place of the right angle" would in fact be at X, not K. (Drawn by the author.)

151 Michelangelo's Capitol with traces régulateurs *superimposed, from Le Corbusier,* Vers une Architecture, *1923.*

of his assistants at the office in the rue de Sèvres, Justin Serralta and Maisonnier (who was then working on Ronchamp), came up with a deft piece of face-saving: a diagram that preserved appearances while removing the geometrical errors.[24]

In both cases memory was the means of insemination. Judged sublime at Ronchamp, the method proved less satisfactory with the Modulor. Take the "place of the right angle" for instance. Some picture postcards were spread on a table. Le Corbusier recalls himself looking at them:

> *His eye lingered on a picture of Michelangelo's Capitol in Rome. He turned over another card, face downward, and intuitively projected one of its angles (a right angle) on to the façade of the Capitol. Suddenly he was struck afresh by a familiar truth: the right angle governs the composition; the* lieux *(* lieu de l'angle droit: *place of the right angle) command the entire composition. This was to him a revelation, a certitude.*[25]

He was not talking about the right angles that we see all over the façade, but the ones we do not see. An accompanying illustration shows three made visible by the superimposition of right-angled triangles subtended from the base, side, and lower cornice, their apexes corresponding to salient points in the composition, the truth of which may have been familiar to Le Corbusier because he remembered seeing the same type of analyses elsewhere—perhaps those published by August Thiersch in 1883 and Heinrich Wölfflin in 1889.[26] Perhaps Le Corbusier vaguely remembered a diagram, laid out on the same lines, that showed how the golden section can proliferate similar ratios in a unique way. He may also have remembered that the simplest geometric construction of the golden section involved a double square, since back in 1936 he had used this very figure to generate a facade study for the Ministry of Health in Rio.[27] It is anyway likely that confused recollections delivered as prophecy sent Hanning off on a fool's errand in 1943. There is only one "place of the right angle" on the long edge of a double square, and that is in the middle where the two squares meet. Le Corbusier was convinced that it should be to one side, as in the Capitol. "My memories have no historical precision," he once admitted.[28] The progress of the Modulor was the reluctant journey away from this mnemonic blur.

Hands

It is not surprising that creation from fomented memories worked better with art than with mathematics. Mathematics, after all, lives on unambiguous exactitude, whereas there are types of art that die of it. Sometimes architecture seeks deliverance from modern science by relinquishing geometric precision, the species of exactitude most closely associated with science and technology. The architect turned artist takes possession of the building by restoring manual control; he lays hands on it; perhaps he lets others leave their mark too: the workmen, for instance. Le Corbusier's unconcern with exactitude in his later buildings is legendary. One of the first things that occurred to him when he saw the steep path up to the bombed chapel was that the new Ronchamp would have to be built with a minimum of mechanical plant because access was so difficult.[29] When it was finished he went out of his way to praise the gang of builders and their foreman, Bona.[30] The chapel is rough to the touch, reminiscent of primitive constructions, and full of the handiwork of artists and artisans. The whole thing gives the impression of having been crafted. Its most commonly applied adjective is "sculptural," and it is often explained as a natural development of Le Corbusier's own sculptural activity, carried out in collaboration with Joseph Savina, which began in 1944.[31]

According to the story of its creation, the laying on of hands endowed Ronchamp with form in the very beginning, then furnished it with textures, colors, and images toward the end. In between, where development is normally most intense, there was, instead, determination to resist modification of the original idea, which had been "born in the mind" but which achieved immediate physical definition with the charcoal wielded by the sketching Master. In other words, it is at once a pure creation of the mind and a pure creation of the hand. Le Corbusier always insisted that his architecture was at the scale of man; if so, then Ronchamp is a giant magnification of his own handiwork made by many other hands. He once referred to it as a vase.[32]

In 1911 Le Corbusier had written an ecstatic description of Slavic peasant pottery seen on his journey to the east—I should say felt before seen, for that is the order he chose: you feel the generous belly, caress the slender neck, then see the fantastic glazes. The eye is the organ of reason, the hand of instinct. The potter's judgment is in his hands: "Their fingers do the work, not their minds or hearts," and therefore: "First and foremost among these men who do not reason is the instinctive appreciation for the organic line."[33] Then, in a foreshadowing of Heidegger on art and peasants,[34] he praises the splendid ceramic curves

> *created as they are on the wheel of the village potter, whose simple mind probably doesn't wander farther than that of his neighbor the grocer's but whose fingers unconsciously obey the rules of an age-old tradition,*

152   Spanish pot, from Le Corbusier,

L'Art décoratif d'aujourd'hui, *1925.*

153 *Hans Erni*, Die neuen Ikarier, 1940.

154 *Ronchamp, east end of roof shell.*

COMIC LINES **283**

> *in contrast to those forms of a disturbing fantasy, or a stupefying imbecility, conceived by who knows whom in the unknown corners of large modern factories; those are nothing but the foolish whims of some low-ranking draftsman . . .* [35]

Even during his most fervent *machinolatrie,* Le Corbusier appreciated pottery. The subject arose 14 years later in *The Decorative Art of Today.* Modern man was now given the alternative of plagiarizing folk art or creating its contemporary equivalent. A deep sense of loss will propel us either into slavish imitation or toward a new resolution.[36] We live on the knife's edge, so the draftsman in the factory can turn into the hero of our times, if only his mind can overcome the sentimental attachment to what his hands used to do: "He no longer works with his hands. His spirit gives the orders. He has delegated to the machine the work of his clumsy and unskilful hands. Freed, his spirit works freely. On squared paper he draws out the daring curves of his dreams."[37] Now it should be obvious from what has been said that the curves of Ronchamp ought to be of the hand, not of the graph. The roof was inspired by a crab's shell (or was it an aerofoil?) and the chapel was curved like a woman's body (or was it a ship's hull?).

Sex

Le Corbusier's sea change was accompanied by increasingly frequent evocations of female form, first in his painting, then in his architecture. This has been established by von Moos, McLeod, Green, and others.[38] Female form is curvaceous, and so is Ronchamp. One writer, noting that it "made the break with the straight line possible and complete," quoted Le Corbusier's own preference to establish a female identity for the building: "'I like the skin of women,' Le Corbusier said and promptly constructed a building like a person."[39] Another describes the forms as "surfacing from a prerational existence charged with eros."[40] To give a chapel dedicated to the Virgin a voluptuous carcass was to risk mixing sacrilege with titillation, especially when a thread lead back in one direction to some well-publicized dirty postcards, and forward in the other to the most hallowed institutions, for Le Corbusier was to name the three towers of Ronchamp, two of them white inside and one red inside, after the Virgin Mary, his mother Marie, and his wife Yvonne.[41]

Stephen Gardiner has proposed the body as a link between the Modulor and Ronchamp. Using an ingenious cross conversion, he claims that the Modulor imposes measured proportions on the human (male) figure, while Ronchamp imposes the "free-flowing forms" of the human (female) figure onto an architectural frame.[42] The body becomes a meeting

155 Ronchamp, view into cowl of red tower.

point for opposites and a medium of exchange: Modulor man to voluptuous woman; regulated measure to free form. Thus the requirement of synthesis is satisfied, but all the ingenuity is in the explanation, not in what it explains.

Every time that the issue of sexual identity obtruded into Le Corbusier's discussion of the Modulor, the male principle won over the female. In one incident, Maisonnier and Serralta decided to change the sex and position of the Modulor figure, opening her legs to fit her into a Vitruvian circle. Le Corbusier attributed this aberration to Serralta's weakness for the ladies, but records a shudder at the idea of a woman 1.83 meters tall.[43] His treatment of Mlle. Elisa Maillard, a mathematician who had written on the golden section, was equally revealing. She was involved in the early stages of the Modulor, tried to help Le Corbusier and Hanning construct their impossible diagram in 1943, put Le Corbusier in contact with René Taton, a prominent mathematician at the Sorbonne, and made a last attempt to modify the ill-fated diagram in 1948. Jerzy Soltan remembers her as enlightened and objective, comparing favorably with Prince Matila Ghyka, promoter of mystification, whom Le Corbusier was inclined to favor as his mathematical advisor.[44]

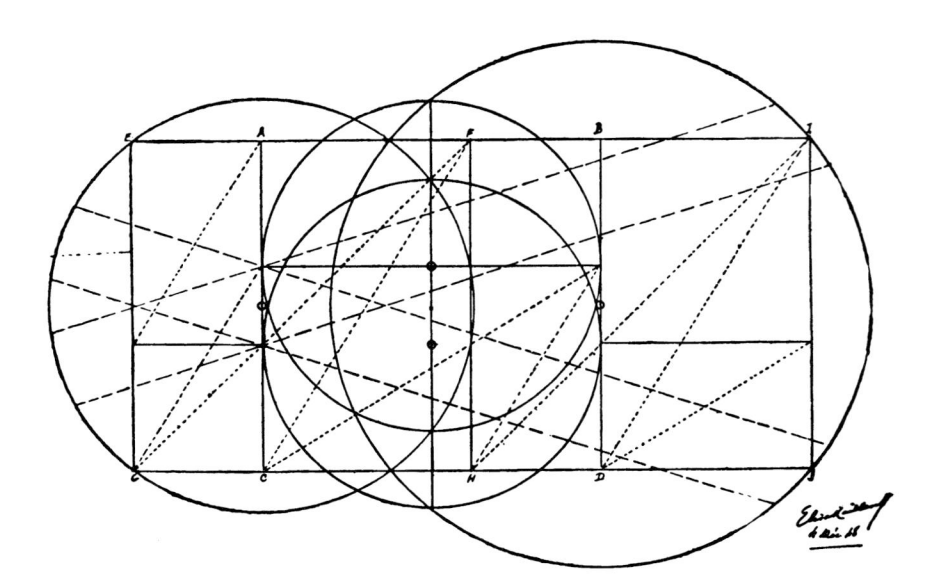

156 Le Corbusier's revision of Maillart's diagram, from The Modulor, 1950.

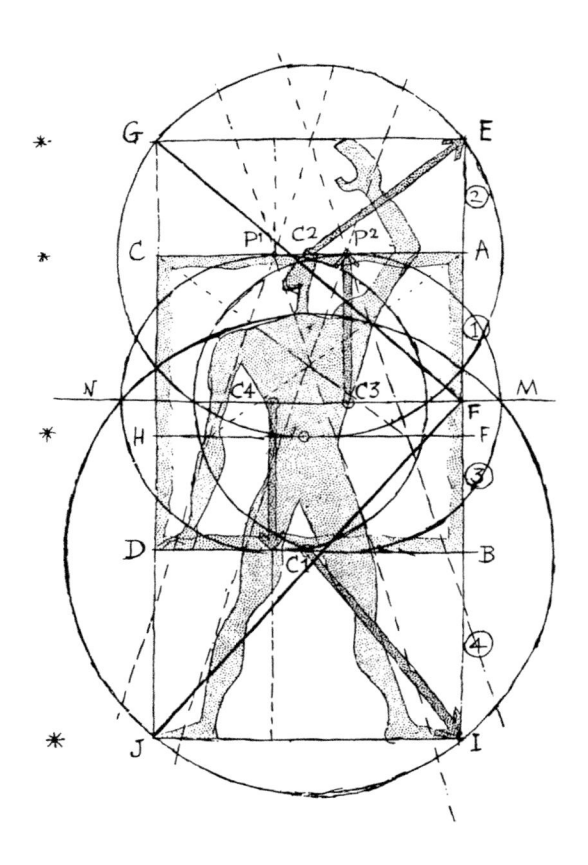

Eight days after Maillard delivered her new answer to the three-square problem, Le Corbusier decided to change it. With no acknowledgment of Maillard's generosity, he published her original with an explanation of its shortcomings so that he could contrast it with his revision. The Maillard diagram was sideways and had too many circles:

> *I set the* recumbent *Maillard drawing* upright *and coloured it. Into it I set the man-with-arm-upraised. I converted the reading of circles to a reading of rectangles and squares. And I wrote: "This sketch closes our investigation of the 'Modulor' by confirming the initial hypothesis." And: "HERE, the GODS play!"*[45]

Modulor I ends with a restoration of the male principle. The Maillard diagram was feminine. A few weeks prior to these derogatory remarks Le Corbusier had written about gender difference, taking his cue from a distinction made by the Catholic dramatist Paul Claudel in *L'Annonce faite à Marie* between the Master of the Compasses and the Master of the Rule.[46] The former is bourgeois, academic, effeminate; the latter, active, constructive, masculine. The circles, stars, polygons, and polyhedra of the compasses were condemned. The straight line and right angle of the ruler were praised. The two instruments gave rise to two kinds of geometry and two kinds of architecture: "In the one, strong objectivity of forms, under the intense light of a Mediterranean sun: *male* architecture. In the other, limitless subjectivity rising against a clouded sky: *female* architecture."[47] Le Corbusier called female architecture "dangerous" because it tended to dissipation rather than concertedness. This was written only two years before the birth of Ronchamp, dedicated to women, masterpiece of subjectivity, reputed destroyer of the right angle and straight line.

Perhaps Le Corbusier accepted the female *subject* for his art but not the effeminate *practice* of his art, in which respect he would merely have been perpetuating a bias ingrained in Western art. This could explain everything, but one or two more contextual events should be recorded. 1948 was also the year in which he published the drawing of Medusa and Apollo combined, done in 1941, which has become the emblem of his late tragic phase.[48] Two profiles make one face, almost. On the left side, Medusa's violent female grimace; on the right, Apollo's serene male smile. I do not think anyone has thought it worth mentioning that in the drawing the sexual identities were switched around; Apollo looks like a woman, Medusa like a man. In 1950 Le Corbusier conceived Ronchamp alone. Was it a virgin birth, or was it Athena bursting forth fully armed from the head of Zeus? From 1947 to 1955 Le Corbusier was working

157 Le Corbusier, **Apollo and Medusa** *cameo, first published 1948.*

158 Le Corbusier, plate from **Poème de l'angle droit,** *1955.*

on the *Poème de l'angle droit* in which Amazons play the part of nomadic, aggressive muses.[49] The assertion of the male principle occurs while Le Corbusier surrounds himself with personae of mixed or uncertain gender, and while he attributes to himself, as architect, the capacity to give birth.[50]

Look closer and the confusion increases. It increases in particular regions, not everywhere. Le Corbusier's agitated mind never comes to rest, but oscillates in a pattern. Compass curves, free lines, or ruled? Androgynes, amazons, or men? And where the intellectual confusion is greatest, the architectural synthesis will be most effective.

Behind the Wall

An attempt must now be made to answer Le Corbusier's challenge. The Modulor is, in fact, not so hard to find at Ronchamp, and still easier to find in dimensioned drawings of Ronchamp, since it is expressed in a handful of specific measurements. Their presence in the liturgical furniture (altar tables, pulpits, offertory boxes, founding stone, niches, ciborium, confessionals) is straightforward, leaving little rectangular islands dotted here and there in a larger swell.[51]

159 *Ronchamp, plan, from Le Corbusier, Ronchamp, 1954.*

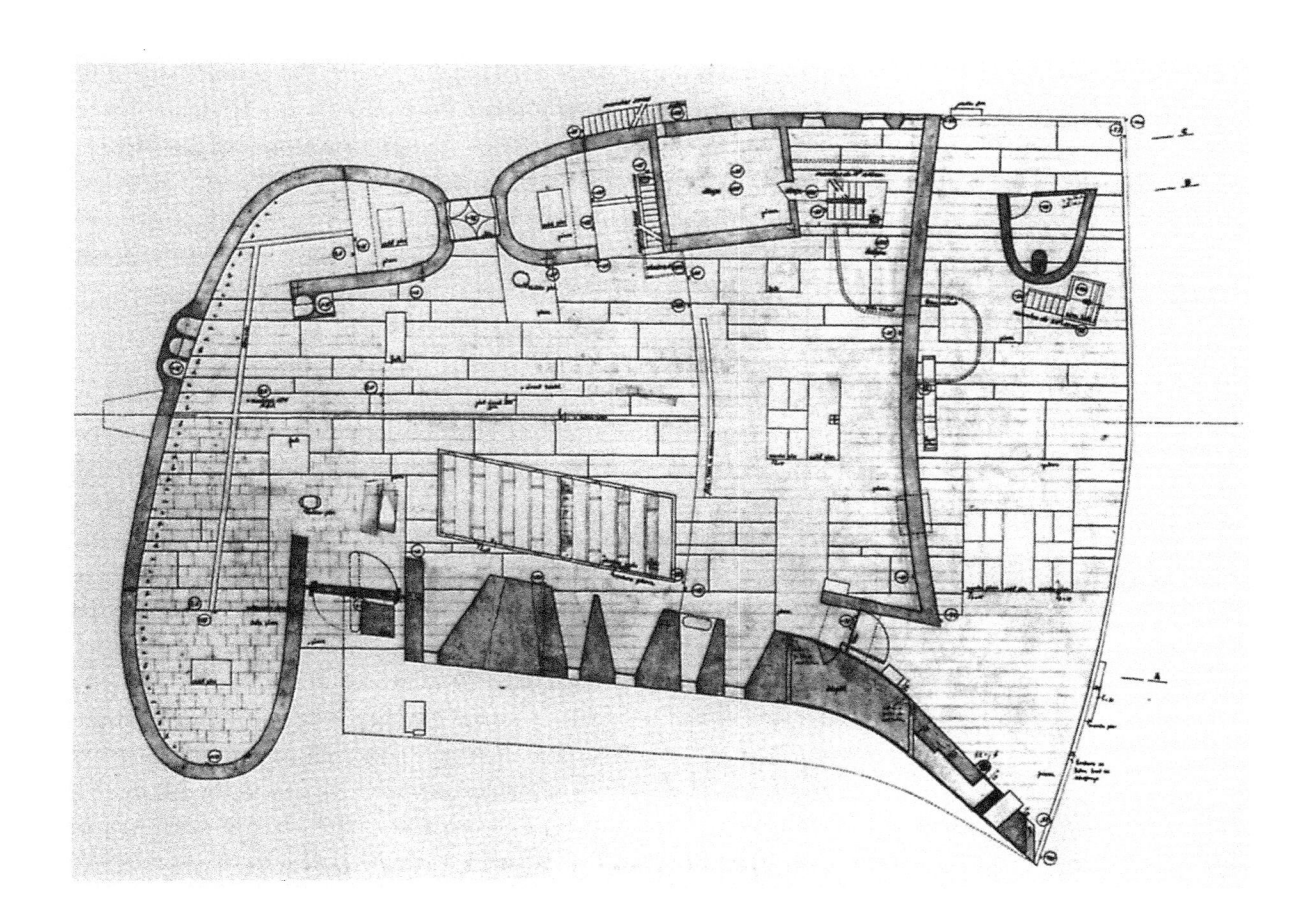

A complex weave of Modulor dimensions is indicated in the drawings for the pavement of the nave and podia.[52] The Modulor paving helps to locate eight isolated points on the outline of the encasing walls. Otherwise, paving and walling lead separate lives. In any case, the Modulor patchwork shown on the drawings cannot be transcribed exactly on the floor, since the floor undulates gently in conformity with the ground, distorting proportions ever so slightly.

The effect of distortion is more considerable in the south wall, not only because it leans and twists, so that proportions worked out on a flat sheet of paper cannot apply, but because the pyramidal window embrasures make an uncoordinated array of anamorphic perspective boxes, collapsing and stretching the perception of relative size. Underlying these shifting surfaces is a net of Modulor dimensions, not quite identical with the ladder of structure inside the hollow wall; not quite identical with the fenestration either. This quasi-structural armature was indicated in the earliest interior sketch, as well as in the final drawings.[53] The cavity between the two roof shells is a Modulor height. The lowest point of the concave roof shell, where it meets the west wall, is a Modulor distance from the floor. The colored glass windows are echoes of the Modulor "panel exercises" done by the assistants at the rue de Sèvres.[54] Some look like gaudy parodies of Mondrian paintings. Often their dimensions are only approximately Modulor. Sometimes a set of related measurements will slip between the Modulor sequence, but maintain a rough equivalence to the golden ratio amongst themselves.

The Modulor also played a part in the design of the enameled steel panels mounted on the south door. Le Corbusier's buildings of the fifties usually incorporate strategically positioned advertisements for the Modulor system, often in the form of cast bas-reliefs of Modulor man. None was so prominently displayed as on the door of Ronchamp, although the message was correspondingly recondite. The brightly colored motifs on either side are a compendium of Corbusian signs (hands, clouds, the serpentine vein, etc.), including the double square, an incomplete pentagram, and a five-pointed star, all of which allude to the geometric construction of the golden section. The diagonals of a regular pentagon intersect in the golden ratio, making a five-pointed star or pentagram. Inside the pentagram is a smaller pentagon, and so forth. All the line segments in this infinitely extensible web can be arranged as consecutive terms of the golden ratio. A geometric series is defined thus: as a is to b, so b is to c. Three terms are necessary. The golden ratio, a special case, requires only two terms: as a is to b, so b is to $a + b$. Only one multiple satisfies the expression. The interlocking similar triangles of the pentagram give a clear visual demonstration of the unique marriage of addition and multiplication. This is the kind of thing that some mathematicians have been prone to call beautiful.[55]

In his outburst at the end of *Modulor I,* Le Corbusier dismissed stars and polygons as planimetric gewgaws for Beaux-Arts academicians. Was he diverting his annoyance from a more obvious target? When "the Divine Proportion" was first made the subject of a treatise by Luca Pacioli at the end of the fifteenth century, when Kepler called it a jewel in the seventeenth, when expounded as mathematical aesthetics in the nineteenth and twentieth centuries, the main topics were polygons and polyhedra, regular and stellated. Le Corbusier's disdain for the polygonal, stellated architecture of compasses therefore also suggests a distinction between the forms proper to architectural beauty and the forms in which mathematical beauty is beheld. Although he never gave up the quest for a geometric construction of the Modulor,[56] Le Corbusier became more and more irritated by the startling ugliness of the various "correct solutions" offered by well-wishers such as Bernhard Hoesli, Hansjörg Meyer, R.-F. Duffau,[57] and of course Maillard. He wanted the beauty of mathematics without the sensibility of mathematicians.

One explanation for the display of the demoted pentagon and star at the entrance to the chapel is that Ronchamp is, indeed, female architecture.[58] Richard Moore argues thus, but it was both affirmed and denied by Le Corbusier. It depends on which way you look at it—literally. Six years after the Maillard revision, the outer leaf of the Ronchamp door was composed on a recumbent pair of circumscribed pentagons, and the inner leaf was composed on an erect version of the same. So female and male signs were put back to back. This is made absolutely clear in diagrams published by Le Corbusier in 1955, except that, again, he superimposed a male interpretation on a female organization.[59] He said that the recumbent schema of the outer leaf had been inspired by a painting. In a rerun of earlier revelations he suddenly noticed what was normally invisible: "I was telephoning—the photograph of the *Rétable de Boulbon* was in front of me upside down. The pentagon hit me in the eyes."[60] The Boulbon altarpiece is a resurrection. In Le Corbusier's analysis of its underlying geometry the rising figure of Christ became a spearhead thrusting up from the center of a pentagon, which, though not regular, and therefore not productive of the golden ratio, was thought by him sufficiently similar to substitute. Freud and Jung sort out their differences on the door; a stiff, phallic Christ takes the part of the animus against the horizontal anima. This hardly represents a synthesis.

To summarize, the Modulor can be identified in three guises at Ronchamp. It can be measured directly in the most obvious locations such as the paving and altar tables; it is symbolized in the iconography of the door; it is buried in the structure. It is conspicuously absent from the superficial carcass of the building, where it only emerges very occasionally as a stranded vertical dimension. It is either behind the wall ("where the Gods play!" according to Le Corbusier) or it is stuck on afterward. It is either displayed as a sign of its own deeper presence, or it is altogether invisible.

Le Corbusier's experiments with proportion may be thought of as polarizations of the classical norm. From the start he had recognized the value of hidden order in classical architecture (as in the Capitol), yet in numerous of his facades from the 1910s and 1920s, the multiplication of similar rectangles on a paper-thin plane pulls the hidden order up to the very threshold of perception, so that with a little concentration the correspondences and iterations may be seen with the naked eye without the aid of the *traces régulateurs*. Try it out on the Villa at Garches.[61] It was in 1928 that he wrote: "For the mechanism of ratios to be effective, the quantities they generate must be perceptible, legible."[62] What was subliminal in the modeled surfaces of classical architecture was made liminal in the purist wall-plane, but then withdrawn altogether at Ronchamp despite all the advertisements for it.

Theories of proportion as traditionally formulated, and as found in the Modulor, are quite inadequate to the task of describing complex shape. It transpires that the Modulor operates at Ronchamp as a fictional wish-fulfilment. It lurks behind the wall as if it were responsible for it; poses in front of it as if an accidental exposure. It acts as if it made all the difference in the world while making hardly any difference at all.

No real difference, except insofar as it was in retreat, because at Ronchamp withdrawal of the Modulor into the foreground and background accompanied a truly radical reversal of architecture's organization. Architectural structure is generally orthogonal, architectural ornament not so. The free curves of ornament (fronds, swags, volutes, drapes, and so forth) customarily cling to the architectural frame like ivy. This has never been thought of as a reversible relation. It is one thing to do without decoration, it is another to switch the geometric identity of decoration and structure around, which is what happens at Ronchamp, where a kind of decorative relief is provided by occasional Modulor blocks and boxes in an encompassing field of warped structural surfaces. Demonstrable episodes of traditional proportion cling to the last vestiges of orthogonal form. Angel Guido once remarked that the *traces régulateurs* on Le Corbusier's purist buildings were "already decoration."[63] At Ronchamp this was true of the Modulor, but the decoration was now helping to cancel one of architecture's fundamental constants. In the first place (classical) curvilinear ornament provided a *frisson*, in the second (purist) decoration was reduced to rectilinear proportions, in the third (Ronchamp) decoration stays rectilinear and the structure turns plastic. Ornament now provides reassurance and the structure provides the *frisson*.

Compare Ronchamp to the church built by Alberto Sartoris at Lourtier, Switzerland, in 1932. The close resemblance between the southeast view of Ronchamp and the photograph of Lourtier from the southwest published in Sartoris's *Architecture nouvelle* in 1948, which Le Corbusier certainly saw, was noticed by Robert Slutzky.[64] Ronchamp is a version of Lourtier with the functions scrambled. Lourtier is a rectified preliminary for Ronchamp. Le Corbusier may well have taken everything from Lourtier except the rectangular framework for architectural proportion; he took everything except that which it was almost impossible to refuse.

160   Church at Lourtier, Switzerland, Alberto Sartoris, 1932, view published in Sartoris, Architecture nouvelle, *1948.*

Reversibility is a feature. Ronchamp turns inside-out. Mass can be performed on either side of the east wall, and the venerated statue of the Virgin, ensconced within the wall, can be cranked by hand to face either direction.[65] Contradicting his earlier insistence that a building should be designed from the inside outward,[66] Ronchamp was designed from the outside inward. The earliest sketches were all of the exterior walls, construed as receptors of the "four horizons" disclosed at the top of Bourlèment hill. It is as if these four landscapes (so distinct to Le Corbusier, and contemplated in colored sketches made on site when he sat down in the office to make his first designs)[67] rushed toward the site to be caught

161 *Ronchamp, interior.*

by the building. Le Corbusier called it "visual acoustics." Distant pressures impinge on the walls and roof, which push in, threatening to evacuate the compressed space inside.[68] Le Corbusier had avoided use of the term space in its heyday. Now, after he had finally admitted it into his vocabulary, he turned the positive bubble into a deflation, as if all the active forces came from without. From within you cannot see much of the outside world, but you suffer its impact and receive rumors of its immense power. Occupation of this resultant cavity induces an anxious premonition of helplessness, which is not surprising, given the almost unconscious translation of architectural interiority into a metaphor of mental interiority. A person alone in a room is like the soul in the body. "Inside: face to face with yourself. Outside: 10,000 pilgrims,"[69] mused Le Corbusier. But Ronchamp, a building born of subjectivity, may also be interpreted as an assault on the ubiquitous representation of the subjective self in Western architecture;[70] a self in control of an interior domain that opens out confidently to the exterior; a consciousness in charge of the doors and windows of the soul, to recall an exhausted literary metaphor not yet exhausted in architecture. It has the hallmarks of a church but it is not like being inside one; more like an aquarium or a street at night, environments dependent on electric light (conspicuous by its absence at Ronchamp), which contracts what can be seen into puddles of peripheral illumination that do not extend to the observer, who becomes an outsider looking in. Blind forces without, an anteroom within, the focus of the interior spreads into the thickness of construction, into the cul-de-sacs of the chapels and into the luminous embrasures. Light has difficulty in penetrating further. Within this thickness, all evidence is trapped and the duality of interior and exterior reduces to an interface.

Xenakis

The last 120 pages of *Modulor II* describe how the system was used in Le Corbusier's own architecture. Unexpectedly, and uncharacteristically, the last 5 pages of the book are devoted to the work of one of his assistants, Iannis Xenakis.

Xenakis is now internationally recognized as a composer. Between 1948 and 1960 he worked for Le Corbusier. He had not trained as a musician, nor as an architect, but as an engineer, obtaining his diploma from the Athens Polytechnic in 1947. His job with Le Corbusier was a bread-and-butter affair at first, doing structural calculations for the *unités* as a member of the technical section of the ATBAT consortium, staying on without much enthusiasm when the consortium was disbanded.[71] His commitment was to music, although the prospect of earning a living from it was remote. But as his music developed, so did his interest in architecture. His first major musical composition was completed in 1953, and in the following year he was entrusted with the monastery of La Tourette by Le Corbusier. There is no doubt that Xenakis was responsible for the brilliant deployment of the Modulor in *pans de verre ondulatoires* (undulatory glazed panels) that were first developed for the south elevation of La Tourette,[72] then used in several later projects. Instead of composing Modulor solids, or Modulor surfaces, Xenakis composed a Modulor line; a ribbon of identical mullions, closely spaced at varying Modulor intervals which gave the impression of continuous dilation and contraction "like accordions."[73] Xenakis, whose first study for the *pans de verre ondulatoires* was a band of graded shading,[74] had a musician's appreciation of the way in which a row of discontinuous elements can give rise to a continuous effect, as notes on a staff give rise to a melody. By collapsing the Modulor into a line, and by collapsing the intervals along the line into a frequency, he brought something new to architecture.

Le Corbusier mentions this, but the place of honor was reserved for Xenakis because of his music. *The Modulor*, which began with music, ends with it. The tables had been turned. Xenakis had used the Modulor in his orchestral composition *Metastasis*. Music, source of inspiration for architecture, was borrowing back inspiration from architecture. It was a good closing shot.

Xenakis was not interested in the scale of man. The reason he had borrowed the Modulor was because, as explained earlier, it exploited a rare equivalence between a geometric series and an additive series. He was searching for some measure of unity in his musical notation, and was bothered by the irrelation between the geometric progression that defined the pitches of notes in the 12-tone scale, and the additive properties that conventionally pertain between note lengths and bar lengths. So he devised his own musical Fibonacci series to articulate the percussion and the string episodes in *Metastasis*.[75]

Jullian de la Fuente remembers Xenakis as the one person in the office who could intimidate Le Corbusier.[76] Both Le Corbusier and Xenakis can be described as strong personalities, although what began as mutual respect in recognition of it degenerated into mutual hostility. Finally, Xenakis was sacked, together with Maisonnier and Tobito, at the culmination of a bitter wrangle over recognition. Prior to his dismissal, Xenakis had forced Le Corbusier's hand, obtaining public credit for his part in designing the Philips Pavilion of 1958. No one else ever wrested an admission of this sort out of Le Corbusier.

162   Iannis Xenakis's pans de verre on-dulatoires, *monastery of La Tourette, 1955–1959.*

But then the Philips Pavilion for the Brussels world's fair, a glorification of technology, appears out of place in Le Corbusier's postwar projects; a shiny, stiff, silver tent of spiky conoids and hyperbolic paraboloid shells. It served as the setting for a sound-and-light experience that lasted ten minutes, including two minutes of electronic music composed by Xenakis, added because the work commissioned from Edgard Varèse, the *Poème électronique*, was too short.[77] The intestinal plan with inlet and outlet to cater for the crowds was Le Corbusier's idea. Xenakis worked for several weeks in October 1956 to bring the plan into three dimensions. It was a tremendously difficult problem for which he devised a typically extreme solution: ruled surfaces. A ruled surface is a curved surface generated from straight lines. The ruled surfaces devised for the Pavilion were an unusually complicated combination for which even Xenakis was unable to calculate the static stresses. Soon two other engineers, Vreedburgh, a professor of mechanics at the Technical University of Delft, and Duyster, from Brussels, became involved. Within a fortnight they had done the calculations and tested a scale model, demonstrating that the Pavilion could be built with a 5-centimeter-thick prestressed concrete shell.[78]

The prima facie evidence suggests that the ruled surface was Xenakis's personal contribution to the project. Obviously, with an engineering background he would have known about such things.[79] There had been a hint of what was to come in the *pans de verre ondulatoires*, which suggested curvature with closely spaced straight lines. Moreover, he had used ruled surfaces three years earlier in his music. Continuous *glissandi* are used extensively in *Metastasis*. To coordinate the 52 different string parts written for the orchestra (far more than any composer would normally contemplate), Xenakis resorted to an unusual method of composition: he drew a graph. The horizontal axis represented time, the vertical axis represented pitch. The rising or falling sounds of each instrument were mapped onto the graph as straight lines. All together on the same sheet they gave a picture of the total orchestral sound, much more directly apprehensible than in conventional musical notation. Bars 309–314 use twisting arrays of straight-line *glissandi* to produce surging curved profiles. These were referred to by Xenakis as "ruled surfaces of sound." Nouritza Matossian, who has described the close relation between the music and the pavilion, thought that since Xenakis was working at the rue de Sèvres he probably got these graphical notions for the composition of *Metastasis* from Le Corbusier in some roundabout way, to which Xenakis's response was to write on her draft: "No!! . . . I just had an illumination! Isn't that possible?"[80] By contrast he was always willing to acknowledge the influence of the Modulor.

163 Philips Pavilion, Brussels International Exhibition, Le Corbusier and Iannis Xenakis, 1958.

164 Iannis Xenakis, ruled surfaces of sound in Metastasis, *1954 (from N. Matossian,* Xenakis*).*

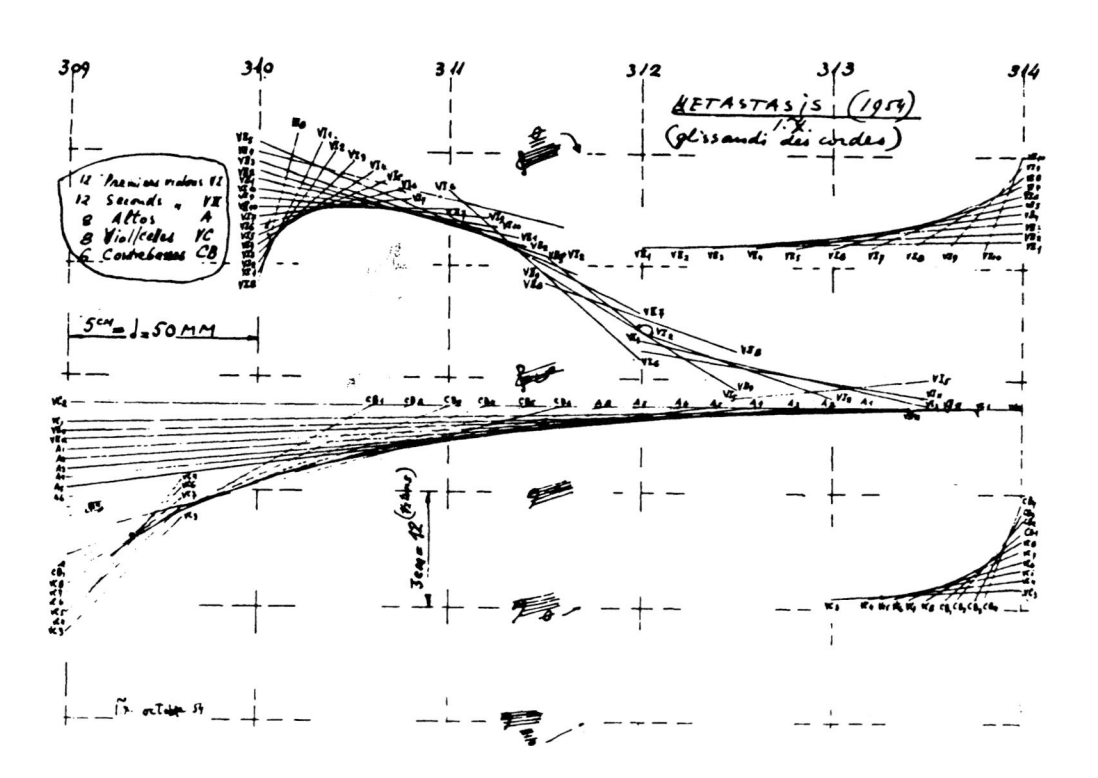

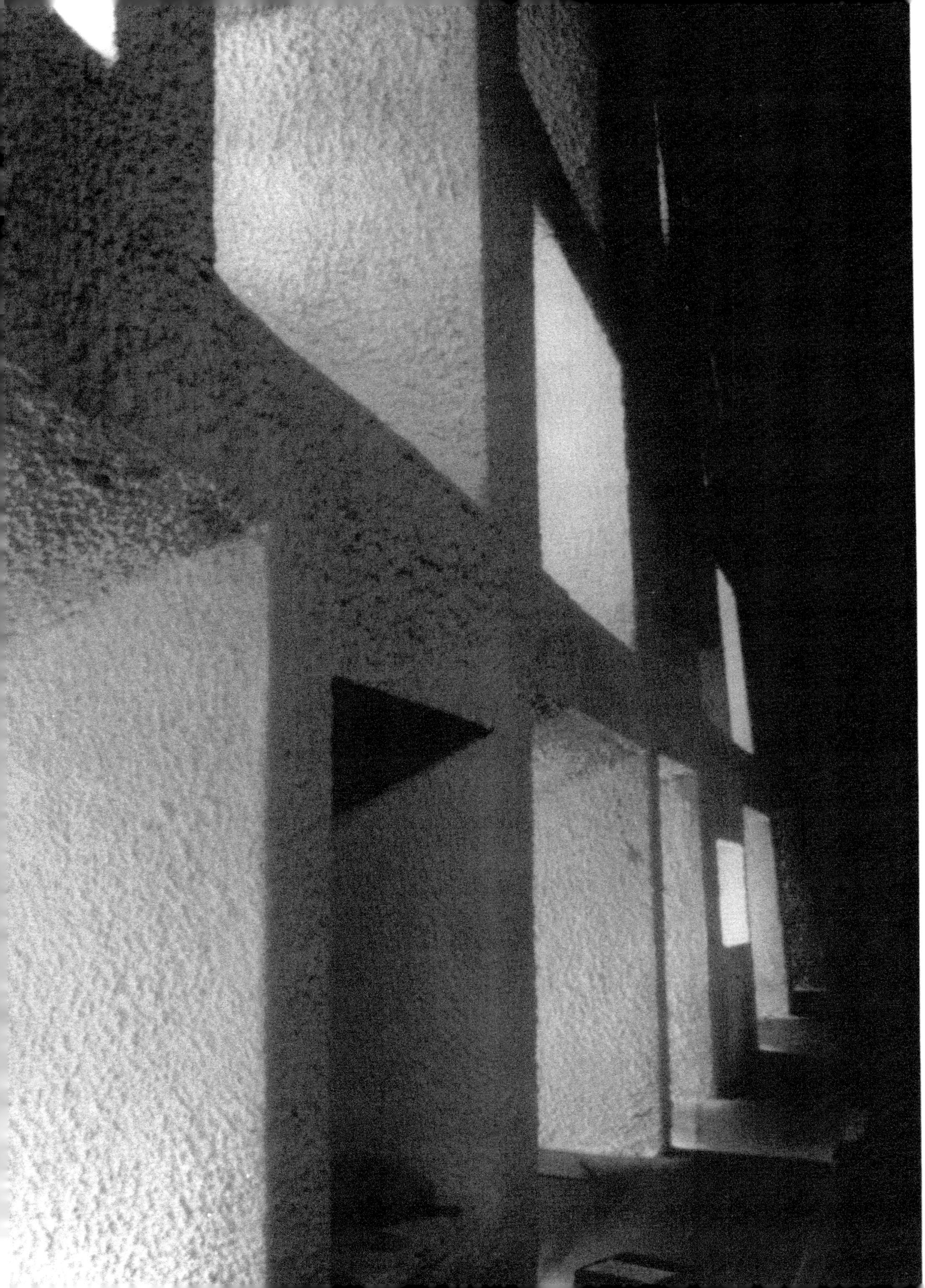

165 Ronchamp, twisted interior surface of south wall.

On several occasions Xenakis has made it clear that his inspiration for the ruled surface was not Le Corbusier, whom he accused of ignorance in technical matters, but Bernard Lafaille,[81] an engineer who had been involved with the ATBAT collaborative at the same time as Xenakis, and who had made pioneering researches into the structural properties of conoid and hyperbolic paraboloid shells in the 1930s.[82] Xenakis said that when Le Corbusier saw his scheme for the Philips Pavilion, "he was so impressed that he immediately tried to show me that my ideas were in the same direction as his sculptural ideas."[83] This is particularly significant because the nonplanar body of Ronchamp had been defined almost entirely by ruled surfaces.

Lines and Bodies

How did the Ronchamp sketches turn into the completed building? There is a working drawing by Olek for the cowl of the south tower in the Fondation Le Corbusier Archive[84] dated 9 December 1953, produced, therefore, after the design was "finalized" and three months after work had begun on site. The top of the south tower, hollow, with a lopsided smooth cusp, preserves to a remarkable degree "the demiurgic experience of the primeval potter," to quote Mircea Eliade out of context.[85] If it is not like a pot, it is like a thumb (an observation made by James Stirling).[86] If it is like an industrialized farm silo, or a nautical vent duct, it would be a silo or duct modeled in clay; an effigy made by hand in honor of the machine-made. The drawing conveys this shape from the plaster model of 1950 by an expedient. It is a good, not an exact transla-

166 Ronchamp, the south tower, drawing by Olek, December 1953 (FLC 7189).

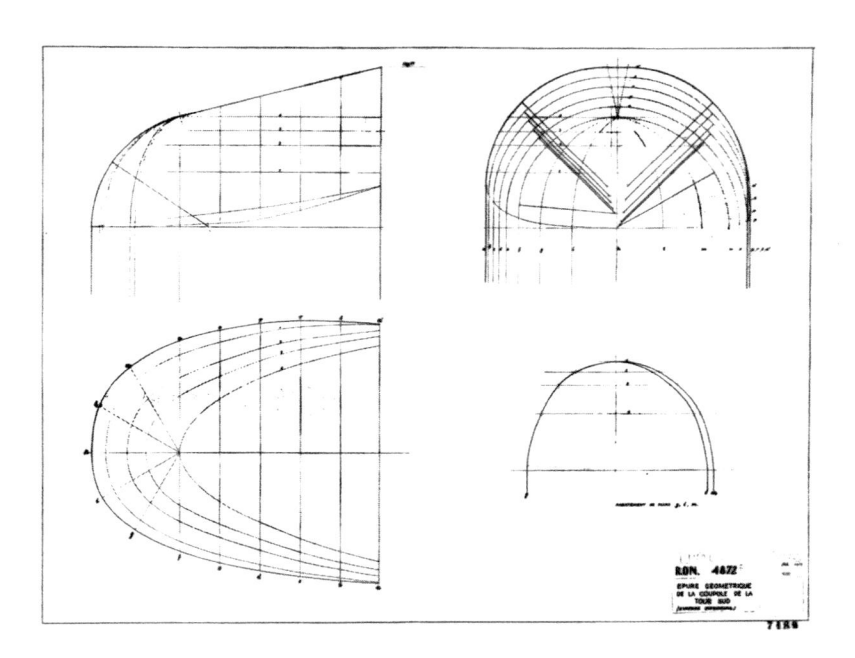

tion. The draftsman creates the effect of a modeled surface as a consequence of the orthographic juggling of circular arcs and straight lines. The most basic elements of geometry are now cast in the lineaments of the form.[87] Through the agency of technical drawing, these elements deftly imposed themselves on the hand-modeled surface derived from the charcoal sketch, altering the form.

André Maisonnier has recorded his version of how Ronchamp was worked up from the sketches. It does not conflict with Le Corbusier's but the emphasis is quite different. Maisonnier, giving Le Corbusier full credit for the sudden and definitive act of creation, writes in a matter-of-fact way of its postnatal development. Like Pauly, he describes the subsequent sketching as a continuous influence until the completion of the plaster model, shown to the archbishop of Besançon in November 1950. This had been carefully cast in molds, but Maisonnier had modeled earlier and smaller versions by hand ("comme une sculpture, sans passer par du modelage").[88] So far, the transition between idea and object was accomplished by a handover from Le Corbusier to his collaborator, much as drawings for sculptures were handed over to Joseph Savina. Le Corbusier would send a sketch to Savina, who then used his imagination to realize it in three dimensions. Maisonnier did the same, though with far less latitude because the Ronchamp sketches were far more comprehensive.

Modification continued even after the plaster model was complete; according to Maisonnier, Le Corbusier was still revising the essential lines of the general silhouette.[89] Another model was then made, not from plaster but of steel wire covered with paper. It was here in the 1:100 wire model that the "géométrie rigoureuse," which would from now on preoccupy Maisonnier, first became evident. The wire model demonstrated how "all the major volumes are defined from curved directrices and rectilinear generators."[90] This is an unaccentuated but clear statement of an important property.

The wire model was crucial to the development of Ronchamp because it provided a rudimentary but effective liaison among sculpture, drawing, and building construction. Undulating surfaces that had gained a second definition in plaster as copies of sketches now gained a third definition stretched as webbing across a wire armature, which gave a tremendous insight into the relation between their possible construction and their drawing. Maisonnier confessed that the next step, orthographic projection, was hard going nevertheless, especially when it came to the southeast corner of the roof, despite having a larger, more detailed model to help him by this time:

167   Ronchamp, wire and paper model, from Le Corbusier, Ronchamp, *1957.*

> *We had studied descriptive geometry a great deal, but at the extremities of the [roof] shell the geometry did not enable us to find the junctions and express the double curvature. I had worked very hard at this enterprise. A well-qualified engineer had translated our plans into reinforcing, into structure, very accurately. At the office there were always engineers who would calculate what was necessary. At Chandigarh as at Ronchamp, we were giving dimensionings that corresponded almost exactly with those established by more accurate methods.*[91]

It sounds as if his hastily tutored efforts paid off, although on another occasion he said that the self-same roof detail at Ronchamp had taken as much time and thought as the whole of Chandigarh. This detail, like the cap of the south tower, was one of the few parts of the building not defined by curved directrices and straight-line generators. That is why he found it so troublesome. As to the rest:

> *The most characteristic geometric elements are the roof shell, composed of two reversed conoids parallel to one another (these are the two concrete skins of the roof), and the south wall composed of two opposed ruled surfaces that start off obliquely from the main door, straightening up as they pass along the plan of the wall and arriving as two verticals at the southeast corner of the building.*[92]

This constitution, already evident in the wire model, lends itself to orthographic projection, and the Archive drawings corroborate what Maisonnier says. The roof is of particular interest. The directrix at the west end is a tilted, shallow V shape with a rounded valley. At the east end it is an inclined straight line that meets the south wall at its highest point.

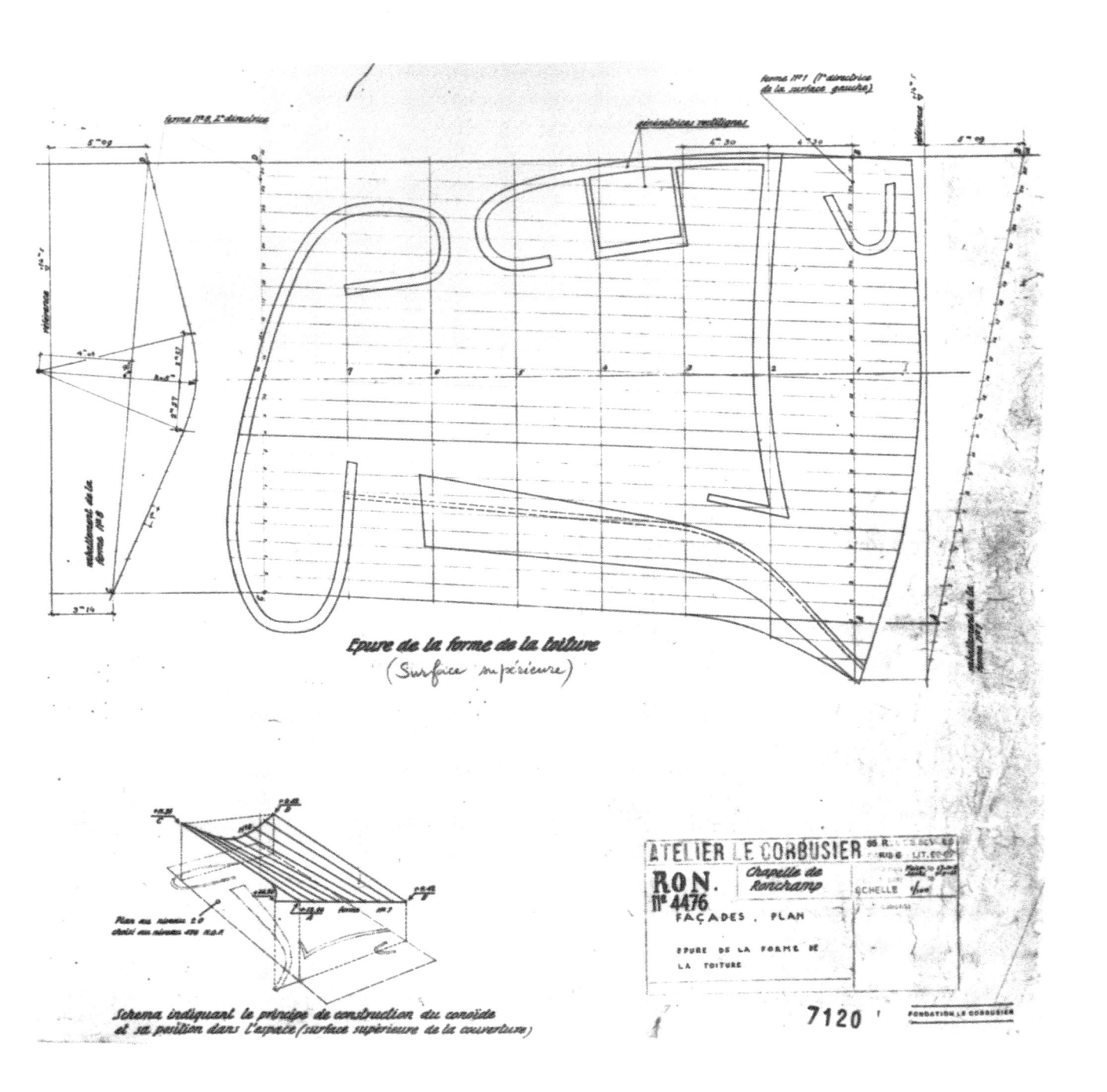

168 Ronchamp, roof shell, drawing by André Maisonnier, September 1952 (FLC 7120).

The directrices, parallel to one another in plan, are joined by equally spaced lines, also parallel in plan but not parallel in space. These lines, the generators, define the surface. It is remarkably easy to find the projections in elevation of any points on surfaces so defined, for example, the line of intersection with the twisting inclination of the south wall—until it curves up at the end into the corner that gave so much trouble.

Le Corbusier likened the roof to a crab shell; Maisonnier likened it to an aircraft wing. If we put both the organic and the mechanical comparisons aside we find another source of similarity. The conoid devised for the roof is a variation of a figure devised for the purpose of architectural projection by the seventeenth-century architect and mathematician Guarino Guarini, which is worth savoring when we read that "the outstanding feature [of Ronchamp], characteristic of the modelled structure as opposed to the board-designed, is the roof."[93]

The polymath Guarini recorded this type of conoid on the two occasions on which he wrote about projection in *Architettura civile* and *Euclides adauctus*.[94] It was Guarini who first published stereotomy in Italy and who first made use of its projective forms in his buildings, despite the fact that he did not usually build in stone but in brick. He had visited Paris and learnt from the French treatises, especially Derand's *L'Architecture des voûtes*. Practically all the examples used by Guarini can be traced to French stereotomy.[95] His conoid was generalized from a surface found in *arrière-voussoirs* (window openings arched on one side and rectangular on the other). A right cone terminates in a line instead of a point. The line is equal to the diameter of the cone's base. In orthographic projection the end elevation of this figure is a triangle, its side elevation a rectangle, its plan a circle. Guarini defines the surface by stipulating that it be made of perpendicular parallel lines in side elevation, which are parallel lines in plan as well. The surface is complex and mobile; its projections are simple and easy to handle. The roof of Ronchamp, although modified in several particulars (it is, in fact, a conoid flanked by two hyperbolic paraboloid surfaces), preserves the projective advantages discovered in Guarini's conoid.[96] There is no direct influence here of course. What I want to stress is that Guarini's conoid and its derivatives are ways of getting three-dimensional complexity from two-dimensional simplicity.

Guarini's conoid is a ruled surface, as were certain prewar aircraft wings, as also is the main part of Ronchamp's roof. Le Corbusier compiled an impressive folio of photographs, published in 1935 as *Aircraft*. The roof models made by Maisonnier were like airfoils. Vladimir Bodiansky, chief engineer at the rue de Sèvres and Le Corbusier's right-hand man in the

169 Four projections of Guarini's conoid (redrawn by the author).

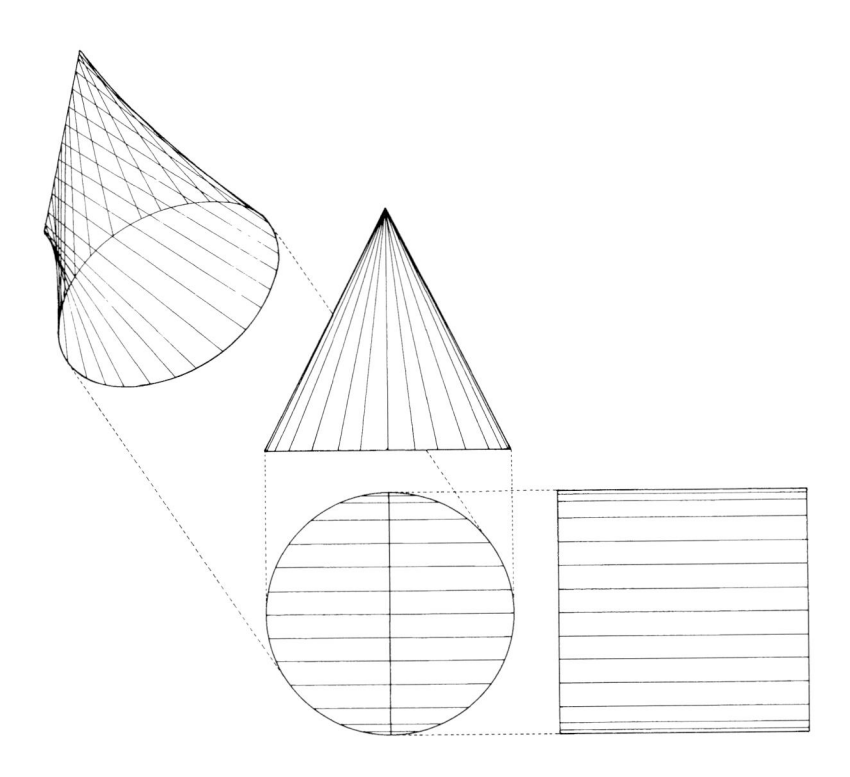

early postwar years, had been an aircraft designer before he turned to architecture.⁹⁷ Aircraft may well provide the missing link between Guarini and Ronchamp, Bodiansky the missing link between the ruled surfaces of Le Corbusier and Xenakis, for neither Ronchamp nor the Philips Pavilion were isolated instances, as we shall see.

Ronchamp's roof does look a bit like an aircraft wing. It also looks a bit like a lot of other things. Apparently it was everything at once that Le Corbusier had thought, a jostling multitude of retrospective images made singular. It resembles many things that were in Le Corbusier's past, and many others that just turn up, like the open umbrella being used to ward off the sun that was caught by an observant photographer in mirror symmetry of the roof shell's east edge.⁹⁸ Ronchamp is not adequately defined by being tethered ever more securely to Le Corbusier's personal stock of memory images, although a great deal of historical research has been directed to that end. These researches may help us to understand something of the creative process, but the extent of Ronchamp's resemblance to other things is exceptionally broad. Nor is it only an issue of extent. More is at stake. The nature of the building alters drastically when what it resembles cannot be traced back to Le Corbusier's mind; it becomes less predictable and its meaning is less stable.

It does not take much to bring out the historians' valuable prejudice that remote events have more effect on local events than we would credit. But historians sometimes understand remoteness only in chronological terms. There is a corresponding tendency in the growing corpus of Le Corbusier studies to emphasize ancient symbols and references, especially in the late work, and to identify arcane content with profundity and humanity—encouraged by Le Corbusier's penchant for recollection. Thus H. Allen Brooks rejects the too frequently voiced opinion that the shell of the Chandigarh Assembly Hall looks like a cooling tower, because it emphasizes the mechanical aspect, whereas he prefers what he calls the poetic likeness to vernacular farm chimney towers in the Jura, where Le Corbusier grew up.[99] Whilst the observation is poignant, promotion of the traditional farm over the modern power plant is only credible if one believes that poetic power increases as likeness diminishes, for the Jura farm towers were wooden pyramids, whereas cooling towers are concrete hyperboloids of revolution, the same shape and material as the Chandigarh Assembly Hall. The hyperboloid of revolution is a ruled surface.

It depends in any case on what one defines as remote. The machine aesthetic brought something familiar to modern experience, but remote from architecture, to bear on modern architecture. By 1950, mechanical analogies were understood to be the basis of Le Corbusier's prewar work. The conceptually remote had by that time been made familiar. Le Corbusier's later work then moved attention to other remote areas. In consequence the mechanical aspect of the late work, far from being played up as Brooks maintains, has been played down. One unexceptional claim would be that even Ronchamp has its mechanical aspect. Its profile is a bit like a dam; its towers a bit like silos; its south and east walls a bit like the acoustic early warning dishes built of concrete in 1928 on the English coast to warn of enemy air attacks.[100] All are wrested from the specious present and transmogrified by the giant potter's hands. The transmogrification of the airfoil was so complete that one critic has interpreted Ronchamp as an expiation of the sin of aeronautical lightness—a postwar reaction against planes.[101] In *Aircraft* there is a picture of the U.S. aircraft carrier *Lexington*. Its prow looms up. The wings of numerous biplanes jetty beyond the hull. The caption reads: "And Neptune rises from the sea, crowned with strange garlands, the weapons of Mars."[102] It looks a bit like Ronchamp.

It is perhaps ironic, certainly interesting, that Ronchamp exemplifies more advanced engineering design than any of Le Corbusier's prewar buildings; not more advanced construction on site, but more advanced

170   The U.S. aircraft carrier **Lexington,** *from Le Corbusier,* Aircraft, *1935.*

design as defined on the drafting table. At the rue de Sèvres the ruled surface became the principal type of the larger scaled curves, taking architectural shape beyond its conventional limits.

After Ronchamp, ruled surfaces appear in the vaults of the High Court building at Chandigarh (1952),[103] where parallel generators slide between parabolic arches and a straight line; in the crypt and "canons de lumière" at La Tourette (1954);[104] in the "cooling tower" of the National Assembly building at Chandigarh (1956); in the Philips Pavilion (1958); in the sketches for the Museum of Knowledge at Chandigarh (1960);[105] in the tower of the chapel at Firminy (1960), producing an improbable continuity between a square plan and a steeply inclined oval roof;[106] in the roof of the Customs Post at the Écluse de Kembs-Niffer (1962); and in the sketches for the Olivetti Computer Center, Milan (1963).[107]

Before Ronchamp, ruled surfaces appear in the pilotis and rooftop nursery of the Marseilles *unité d'habitation* (1947–1949),[108] clearly shown in drawings from the ATBAT technical section under Bodiansky. Prior to this there are no examples of ruled surfaces in Le Corbusier's built work, but premonitions occur in sketches of the roof shells of the "House for Myself" (1929), influenced, observes von Moos, by Freyssinet's engineering;[109] also, more suggestively, in the cantilevered roof structures designed for the French Pavilion at the Exposition de l'Eau, Liège (1937),[110] which cannot be defined as ruled surfaces, but which are very close relatives of Ronchamp nevertheless.

So Xenakis was not the initiator. His use of ruled surfaces was the most daring, the most technically challenging, and the most explicit, but not the first to come from the atelier.[111] Ronchamp was the first sizable structure to incorporate ruled surfaces. Since it was preceded by modest applications devised by the technical section of ATBAT, a reasonable interpretation of events would be that Le Corbusier's untrameled creation, which spilled out in sketches, was afterward successfully trussed into ruled surfaces by the engineers in collaboration with Maisonnier. They had found a convenient way to dress the naked architectural body that was no more integral to Le Corbusier's creation than a corset to the human figure.

There are two tantalizing bits of evidence to set against this. In Le Corbusier's notebook D17 from the summer of 1950 there are several pages of designs for Ronchamp, including the earliest sketches made from the Paris-Basel train. Following them is a scribbled diagram, apparently un-

171   Le Corbusier, diagram from sketchbook D17, 1950 (FLC 275).

related. It may have nothing to do with Ronchamp, but I think it has. A lenticular shape is divided by a horizontal line. A few verticals and diagonals are roughly inscribed within the lenticular outline. Next to it is a long horizontal triangle also divided by a lengthwise line.[112] These are presumably the end and side elevations of something, very likely a roof, because the end elevation is identical to one variant of the aforementioned roof design for the French Pavilion at Liège of 1937.[113] No final pronouncements can be made, but the simplest explanation is that the diagram shows Guarini's conoid with a segmental rather than a circular base. I think that this was Le Corbusier's first idea for the Ronchamp roof shell. If so then the ruled surface was integral from the start.

That would mean that the origin of the ruled surface was structural, yet there is also an indication that it came from art. In 1947 Le Corbusier wrote a few words for an exhibition catalogue of Antoine Pevsner's postwar sculptures made from welded metal rods:

> *For a long time I have thought that in certain places I call "acoustical" (because they are the foci that govern spaces), great forms with warped surfaces derived from an intelligent geometry could occupy our great buildings of concrete, metal, or glass. I have looked for the man who, like the old ship builders, could join framework and boards to make formwork in which unexpected concrete statues could be cast. Before the buildings, from their sides or from their facades, the forms would invoke space. Sculpture invoking space . . . it seems to me that this is in line with these plastic tendencies. Architecture and sculpture: the masterly, correct, and magnificent play of forms in light.*[114]

It only needs sculpture to be pushed a little further into architecture and this would read like a prospectus for Ronchamp, ruled surfaces and all, for Le Corbusier spells out the possibility that Pevsner's straight metal rods could be models for timber shuttering.[115]

Since 1941 Naum Gabo and then his brother Antoine Pevsner had been making "linear constructions" by stretching threads or wires over a rigid formwork to imply continuously warping surfaces. Gabo had trained as an engineer at the Munich Polytechnic. Was his sculpture science or art? He said he had looked to science but had been unable to find a form that satisfied his inner vision of space:

> *You cannot reason in a work of art. Looking at a work of art of mine, this statement may sound paradoxical, since at first glance it may seem that I am using forms, shapes and lines taken from science; but those who think so, forget that all forms, all shapes, all lines, elementary as well as complicated, geometrical as well as so-called "free", are neither the privilege nor the invention of science. The square and the circle and the triangle and all the rest of them were present in the human observation and served as an expression in works of art of the earliest time. As a matter of fact, it can be asserted that it was actually from the artist that the scientific mind borrowed them and then started to use them as a means of investigation and calculation.*[116]

These were artists playing with borrowed equipment. Gabo declared it and then exonerated himself by claiming that science originally had done the borrowing from artists. Le Corbusier did not declare it, and managed to convince his public that the forms he obtained were prescientific. Gabo retained the dematerialized form of the lines (as if drafting in space); Le Corbusier smoothed the lines into surface and solidified the surface into mass. Pevsner's welded rod constructions lay in between.

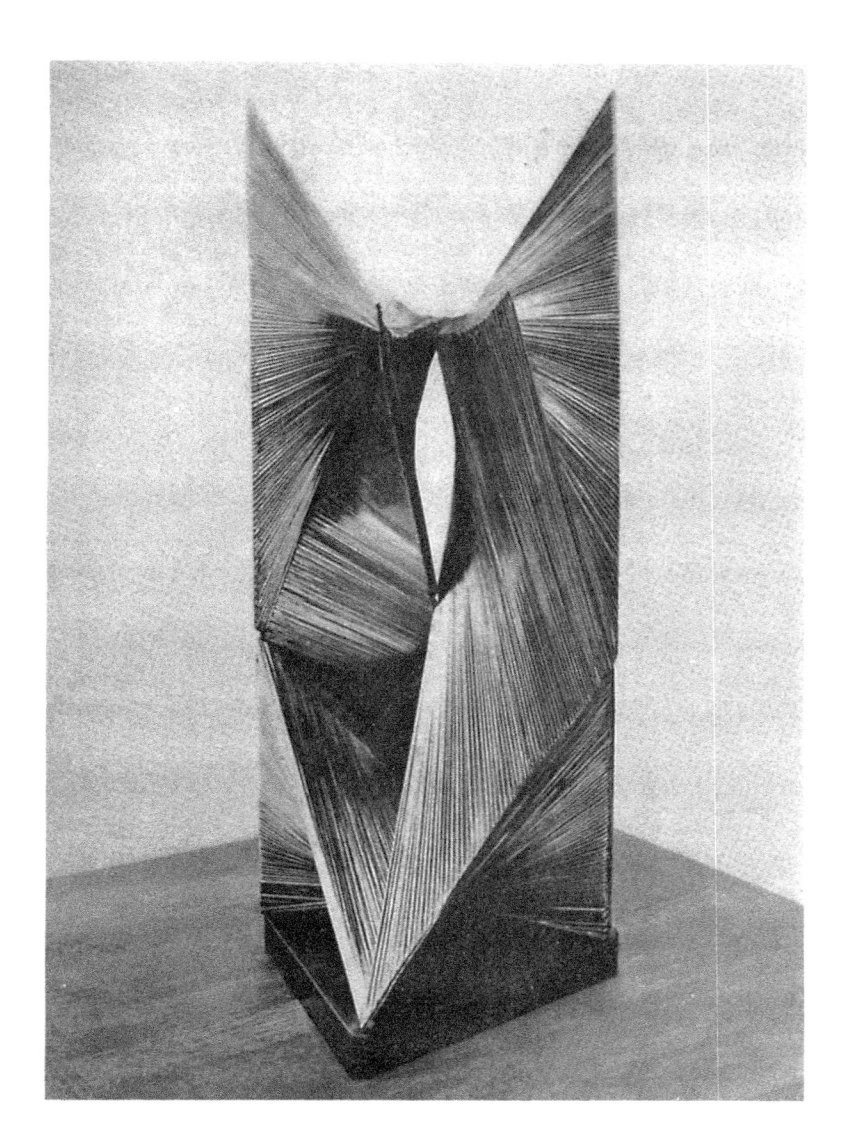

172   Antoine Pevsner, Construction dynamique, *from René Drouin catalogue* La Première exposition des oeuvres de Pevsner, *1947.*

Just as the form of Ronchamp may be traced to an engineering source or a sculptural source, so also were there two kinds of advocacy for the linear body. In art theory the linear body was invoked as a way of increasing the scope of the imagination, and was associated with a school of thought from Michelangelo to Hogarth that insisted on the artist's freedom. Hogarth, actively hostile to the straight line and generally antagonistic to geometry, recommended that we think of solid bodies, especially human bodies, as thin shells, "made up of very fine threads, closely connected together, and equally perceptible."[117] Dressing the object in lines enables us to understand contour as a three-dimensional property and not mere drawing on paper. It helps us to envisage the solid form within.

173 André Maisonnier, elevation of the Ronchamp roof as a ruled surface, March 1951 (FLC 7269).

174 Interior of Ronchamp showing underside of roof shell.

It also helps us to memorize complex, unsurveyable forms so that "they will be as strong and perfect (the images remembered) as those of the most plain and regular forms, such as cubes and spheres."[118] Compare this with the modern engineer's statement that "more complicated geometric forms may also be made comprehensible through lineament. The lines give a geometrical frame. However, in each case it is essential that the form is not capricious, but follows a certain geometric law."[119] Ronchamp may be regarded as a hybrid of these two kinds of advice. The free shapes captured in the imaginative technique were brought into being via the engineering technique.

Ruled surfaces were commonplace in aeronautical, naval, and civil engineering. Since the 1930s they had appeared in modern architecture in the easily recognized form of hyperbolic paraboloid saddles.[120] As such they were symbols of technological advance, badges of contemporaneity. In certain hands they demonstrated the triumph of a new architectural beauty, based on calculation yet transcending the exhausted dreams of art; the kind of thing that Le Corbusier used to write so much about. But Le Corbusier deferred exploitation of the ruled surface until after his love affair with technology was over, and then, when he used it, he neither cared to exhibit it like a badge nor did he use it in the same way as engineers used it, or for the same reasons that they did.

Felix Candela, whose graceful thin-shell structures were becoming known in Europe during the 1950s, dismissed all surfaces except the hyperbolic paraboloid as stylistic sensationalism, his reason being that the hyperbolic paraboloid was the only compound surface that could be easily calculated. The stresses in the shell could be analyzed; its structural behavior was therefore predictable: "This is its real justification and a far more valid one than the beauty of its form," he wrote in 1958, the year of the Philips Pavilion, which had certainly not been designed for ease of calculation any more than had Ronchamp.[121] The functionality of engineering form—the thing that makes it work so well in the outside world—has always been closely tied to the introverted issue of calculability. Le Corbusier's ruled surfaces were adapted to expressive ends in flagrant disregard of such criteria. Only the Chandigarh hyperboloid of revolution could be described as anything like a textbook example, while the most original results were obtained by manipulating wire and string models for Ronchamp, the Philips Pavilion, and the Firminy chapel until they looked right.[122]

Not Modulor measure but the rigid lines of ruled surfaces and translated arcs lay behind the free form of Ronchamp. The Modulor, good fable

COMIC LINES **313**

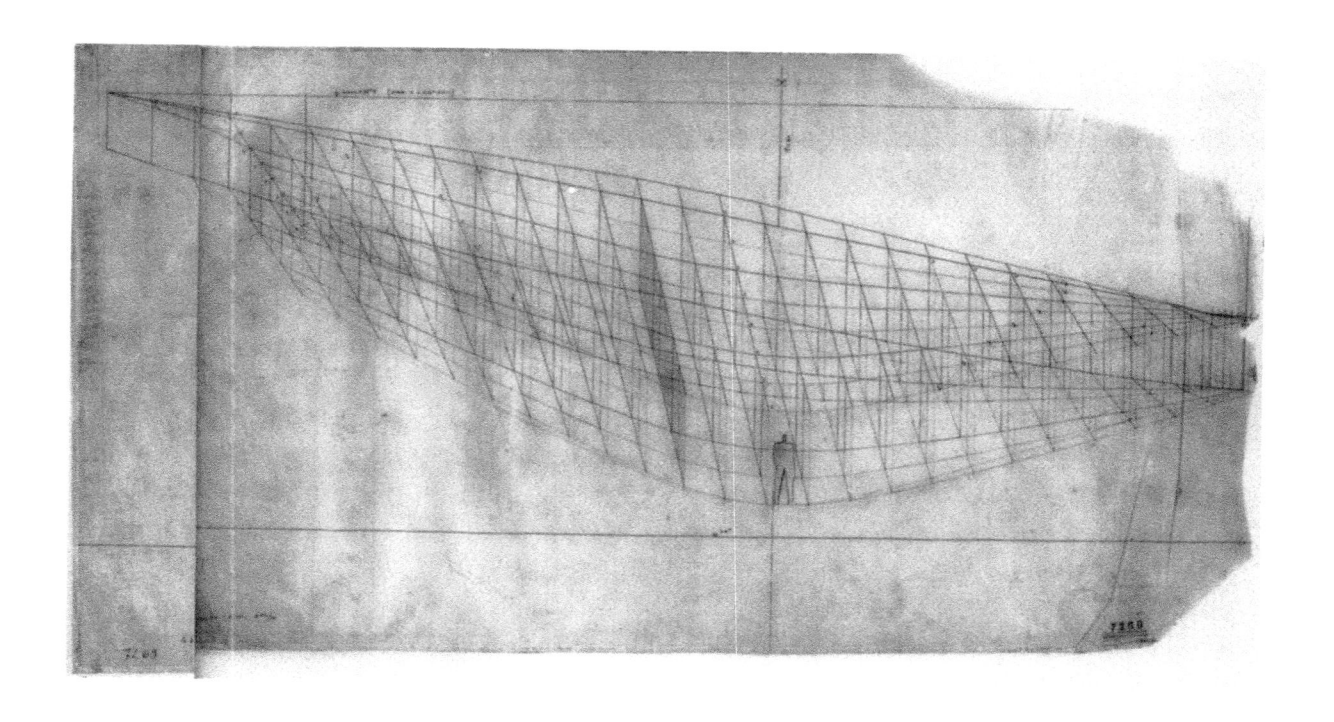

that it is, tells a story about a hidden regulating agency, but is not that agency itself. Le Corbusier's challenge to find the Modulor with the naked eye could now be converted into a challenge to find the ruled surfaces, which are fairly well camouflaged at Ronchamp, in contrast with Xenakis's Philips Pavilion where the generating lines are expressed in the construction, and in contrast with the Chandigarh Assembly Hall, the shell of which is a characteristic, recognizable example of the type. On the underbelly of the Ronchamp roof, timber formwork for the concrete shell was laid across the generators, emphasizing lines of increasing curvature. Consequently it is impossible to discern the geometry of the surface on the interior, where it is most extensively exposed, even with prior knowledge of its generation.[123]

Secrets

Why did Le Corbusier go to the trouble of setting up such an elaborate false trail in search of the mathematics behind the walls of Ronchamp? Immediately after declaring that the Modulor was everywhere at Ronchamp, he made the blank statement "Curved volumes governed by rectilinear generators," which indicates the presence of ruled surfaces but implies that they were somehow governed by Modulor lines.[124] A draft of the same passage read:

> *Modulor reduced to 4.52 m = 2 × 2.26 m. It defies; I defy the visitor to discover it for himself. If this had not been stretched like the strings of the bow [chords of the arc], the game of proportions would not have been played!*[125]

This version, which he did not publish, suggests in an oblique way that he understood the Modulor was barely operative on the shell of Ronchamp.

Le Corbusier made no reference to rectilinear generators in the two *Modulor* books. The nearest he came to it was when he illustrated the interference patterns that suddenly became visible when two sheets of Zip-a-tone lines were overlaid—the kind of thing to be found, he concluded, "dwelling only in privileged places, inaccessible to modest folk."[126] The Zip-a-tone illustrations were reproduced without explanation in his book on Ronchamp.[127] In one typically obscure section of the *Poème de l'angle droit*, he refers to a modern cathedral of *brut* concrete arising from the ability "to think the marriage of lines / to weight the forms."[128] Only in these hinting allusions is there any indication that the ruled surface had usurped the Modulor. It was a secret; a secret not meant for modest folk.

But there is something wrong. It is not the right sort of secret; not that of Pythagoras, Hermes Trismegistus, Paracelsus, or the Cathar heretics, nor that of the alchemist's laboratory or the medieval masons' cutting floor. Those are the cultural secrets we expect to find. They too were provided at Ronchamp, which can be interpreted, quite legitimately at one level, as an alchemical rebus[129] just as the Tour St. Jacques had been a "giant rebus worked out on the basis of the Cabala," according to Le Corbusier.[130] Revelation needs to unfold at a certain pace. Secrets are necessary as retardants. We are drawn into cultural recollection by the artifice of secrets, the deciphering of which will always provide an echo of the excitement of original discovery; the impenetrability of which will remind us of how mysterious things are. If we are surprised at such secrets it is by convention, and without anxiety, for they are quite reassuring by now.

Le Corbusier refers to secrets in the first part of *When the Cathedrals Were White*, written in 1935–1936, where he dreamt up what he later called "the Middle Ages of the mind,"[131] which is a protomodern Middle Ages, all whiteness and order, springing from the debris of antiquity. People were "raw and frank"; there were no academies, no regulations. Architecture was an expression of participation, liberty, and unanimity, and yet there were secrets: "Human beings observed the hermetic rules of Pythagoras" in their search for the laws of harmony. "The law of numbers was transmitted from mouth to mouth among initiates, after the exchange of secret signs." He added a footnote to explain why:

> *Books did not yet exist. These rules of harmony are complicated, delicate. To understand the reason in them you have to have a spirit of some sensibility. Speak of them openly? That would be to put them in danger of errors of fact and of understanding; after three generations they would have become grotesque and the works constructed in accordance with their law would have been caricatures. They must be absolutely exact. . . . When books became one of the most precious instruments of knowledge, the secret of the rules of harmony no longer had any justification.*[132]

The book will destroy the building unless architecture keeps something within its own domain. By the time of Ronchamp's construction, the Pythagorean secret of harmony had been blown by publication of the *Modulor*, leaving the "space which cannot be described in words"[133] and the secret of the ruled surface, which is the kind of secret that creates an embarrassed silence.

But, as secrets go, Ronchamp's secret was hardly well kept by Le Corbusier. Although it was not presented on a plate, nothing much was put in the way of anyone noticing it either. Perhaps no one wants to notice it because it lowers the tone. Is its exposure a disservice? Does it turn a masterpiece into a caricature? We are left with two answers. First: that it is the critic's duty to try and take the mystery from buildings, even if for low-minded motives and even if it may deprive the building of its means of support. Le Corbusier once gave some excellent advice to students: "One must always say what one sees, above all one must always, and this is more difficult, see what one sees."[134] This epigram was repeated at the beginning of Jean Petit's *Livre de Ronchamp*. To see what one sees at Ronchamp is not to see what is said about it, but the effort of sight spends the building because every resource is exhaustible. On the other hand—and this is the second answer—exposition of the secret may add more than it subtracts, which I believe to be true in this instance.

Caricature

Ronchamp is androgynous, it is dangerous, and it is funny. The creation of great architecture is invariably depicted as a serious undertaking; tragedy and epic are its appropriate genres, and Le Corbusier, who was described by Rudolf Wittkower as a "cross between a prophet and a salesman of rare ability,"[135] knew how to present himself in the appropriate roles. But a number of commentators have noticed anyway that Le Corbusier's tragic muse was occasionally Quixotic, even risible.[136] No one, it seems to me, was more keenly aware of this than Le Corbusier himself. His postwar writing is full of episodes of pomposity interlarded with ironic self-deprecation. One day he came across his young assistant Jullian de la Fuente reading Rabelais at his desk. Instead of berating him for time-wasting, as expected, he confided that this was indeed a great book well worth the reading. Only two others bore comparison: The Bible and *Don Quixote*.[137] Consider the heroes of these three works: men larger than life whose presence in the world is predicated on some grand and absurd delusion: tragic, comic, or grotesque. The autobiographical inference is irresistible. On another occasion, Le Corbusier referred to himself as an acrobat, unappreciated by a public who could not understand why he should go to so much trouble to do tricks that only made them feel uncomfortable.[138] Ronchamp, he once said, was a "problème de robinets."[139] Works of architecture do not necessarily reflect the personality of the architect, but Ronchamp was as comical in its self-consciously tragic pose as Le Corbusier, and, as with the man, it is as hard to tell whether the humor in this modern cartoon of something antediluvian was intended or not.

Ronchamp nevertheless has been taken very seriously. Incessant commentary has rendered it universally respectable. This is quite an achievement when so much about it borders on farce. Slight shifts in interpretation can make a lot of difference. When Bona, the foreman, carried the cross to install it behind the high altar the workmen were struck with awe. To cover their embarrassment they began to tell jokes. That is how Le Corbusier saw the event.[140] Is Ronchamp reminiscent of the Ark left stranded on Ararat, or like "bits of broken china thrown on top of the hill"?[141] Is the roof like a bird's wing, or does the whole edifice look like a decoy? Is it an alighting dove or a sitting duck? Good taste bids us suppress the latter in favor of the former, although the latter is as easy to see.

According to the classical formula, art is imitation. It is also said that art imitates art, architecture imitates architecture. Ronchamp imitates practically everything except good acceptable architecture. It copies everything before and after the fact of its own creation indiscriminately, like Woody Allen's *Zelig*.[142] At Ronchamp, imitation is caricature, which makes it wildly unpredictable, eschewing iconographic stability to the same degree that its twisted carcass eschews visual stability.

Writing of the comic element in forms, Henri Bergson arrived at this formula: "Something mechanical encrusted on the living." Laughter is only prompted by human beings or representations of human beings, thought Bergson, but given the anthropomorphism of Ronchamp, let us apply what he said about the art of caricature to the building. Bergson continues: "The more exactly these two images, that of a person and that of a machine, fit into each other, the more striking is the comic effect, and the more consummate the art of the draftsman." Then he reflects that "this view of the mechanical and the living dovetailed into each other makes us incline towards the vaguer image of *some rigidity or other* applied to the mobility of life, in an awkward attempt to follow its lines and counterfeit its suppleness."[143]

This is so nearly perfect a description of Ronchamp that we might easily overlook that the rigidity in question (the ruled surface and the translated arc) is the covert cause of its suppleness and is by no means awkward. Only when publicized does it become so. To point out the rigidity within the supple surfaces is therefore to express what is either laughable or inappropriate to serious criticism. The building's humanity is compromised, the aura of timelessness dispelled with this untimely reminder of the present, and we are confronted with a work that is doubly reactionary, using technocratic means for pietistic ends. On the other hand, what better illustration could we find of Le Corbusier's old maxim that "les techniques sont l'assiette même du lyrisme" (technique is the very basis of lyricism).[144]

However, by far the most important property of the corrosive ripple of laughter that circulates around Ronchamp is its even distribution; it does not all rebound on the architect, nor is it all directed at the architectural critics, nor engineers, nor the pilgrims, nor the church, nor art, nor the public, but prompts some discomposure in each. Its pressure is constant in all directions.

The trouble is that laughter will seem illicit when discharged in an atmosphere of humble faith. It is said that, in the face of his own unbelief, Le Corbusier focused on the touching faith of others. Several photographs of pilgrims are included in the various editions of his monograph on Ronchamp. The most prominent is a blow-up centered on a middle-aged woman in the attitude of prayer, her expression full of naive wonder as she regards the icon.[145] Magnification turns her into an icon as well, and what she represents, surely, is the tremendous resemblance between

175 *Woman at prayer, from Le Corbusier,* Ronchamp, *1957.*

176 Plate from Le Corbusier, Aircraft, *1935. His caption reads: "The most exact laws of acoustics will help in aerial defense. Like the ear of a dog or of a horse, the three sounding conches turn their tympana to various quarters of the horizon...."*

religious awe and architectural appreciation. The holy alliance between architecture and religion is more than just historical. The means of architectural expression have their origin in religious buildings; there they were developed and perpetuated. A great achievement over the last three centuries has been the divorce of the means of architectural expression from the original things expressed. Secularized architecture dwells on the difference between apparent and real to keep us alive to our own constructive perception, while sacred architecture dwelt on the same difference to keep us alive to the fallibility of our senses. Choice between these two ways of seeing is normally aided by context, but at Ronchamp they are elided in such a way that it is impossible to tell whether architecture is exploiting a residue of popular faith in furtherance of its now secular aims, or whether faith draws architecture back into its service again. It depends, I think, on the balance of laughter and awe. In the same photograph, beside the middle-aged supplicant, to the left, not so prominent, is a girl laughing. *Modulor II* contains an extensive quotation from *Gargantua and Pantagruel* about a temple for drunkards.[146] An unimpressed English translator of Rabelais ascribed these passages to another author on the grounds that "Pantagruel could hardly have thought it worth while coming so far merely to inspect an early example of a building lit by indirect lighting and constructed on the plan of a heptagon set in a circle, to demonstrate some mathematical formula which is never properly explained."[147] The Temple of the Bottle simply was not funny enough, but Ronchamp is, at times. It is not consistently so, but it slips in and out of parody, in and out of sublimity, as the aspect changes—a few paces this way or that is enough.[148] In between these states it is neither awesome nor comic but odd.

And it is dangerous too, because of its prodigal rejection of so much that has sustained Western architecture for so long, including unstated assumptions of male sexual identity, subjective interiority, and, of course, because it looks funny, which may be why Le Corbusier, and others since, kept trying to provide it with a firm base on the atavism of modern art, on the unerring intuition of genius, on the Modulor, and even on an idea of Swiss peasant authenticity that may have derived from Rousseau.[149] Insofar as it exhibits the correct signs of a Great Work, it is not to be trusted. But insofar as these substantiating efforts give way, insofar as it remains ungainly and not quite respectable, it deserves our continuing respect and admiration.

Chapter Eight **Forms Lost and Found Again**

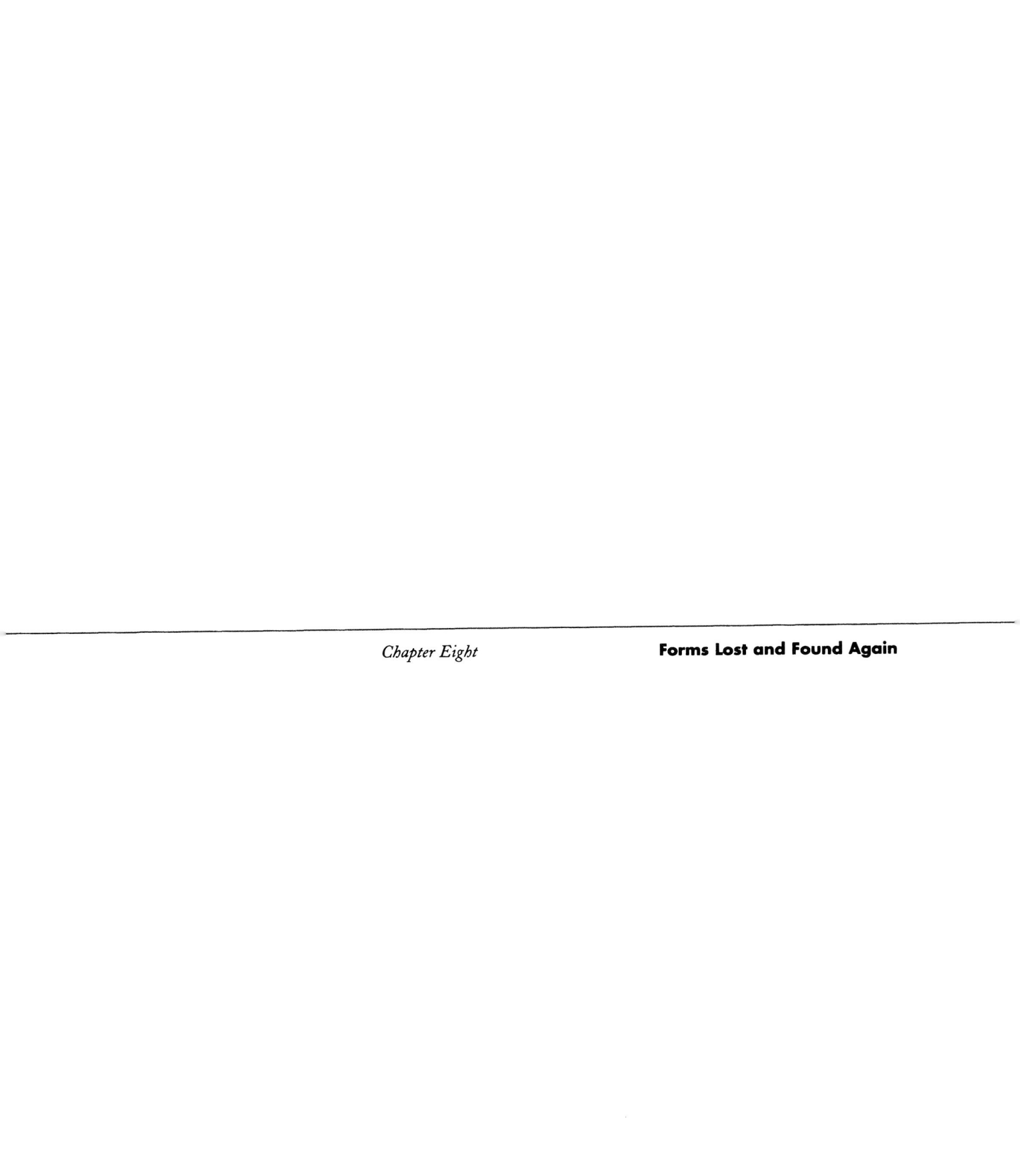

The secret of Ronchamp's geometry was not an edifying cultural secret but the secret of the École Polytechnique. Ten years prior to the Ronchamp commission, in a rather different mood, Le Corbusier had described the École as "a national institution of which we may well be proud," because it used mathematics as a means of mental training for the country's leading administrators.[1] The school was founded in 1795, just after the fall of Robespierre. The royal academies had been suppressed by the Revolutionary authorities, but if the intention was to supersede royal patronage, a new kind of professional institution would need to be devised. The École Polytechnique, free and egalitarian in principle but run on military lines, was conceived as the apex of the new republic's educational system. It was a school of civil, industrial, and above all military engineering, with a tremendous concentration on mathematics because mathematics was understood to be the basis of theoretical understanding and of practical advance in all branches of construction and production. The mathematical emphasis came from the school's cofounder Gaspard Monge, whose earlier attempt to establish a Revolutionary technical school for *sans-culottes* had foundered largely because of their mathematical illiteracy.[2]

Gaspard Monge was a gifted administrator and a gifted mathematician. He had trained as a military engineer at the academy in Mézières, where he began work on what he and his followers would regard as a new branch of geometry, called descriptive geometry. His work of that title, first published in 1799, was a radical restatement of solid geometry. For Euclid, the third dimension was a rather difficult final conquest. For Monge, all geometry was about three dimensions, though practiced in two. His descriptive geometry defined configurations in space by their orthographic projections on two fixed "reference planes" perpendicular to one another. The line of intersection between the two reference planes was treated as a hinge, or fold, which then opened out into a single flat surface. Monge used the fold line in a variety of inventive ways to obtain information about the relation of bodies in space, otherwise more or less unobtainable by graphic means.

Descriptive geometry was a mathematician's generalization of architectural drawing, the powers of which were vastly increased in some respects and reduced in others. Pictures became more abstract, losing much of their illustrative character as solid bodies dissolved into a nexus of trace lines. Ageometric bodies could not be represented, but on the other hand a wider range of more complex geometric figures and their intersections could be represented with much greater ease. In Monge's *Géométrie descriptive,* as in the host of publications following in its wake to this day, the intersections of plane, conic, spheric, and ellipsoidal surfaces, freely oriented in space, are shown penetrating, cutting, intercepting, and touching each other at any angle and in any combination.

Monge provided the basis of modern engineering drawing, not only in the development of descriptive geometry but also by establishing its institutional setting. Regarded as valuable enough to classify as a military secret at Mézières, descriptive geometry was first taught publicly by Monge at the École Normale Supérieure in 1795, and was made one of the essential pillars of the curriculum of the École Polytechnique.[3] However, the reputation of Monge and his school is nowadays preserved less for this practical achievement than for the advances made in pure mathematics, especially projective geometry which had lain dormant since Girard Desargues's discoveries in the early seventeenth century. In the always progressive histories of mathematics, Monge makes no fundamental contribution himself (although he was regarded as having done so in his day), but paves the way for his students Charles-Julien Brianchon (the principle of duality) and Jean-Victor Poncelet (the principle of continuity), who do.

177 Developing the surface of a cone with any base that has been truncated by a sphere, from Gaspard Monge, Géométrie descriptive, *Leroy edition, 1838.*

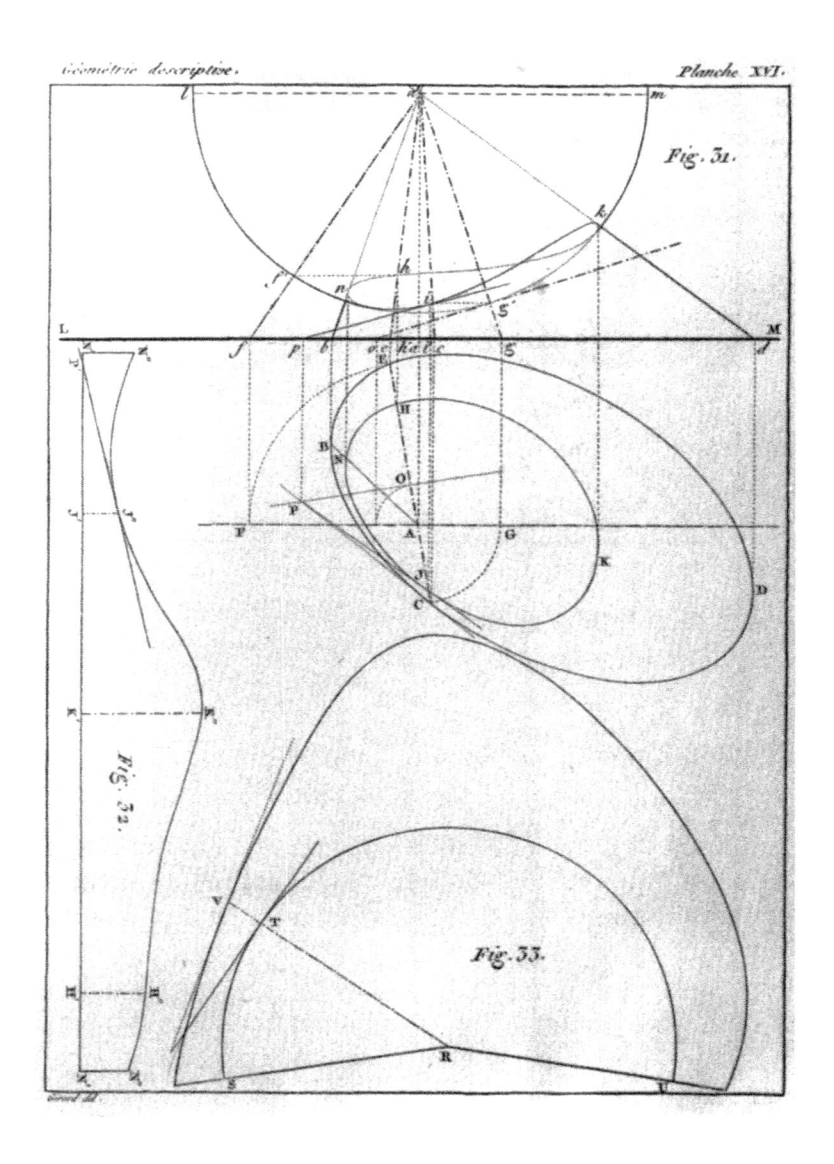

Into this heady, innovative, intellectual atmosphere, and onto the list of scientific celebrities that taught at the École, add the professor of architecture, J. N. L. Durand, who held that position from 1795 to 1830. Durand has the reputation of an uninspired but influential pedagogue. Like Monge, he published his course, *Précis des leçons d'architecture* (1802–1805), which was widely disseminated and assimilated into many other schools. Architecture was a small fraction (about $1/20$) of the foundation course at the École Polytechnique,[4] and since the ambition to establish a mathematical basis for all practices was so clearly stated in the school's prospectus, the mathematician's shadow is frequently seen lurking behind its architectural program.[5] Durand's advocacy of a universal planning grid in which walls and columns are rarely more than emphases of reticulated lines and their coordinates, his penchant for methodical tabulation, his reduction of the classical orders to the numbers 1 to 5,[6] his arguments for economy and utility, and the fact that almost all the illustrations in the *Précis* are orthographic projections[7] have been observed as indications of Monge's impact. The recurrent fear of architecture's being incorporated into the blind operations of technical rationality, subordinate to mathematics, is thereby confirmed.

Durand's teachings do tend toward deracination, but what was he deracinating? He may well have been induced to borrow certain emblems of mathematical rigor from his colleagues. The gridded planning may (but need not) owe something to the coordinate system used to map analytic functions (the École Polytechnique was the theater of development for analytic geometry as well as projective geometry), just as the suppression of perspective may nod toward descriptive geometry.[8] If so, then we

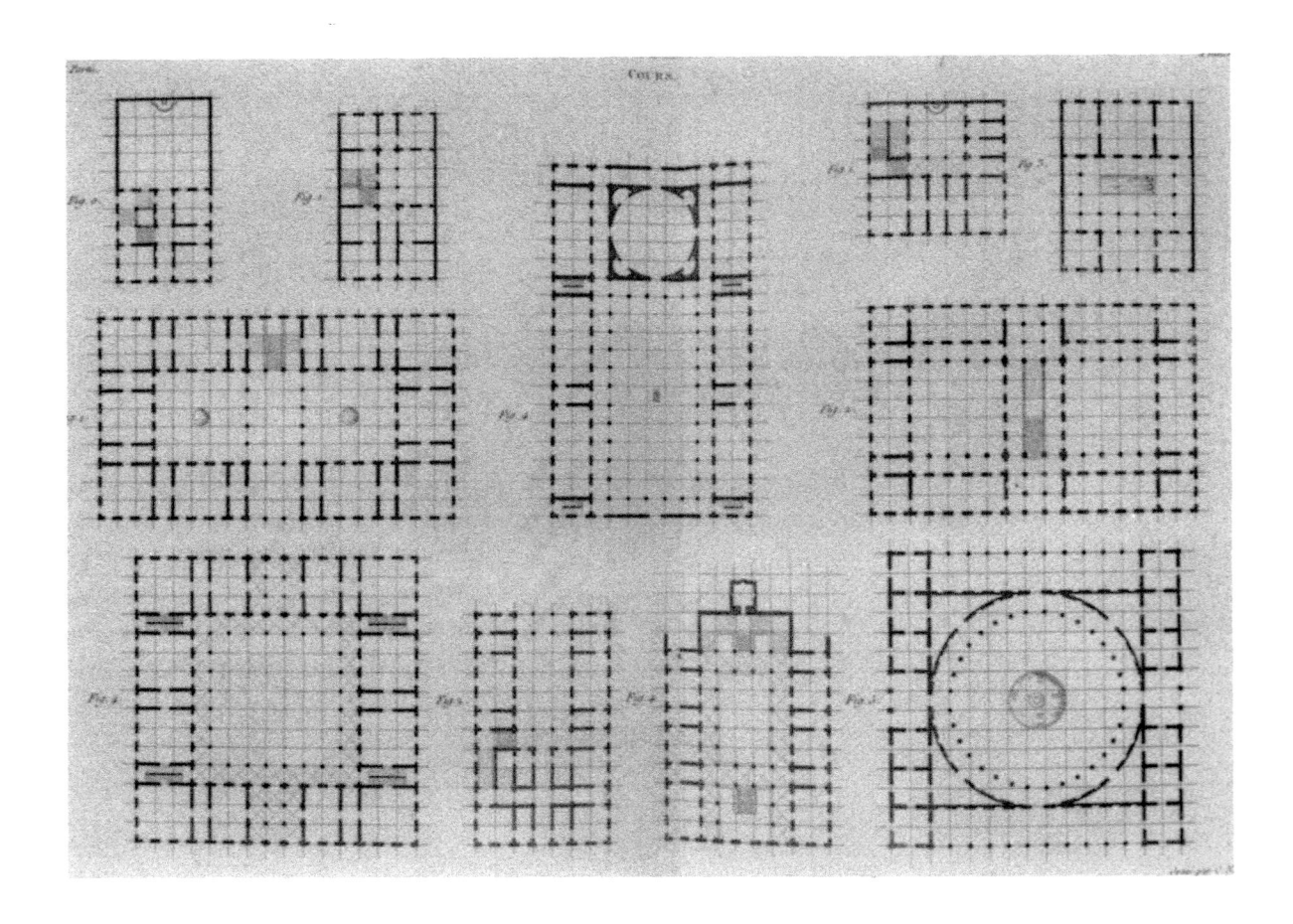

178 Courtyards, from J. N. L Durand,
Précis des leçons d'architecture, *1802–1805.*

should not be fooled by such shallow mimicry, for in neither case did resemblance extend very far.

Durand's grids and his orthographic projections have exactly the opposite tendency to those of the École Polytechnique mathematicians. Cartesian coordinates were used by the mathematicians to define a great variety of curves, an advantage that had been recognized by Descartes himself (although he did not specify rectangular coordinates) and was being actively pursued by Monge while it was as actively inhibited by Durand, for whom the shape of the grid was, as far as possible, identified with the shape of the item mapped into the grid. Nothing could be less mathematical in spirit.[9] And likewise, while descriptive geometry encouraged the free orientation of forms in relation to one another, Durand's orthographic projection was used to enforce the frontal and rectilinear. So whatever the defects in Durand's teaching, Monge's geometry can hardly be held to account for them.

If Durand's methods are to be described as scientific, rational, mathematical, geometrical, or even methodical,[10] we should recognize that they are degenerately so. And if we are prone to see in these methods a devaluation of art, we ought to acknowledge that they are also a devaluation of science and mathematics. From whichever point of view Durand is inspected, he appears to be marching backward.

The degree to which Durand was either oblivious of or resistant to Monge's geometry, which, for good or ill, *could* have been incorporated into a program of architectural instruction, is notable. Its small influence was felt only in stereotomy and shadow projection. I have described elsewhere the surprising result of its application to the representation of shadows on classical capitals and bases.[11] But the impact of descriptive geometry on architecture, as taught or as practiced, was never more than slight. Perhaps that was just as well; we do not really know. We do know that its impact was far more profoundly felt in mechanical engineering.

Industrial technology is extensive and multifarious. Perhaps one of the more pervasive disappointments of our time is that its products nevertheless tend to converge toward a single, universal, unexceptional norm. A cigar box, a milk crate, a filing cabinet, and a block of flats are all examples of packing-case rationality for which similar arguments of expediency can be adduced. When a previous generation of architectural historians saw Durand's lessons as a step toward modernity, it was a step in this direction.[12] Yet when the futurists, constructivists, suprematists, purists, and precisionists, together with early modern architects, exhorted their contemporaries to open their eyes to the sublimity of engineering, other forms were brought in view as well. With the trusses, grids, and compartments came the convulsive beauty of engine bodies, retaining structures, hydrodynamic surfaces, and aerodynamic surfaces—forms produced with Monge's descriptive geometry, the applied (as opposed to the pure) part of which was being taught in engineering courses and trade schools throughout the Western world during the second half of the nineteenth century.[13]

It is well known that the groundwork for Monge's achievement was laid in the eighteenth-century treatises on stereotomy, notably Amédée Frézier's *Théorie de la coupe des pierres* (1737)[14] which is regarded as its immediate precursor. Frézier, architect and military engineer, had divided his treatise into theoretical and practical parts, stereotomy being a more general and abstract science that might be applied to numerous fabrications, of which *timotechnie,* stonecutting for vaults, was but one. Monge completed the task of generalization instituted by Frézier.

I have shown in chapter five that attempts to justify architectural stereotomy as a useful skill were consistently belied by results. It was an ingenious technique appreciated for its difficulty and its beauty. Consider then the peculiar course of events thus far: the smooth, svelte, cylindrical, spherical, spheroidal, conic, and conoidal surfaces, with their precise intersections in graceful groins, characteristic of the sensuous, floating masonry of eighteenth-century stereotomy, were expropriated from architecture by Monge, who was not guilty of adulterating architecture by foisting upon it technical methods of design devised for military and industrial purposes, as he is so often accused of doing. What he did was lift a virtuoso technique from architecture, adapt it, and hand it to a broader community of engineers.

Greater ironies were yet to come. As this family of forms died out in stone, it was reborn in metal; as it began to disappear from architecture it began to reappear in steamships, locomotives, and machines. So that when a new generation of architects, wishing to purge architecture of its past, looked on these awesome feats of engineering with unnaturally innocent eyes, they saw in them a deliverance and sometimes tried to emulate them in a humble way; one aspect of the machine aesthetic being the taut curve in the wafer-thin wall.

As if this were not enough, we now find that Le Corbusier's escape from his earlier *machinolatrie* was effected by recourse to geometric methods of representation and geometric forms that can be traced back along this

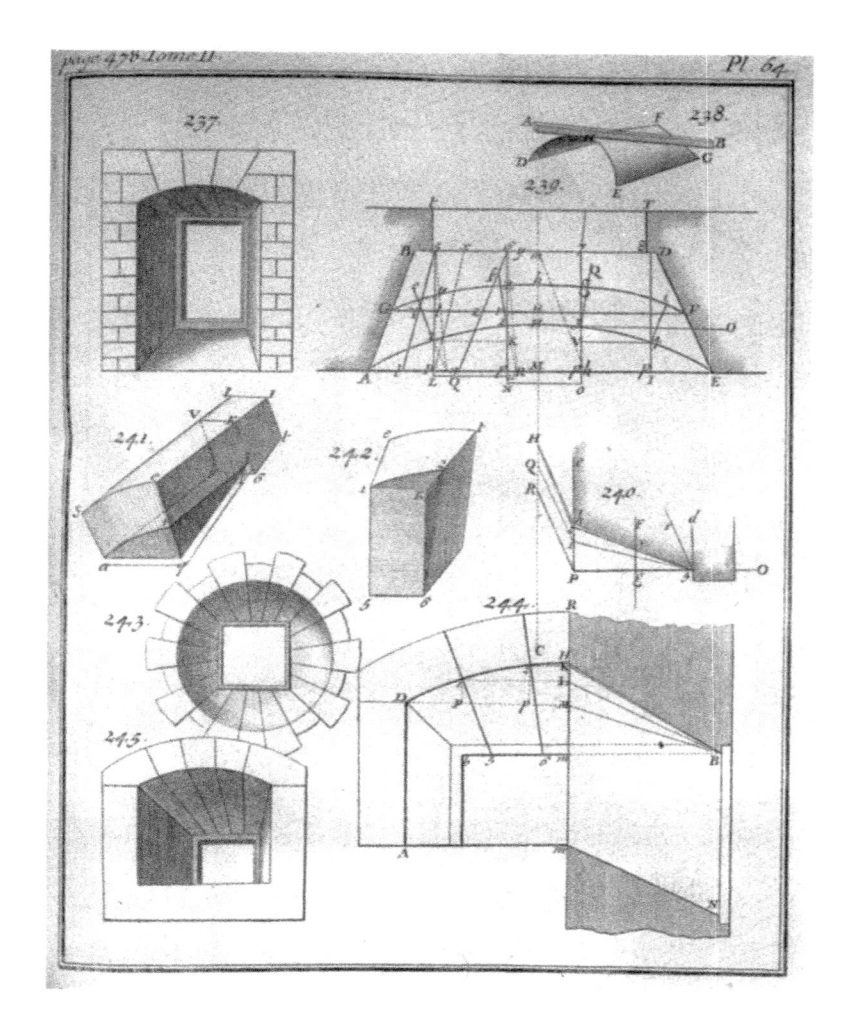

179 A. F. Frézier, *conoidal* arrière-voussoirs, *from* La Théorie et la pratique de la coupe des pierres, *1739.*

same genealogical line through industrial engineering back to Monge, and beyond Monge to stereotomy. Such was the pedigree of the ruled surface in this comedy of errors (in interpretation) and forgetfulness (in practice).

I have shown that Ronchamp's ruled surface roof was similar to the conoid described by Guarini, which can be traced back to Derand and seventeenth-century stereotomy. Frézier, who later described the same figure, gave the whole group of curved surfaces made with lines a generic title: *corps régulièrement irrégulier*. Without defining them further, he noticed that such bodies received no mention in geometry.[15] But it was Monge who first used the term ruled surface (*surface reglée*) as a classification. Not all curved surfaces generated from straight lines belong to

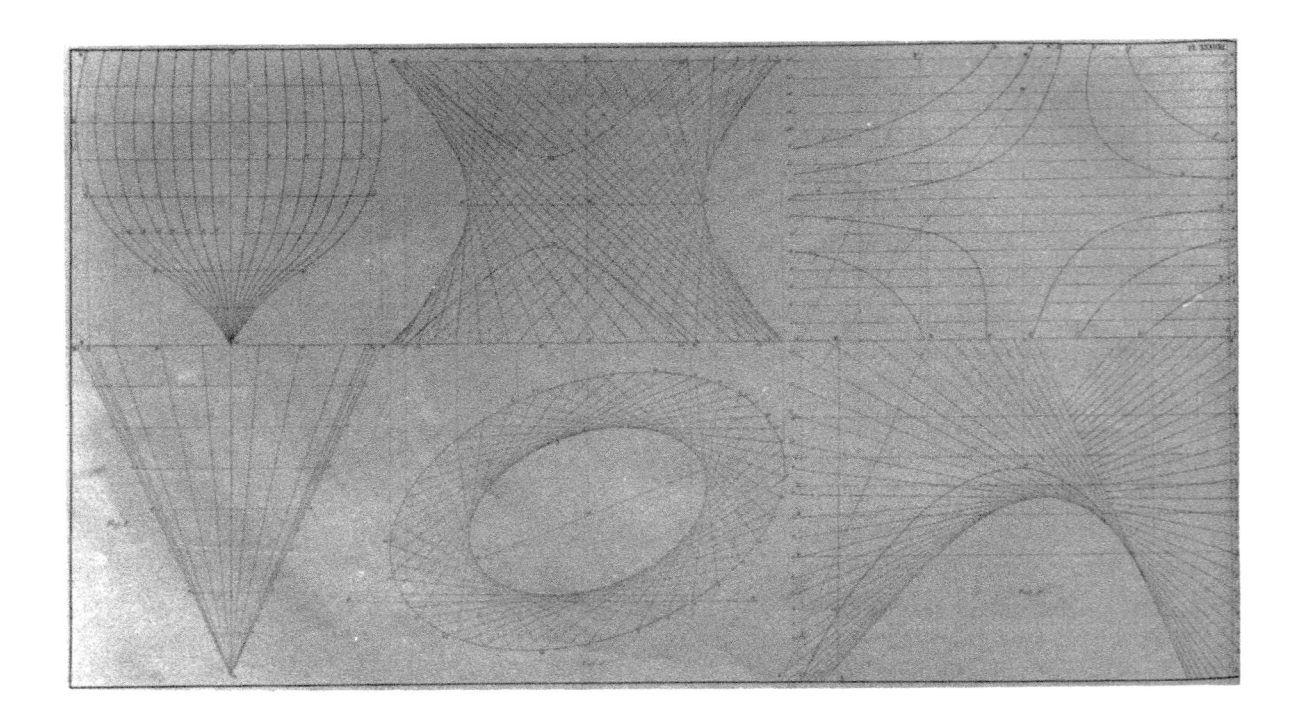

180 Ruled surfaces, from Thomas Bradley, Elements of Geometrical Drawing, *1862.*

this class, only those that are not developable, that is, those that cannot be unrolled into a flat sheet.[16] Monge and his collaborator Hachette introduced a brief treatment of ruled surfaces in later editions of *Géométrie descriptive* (1811).[17] Monge had already dealt with them elsewhere—in algebra.[18] He undertook the spatializing of geometry on two fronts, advancing analytic (algebraic) geometry in the same direction in which he had advanced synthetic (graphic) geometry. Algebraic functions with three variables (x, y, z) could be described as surfaces in a three-dimensional coordinate space. Thus the expression $x^2 + y^2 + z^2 = R^2$ represents a spherical surface with radius R; $x^2 + y^2 = z^2$ represents a cone; $x^2 - y^2 = z$ represents a hyperbolic paraboloid saddle; and $x^2 + y^2/z^2 = 1$ a conoid.

Monge developed the two kinds of geometry side by side as equally valid varieties of space description that became mutually enlightening. His algebraic-cum-graphic treatment of the ruled surface is a case in point. Some ruled surfaces, such as the elliptical hyperboloid, maintained their place at the more advanced levels of drafting instruction during the nineteenth century as examples of this cross fertilization of analytic and synthetic geometries, although they were of no proven practical value at the time. Others, such as the conoid, were related back to architecture and shipbuilding.[19] Practical applications steadily increased, and finally conoid and hyperbolic paraboloid surfaces found their way to Ronchamp.

181 Stern of the Great Eastern, *photograph by Robert Howlett, 1857, Victoria and Albert Museum.*

When architects attempt to escape from the tyranny of geometry, meaning by that the tyranny of the box, where can they escape to? Either they must give up geometry altogether (which would be exceptionally difficult), or they escape to another, always more complex and demanding geometry, or they do the last while giving the impression of having done the first, as at Ronchamp, which is why Le Corbusier's attitude to Gaudí's architecture during his last years is so revealing.

At an operational level, Gaudí's late work on the Güell Chapel and the Sagrada Familia in Barcelona was considerably more radical than Le Corbusier's at Ronchamp, while several earlier works such as the Palau Güell, Casa Batlló, and Casa Milá are similar to Ronchamp in that they include free-flowing forms that required a highly sophisticated geometric prescience. The street facades of the last two, for instance, are impressive examples of stereotomy.

Le Corbusier first saw Gaudí's architecture when he was invited to Barcelona by J. L. Sert in 1928. He made one sketch of a Gaudí building but did not write a word about him for decades, until in 1957 he did a short essay, "On Discovering Gaudí's Architecture," which later appeared as the introduction to a monograph.[20] It turns out that he had always admired the Catalan's sculptural genius, even when he himself had been a "soapbox" architect. Salvador Dalí remembered things

182   Casa Batlló, Barcelona, Antoni Gaudí, 1904–1906, detail of masonry.

183   Sagrada Familia, Barcelona, Antoni Gaudí, 1884–1926, finial.

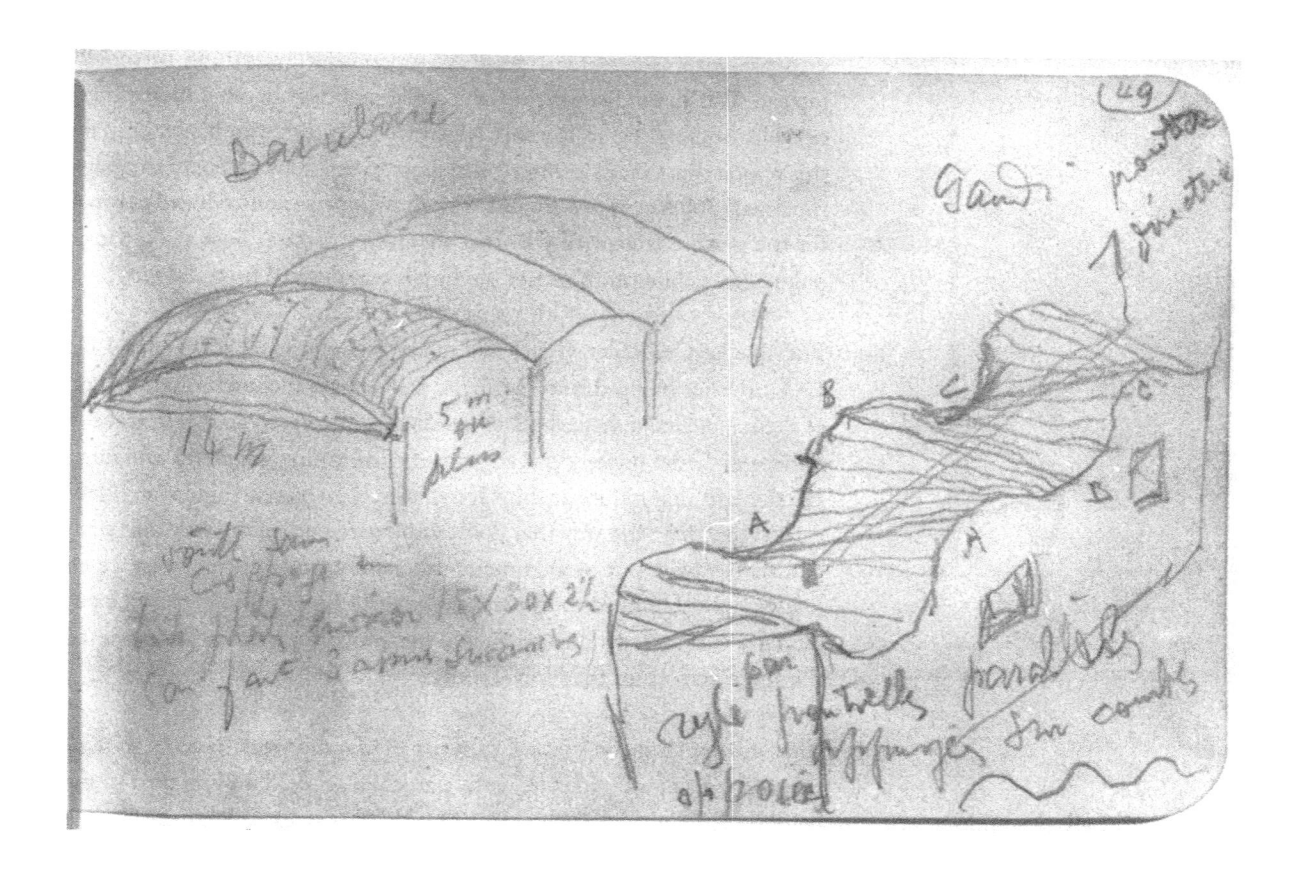

184 Le Corbusier, 1928 sketch of Gaudí's 1909–1910 roof for the Parochial School of Sagrada Familia (FLC 700).

differently. According to him, after touring the major works, Le Corbusier announced that Gaudí was a disgrace to Barcelona.[21] Be that as it may, in the interval Le Corbusier had probably become more appreciative because he recognized the close liaison between modern technical sophistication and "pretechnical" form in Gaudí's work. That is the implicit message of his essay.

In fact Gaudí tried to go beyond this. His later work was an attempt to escape from all design geometry, even this more sophisticated sort, which meant getting rid of the projective drawing as mediator. In the funicular model of the Güell Chapel, which was almost a building in its own right, and in his ceaseless labor on site at the Sagrada Familia he nearly did so, although his biographer and one-time assistant, Cesar Martinell y Brunet, recalled Gaudí's saying that no architect yet possesses "the wisdom and infinite intuition of the angels, who alone are capable of building a cathedral without planning."[22] Gaudí spent forty years on the Sagrada Familia, his final decade devoted exclusively and obsessively to modeling, building, and carving it, yet at his death the portals and towers of one transept were barely complete. In the later stages, he had been able to dispense with architectural drawing, supervising everything personally instead. But as the towers rose, the aging architect became increasingly nervous of climbing to the top. Martinell y Brunet says it was only for this reason that Gaudí resorted once more to the drawing board to design the finials 300 feet above ground. The drawing was reintroduced as proxy for the person. Martinell y Brunet then demonstrates how the intrinsic polyhedral geometry in one of the finials was derived from drawing.

But, like Le Corbusier, Gaudí had mixed feelings about polyhedra. Like Le Corbusier he used ruled surfaces, and he was inclined to put certain of them—helicoid, hyperboloid, and hyperbolic paraboloid—in place of either polyhedra or the right angle as the informing geometry of nature. He thought that nature anyhow took its revenge on rectangular architecture in that way; the way that doors and boards warp, the way that walls and fences twist out of alignment. He even gave each ruled surface an iconography. He liked to think that the three lines of the hyperbolic paraboloid surface (two directrices and one moving generator) represented the three persons of the Holy Trinity.[23]

The one sketch made by Le Corbusier of Gaudí's work in 1928 showed the undulating, warped roof of the parochial school building next to the Sagrada Familia, constructed in 1909–1910. It is a ruled surface.[24]

Chapter Nine **Rumors at the Extremities**

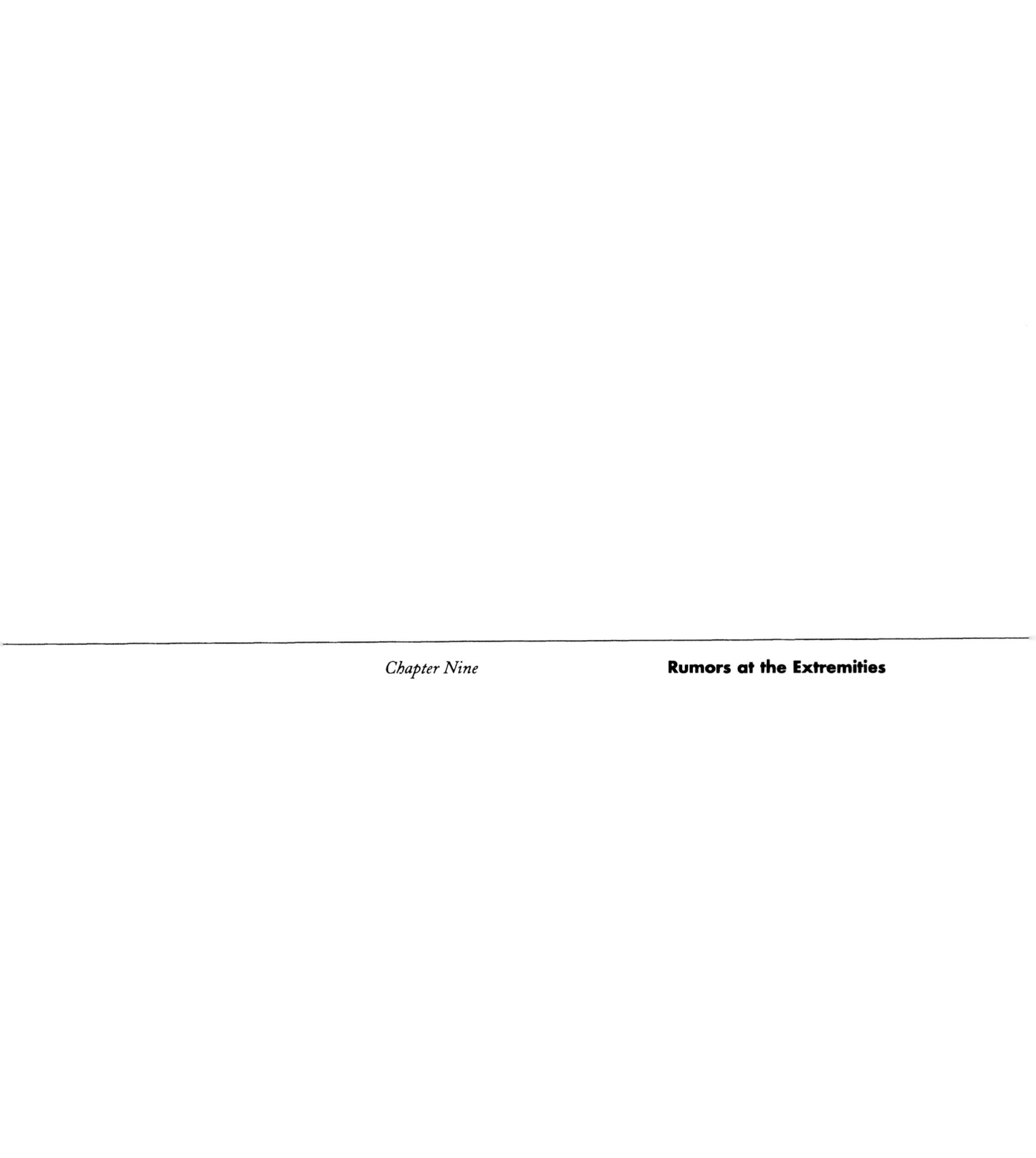

I claimed earlier in this book that the classical heritage of projective drawing went unchallenged in modern architecture. It persists, but not unaltered. Although there has been no radical alteration of practice in relation to drawing, two distinct shifts of emphasis have occurred during the twentieth century. First, the sketch has obtained greater prominence. It is more often dwelt on as the source of originality, and greater value is accorded to it as a means of investigation. Second, axonometric drawings have been slid in between the perspective and orthographic projections as an expeditious way of representing the third dimension without sacrificing the scale measure of plan, elevation, and section. Neither the reliance on the sketch nor the addition of the axonometric are universal, nor are they unprecedented. They are at the opposite extremes of expression, but they are both widely observed tendencies that often occur in the same works. The sketch is indefinite and often amorphous, the axonometric exact and often rectilinear; the sketch is synthetic, the axonometric analytic; the sketch is without obvious geometry, the axonometric full of it. Are we to see them as indicative of a broadening of architectural representation, or as incompatible tendencies pulling in opposite directions? The issue, once again, is how particular types of drawing participate in a general economy of representation, enabling and also limiting vision. I want to begin discussion of this question (and it can be no more than a beginning) by considering two famous examples while continuing the investigation of projection—which extends further than usual in these two examples—permitting us to see how a taste for the outlandish and immeasurable has led to a new employment for geometry in architecture.

In 1923 Theo van Doesburg collaborated with Cornelis van Eesteren, who was then an architectural student, on a *maison particulière*. Although van Doesburg had little involvement in the design, he afterward returned to van Eesteren's axonometric projections of the house and traced a series of momentous variations. Their subsequent influence on modern architecture is well attested, but small errors in the drawings show that van Doesburg's grasp of axonometric projection was none too firm at that stage (some lines indicating the thickness of the plane surfaces are at the wrong angle).[1] The set of tracings, some in color, some in black ink, were given the generic title of *counter-compositions*.

Yve-Alain Bois regards van Doesburg's axonometrics, together with El Lissitzky's, as the first attempts to use this kind of projection in the service of aesthetic aims rather than practical ones.[2] These painters with

185 Theo van Doesburg, Counter-Compositions V, *1924, Stedelijk Museum, Amsterdam.*

architectural ambitions brought about a change of sensibility by exploiting a visual ambiguity that causes the viewer to lose his bearings. At the same time, they had found a way to limit dimensional deformation in pictures. Certitudes have been exchanged. Observers, who knew their place in perspective, are unmoored in the axonometric. It is as if we are floating. Van Doesburg enhanced the effect by removing all indications of orientation or enclosure in the counter-compositions, and also by turning some of the sheets on which they were drawn through 45 degrees. A similarity is suggested between the floating observer and the floating planes. If we all float, we all float in something: space, the space of the twentieth century, says Bois. Undeniably, something new and portentous had been found here, but let us try to define what more exactly.

Let me recall what was noted some time ago by Reyner Banham: the space represented in the counter-compositions is decidedly classical.[3] Axonometric and isometric projections exhibit the three-dimensional Cartesian space explored so methodically by Gaspard Monge and the geometers of the École Polytechnique more emphatically than any other type of drawing, because metric equivalence is preserved, frequently along all three axes. Even more significant than the metric properties of the mapping are the properties of the items so mapped. I have been at pains to show that Monge and his followers conceived the rigid grid of Cartesian coordinates as a conceptual net within which to catch curved and inclined surfaces and the oblique, twisting lines of their intersections. By complete contrast, the modern architectural axonometric takes its own rectangular measuring system as its characteristic content (as had already begun to occur in Durand's planning). The metric framework is the form. Even early telescopic perspective compositions were less completely defined by their method of graphic construction. What we see here, as we see also in the axonometrics of Sartoris, Eisenman, Albers, Held, and many others, are concretions of the measuring grid's reticulated structure. There is nothing intrinsic to axonometric/isometric projections that restricts their content thus, so what holds them in that form? Not science, not technique, but architects and artists. They are the tyrants.

Manuel Corrada has shown how modern artists got hold of the odd idea that the fourth dimension had something to do with color added to space.[4] This is probably what lies behind van Doesburg's title for one of the counter-compositions: *Construction des couleurs dans la 4ème dimension de l'espace-temps*. The confusion arose from Riemann's citing color as a rare example of a three-dimensional manifold (a mathematical "space" defined by combinations of three numbers, which would refer in the case

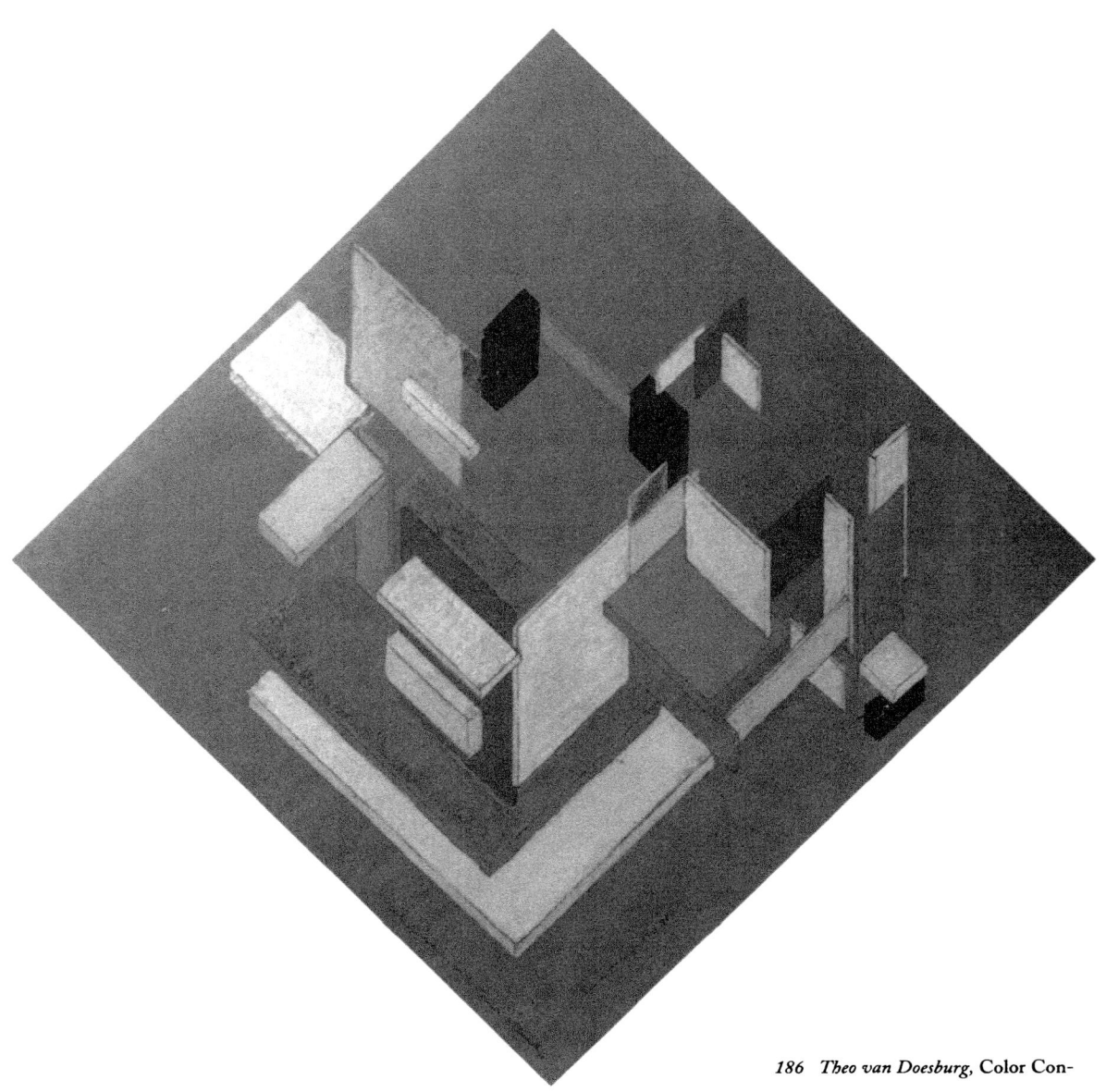

186 Theo van Doesburg, Color Construction in the Fourth Dimension of Space-Time, *1924, Stedelijk Museum, Amsterdam.*

of color to positions along axes of black-white, red-green, and blue-yellow). The only other one Riemann could think of was physical space. Charles Howard Hinton, who was regarded by van Doesburg with some reverence, had published an attractive but somewhat opaque illustration of the successive appearances of a four-dimensional cube using a color-coded space in 1904.[5] The fourth dimension was also identified with time in recognition of Herman Minkowski's space-time continuum, incorporated into relativity theory, which worked at a somewhat larger scale than domestic architecture. Was it therefore a risible pretense for van Doesburg to call three of the counter-compositions space-time constructions, and one a color construction in the fourth dimension? Was it not typical of modern art that something borrowed and misunderstood should be claimed as the very tip of the advance guard and that it be done in such a way as to confound criticism? I do not think these are very interesting questions. We can take it for granted that the answers are affirmative, so we can move on. The question to ask is how these notions affected twentieth-century architecture, not whether they were accurately represented in it.

Van Doesburg pulled van Eesteren's architecture back toward painting, and created an expansive openness where all the orthogonal planes, free of each other and yet more orderly, seem overwhelmingly dominated by an invisible orienting agency in the intervening space itself; an orienting agency far more powerful than the gravitation that is no longer in evidence. It may seem strange that abstracted tracings from architectural drawings would prove to be of much greater interest to architects than the architecture from which they were traced, but it is more easily understood if one considers the counter-compositions as incitements to an effect, still painterly in character, that was exceptionally difficult to obtain in buildings.

Van Doesburg later used a representation of the four-dimensional cube, or hypercube, to emphasize the sense of expansion from enclosure that is such a strong feature of the counter-compositions.[6] Pictures of the hypercube—almost always axonometric or isometric projections—had been in circulation for over forty years. Most of them exploited the implicit ambiguity that was later to become so prominent in architectural axonometrics. The hypercube is an object that cannot be constructed in real space. Its properties can nevertheless be defined by extrapolating from the three dimensions represented by the Cartesian coordinates x_1, x_2, x_3 (or x, y, z) to the four dimensions represented by x_1, x_2, x_3, x_4. Thus operations defining transformations between two and three dimensions can be extended to four and more. From such investigations the hyper-

187 Theo van Doesburg, version of the hypercube, from De Stijl, *7 (1927).*

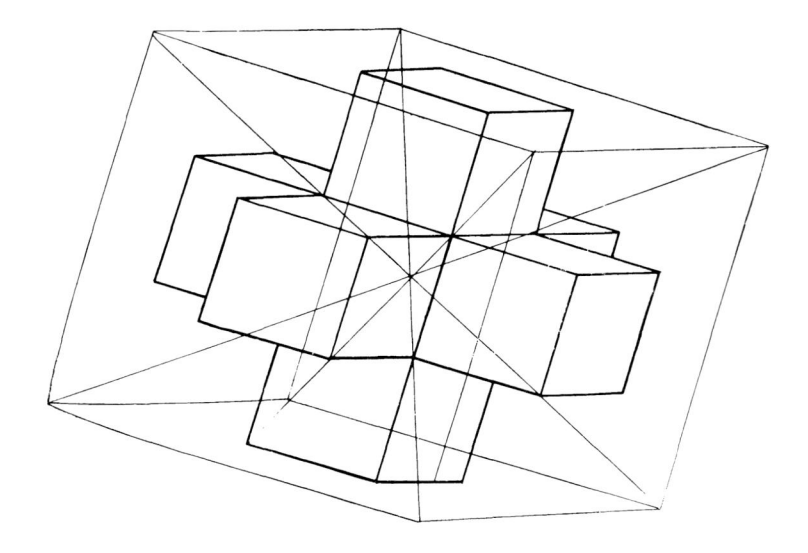

cube is found to possess 16 vertices, 32 equal edges, 24 square faces meeting at right angles, and eight cubic cells. One way to describe it is as a set of eight cubes, each face of which is common to two cubes (just as the twelve linear edges of a three-dimensional cube are each common to two faces). The picture of this figure favored by van Doesburg shows a central cube surrounded by six others extending from it. An imaginative recombination is now required. A larger, encompassing cube represents the eighth cube, which we have to think of, simultaneously, as the same size as each of the seven cubes it contains. In other words, the six outer faces of the smaller cubes extending from the center must somehow join without distortion, and the sides of these same six cubes must coalesce. If we can perform this mental contortion we will have gone some way to envisaging the hypercube.

Other axonometric illustrations of the hypercube made even more effective use of ambiguity, showing the figure more complete and more continuous. It seems to me likely that one such, the most widely reproduced of all, had helped El Lissitzky toward his ingenious axonometric representation of the Proun Room, where all six of its interior surfaces are shown in apparently consistent projection with minimum rupture, although the figure is impossible in three-dimensional space.[7] Both room and hypercube demand that we envisage ourselves looking at them from several directions at once: at the room from above left and below right; at the hypercube from in front, left *and* right (either above or below, whichever way round it happens to be).

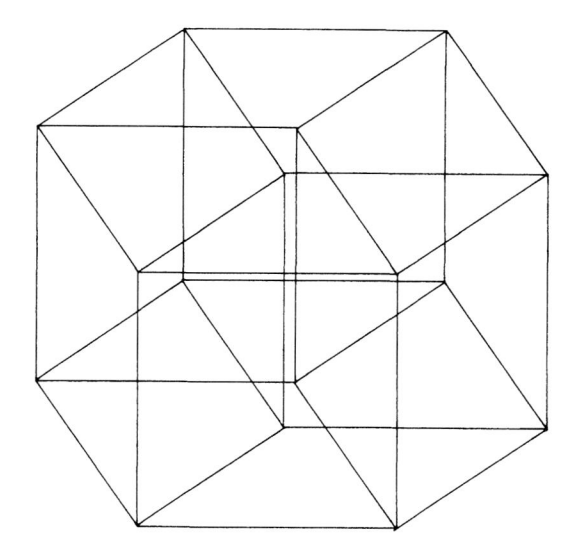

188 Alternative axonometric representation of the hypercube (drawn by the author).

189 El Lissitzky, axonometric of the Proun Room, 1923, from the Kestner Portfolio.

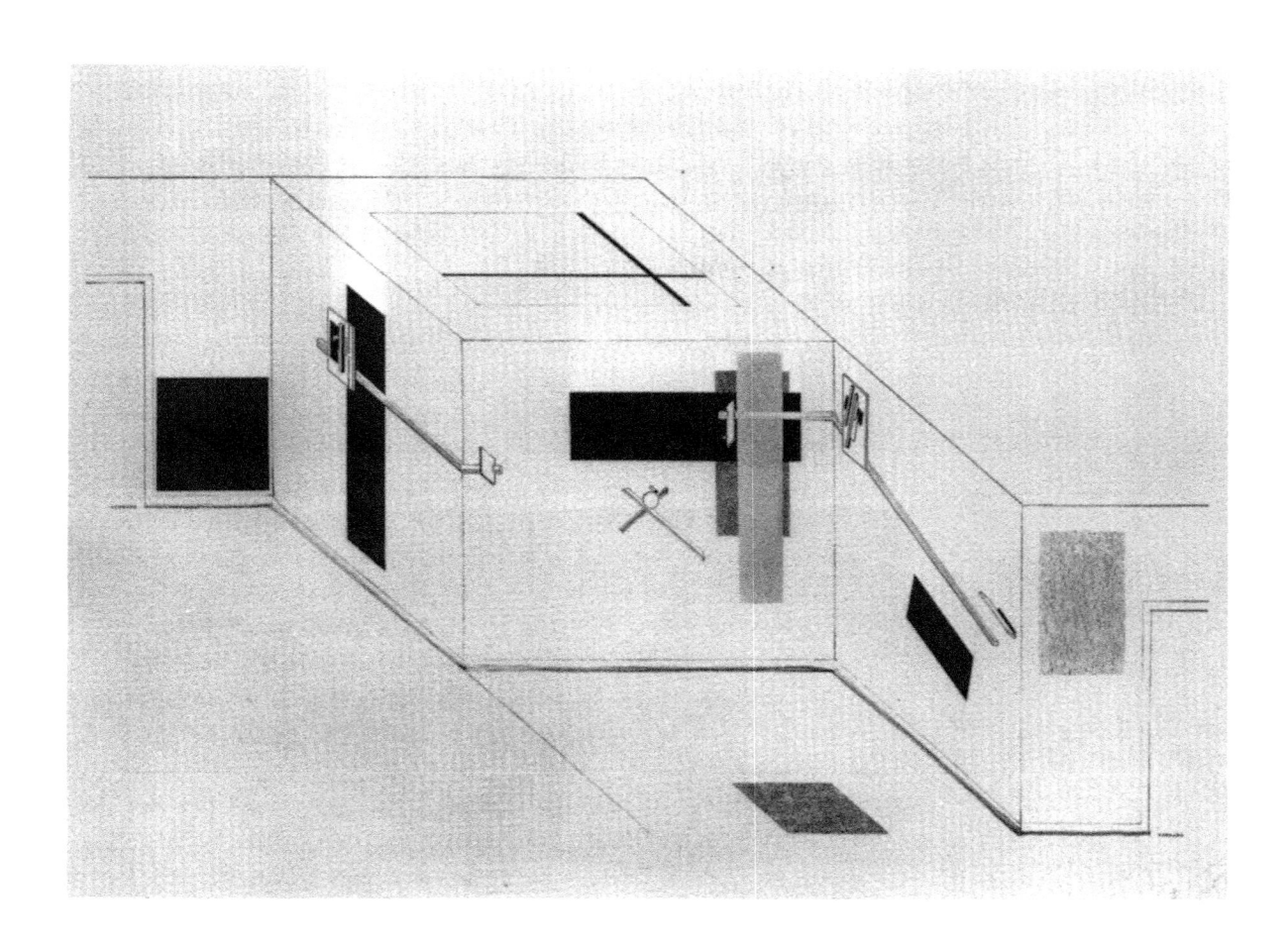

But van Doesburg's choice made the hypercube look expansive. The cube moves out of itself in all directions, relieved of substance, unrealizable yet conceivable, as are the counter-compositions. To get the full effect of the fourth dimension, van Doesburg retreated into two dimensions. There is no paradox involved. The fourth and second share the advantage of not being the third, where fluctuating ambiguities about space are harder to maintain (or at least, harder to maintain if produced through a graphic agency). Projection enabled this advance by retreat, but from the architectural point of view it also raised a difficulty. The attractive feature of axonometric and isometric, despite their imprisoning restraint, has been the wonderful ease with which they produce alternating spatial registration. But the alternation is locked into the drawing. Next to none of the fluctuations in the counter-compositions were exportable into the three-dimensional world they represent. That is why axonometric illustrations of the hypercube work so well.[8] At the same time, we can appreciate from this example how strong the intimation of space can be; the counter-compositions are just flat sheets of paper.

If we move from art to science, we find similar means employed to opposed ends. Van Doesburg wanted to augment ambiguity, Einstein wanted to eliminate it from the discussion of space, but he also used projection as the mediator between the "spherical" Riemannian space he wished to explain and the "flat" Euclidean space that everyone understands.

In the second half of a paper on "Geometry and Experience" published in 1922, Einstein tried to show what was meant by the claim that our universe is finite rather than infinite. On the evidence so far, he was inclined to think it boundless but finite, spherical that is. "Now this is the place," he writes, "where the reader's imagination boggles. 'Nobody can imagine this thing,' he cries indignantly. 'It can be said but it cannot be thought. I can represent to myself a spherical surface well enough, but nothing analogous to it in three dimensions.'"[9] It all hinged on the measurement of distances, and he set out to enlighten the perplexed reader with a simple illustration. The universe appears infinite to us because we imagine that measured distances continue to correlate always and everywhere as they do locally, according to Euclid. But what if they did not?

Einstein resorted to analogy, as physicists are prone to do. The analogue he chose was a stereographic projection that maps a sphere onto a plane surface. This projection, a type of perspective recorded more than a thousand years before Brunelleschi, transforms any circle on the sphere into a circle on the plane (a rather surprising result). Think of a transparent globe with a light shining from one pole that casts shadows onto a plane of projection tangential to its opposite pole. We paste lots of little disks, all the same diameter, each touching the next, over the surface of the globe. On a plane surface every disk would be surrounded by exactly six others, but on the spherical surface there will be fewer than six around any one.

The shadows of all the disks on the globe will project as circles on the plane. If we go round measuring the shadows, they will get bigger as the distance from the point of contact with the globe increases. But if our measuring rod behaved in exactly the same way as the shadows from the globe, enlarging as it moved outward, then its user would survey a spherical surface, not a Euclidean surface. To convert the two-dimensional analogy into three dimensions, we envisage rigid spheres whose metric behavior in space is identical to the metric behavior of the circular shad-

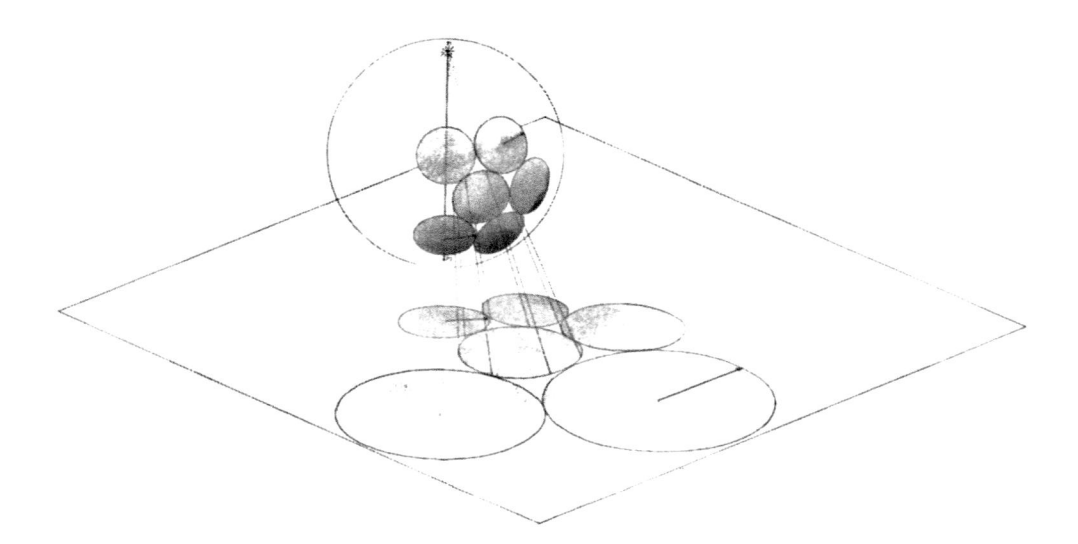

190 *Einstein's analogy of spherical space, using stereographic projection. P1 is a luminous point casting shadows from the disks on the sphere onto a plane of projection tangent to the sphere at P2. (Drawn by the author.)*

ows on the plane. Less than six identical spheres would fit round any other in any plane, though all are the same size. We now have a three-dimensional spherical space, much like the one we inhabit, except that measurements do not add up quite as we expect. Einstein ended by saying that his aim had been "to show that the human faculty of visualization is by no means bound to capitulate to non-Euclidean geometry."[10] He suggests not only that the Euclidean description of space was an indispensable precondition of relativity theory, but that we must return to it to envisage the effects of relativity. Modern physics becomes a host of homely metaphors about uncanny truths.

Turning back to architecture, Erich Mendelsohn's Einstein Tower (1919–1921) is another homely metaphor of relativity. It housed a solar telescope and spectrometer built by the physicist Erwin Freundlich to test the effects of gravitation on light, predicted by the general theory of relativity formulated by Einstein during the Great War.[11] Architect and physicist became cosponsors, raising money to build the observatory at Potsdam, where it still stands, streamlined by an invisible wind. In 1948, making much of his "frequent presence at Einstein's discussions with his collaborators,"[12] Mendelsohn recalled his informing idea of architectural dynamics, which he believed to derive from the new physics: "I define 'architectural dynamics' as the expression of tension innate to elastic building materials, of movement and counter-movement within the innate stability of the building itself." A more obvious source for this idea was early twentieth-century German aesthetics, and the language in which it was couched was that of nineteenth-century mechanics. Despite his meetings with Einstein, Mendelsohn thought of relativistic physics as irrational, if compared, say, to industrial manufacture. According to him, his hat factory at Luckenwalde was rational, whereas "the mystique around Einstein's universe produces a piece of architecture which even its author cannot fully explain by retrospection."[13]

Yet since relativity dealt with motion and the "deformation" of rigid bodies, since it required that space be conceived as a continuously altering field, affected by varied distributions of matter, and since matter and energy could now be regarded as one and the same,[14] the Potsdam Tower could be said to provide quite a good metaphor of the undulations of this continuum—in one sense more apt than Einstein's own, which presupposed an improbably regular distribution of matter. The great difference—and here we have not moved since the Renaissance—is that, unlike the precise analogy offered by Einstein, Mendelsohn's allusion is vague and suggestive. It may lead anywhere, bringing to mind just as easily the tower's diurnal flight round the axis of the earth at about 640 miles per hour. And it could have come from almost anywhere. It was not dependent on relativity theory, which, judging from Mendelsohn's previous projects in the same vein, was more the pretext for the tower than its inspiration or basis.

The Einstein Tower was a built sketch, as Ronchamp is reputed to have been. But, a bilaterally symmetrical shell over an upright steel truss, it remains a far more conventional building in structure and organization than Le Corbusier's chapel. Difficulties nevertheless arose because brick and concrete resisted being turned into calligraphy. There are a wide variety of sketch types employed in the generation of architecture, from

191   Einstein Tower, Potsdam, Erich Mendelsohn, 1919–1921.

cipherlike doodles to indecipherable blurs of graphite. Often it is hard to tell whether sketches are, strictly speaking, projective drawings at all, but in Mendelsohn's case there is no doubt: they are little perspective pictures. He began his heavy investment in the sketch during the Great War, before building anything. He recalled that in the trenches he would draw tiny freehand designs for giant streamlined edifices on small scraps of paper. It is sometimes remarked that the Einstein Tower, his first commission, was the only thing he ever did in such free form. Afterward he would work from sketches, but sketches with the liberty of double-curved surfaces edited out. At Potsdam, far more than at Ronchamp, the building was forced to follow the alien dexterity captured in the architect's first graphic definition; thus they are part of a cosmogony more familiar than those of contemporary astrophysics, expressing the architect's will in a sweeping gesture of creation.

Architecture's Third Geometry

The flirtation with higher mathematics has also led to a marginal but significant change in the relation between architecture and geometry. It is said, sometimes with regret, sometimes in triumph, that modern geometry, from the early nineteenth century to the mid-twentieth, was unconcerned, or at least far less concerned, with visualization. Certainly the reascendancy of synthetic geometry, promoted by Monge, proved brief, and by the end of the nineteenth century it was common knowledge that the space of modern mathematical geometry was no longer confined to the space of ordinary human experience.

Linda Dalrymple Henderson has shown in some detail how artists during the 1910s and 1920s transferred rumors of *n*-dimensional and non-Euclidean geometry into their work, often as if its very invisibility was evidence of transcendence, although this was hardly more than a willful misunderstanding.[15] In architecture it was much the same. Lissitzky, van Doesburg, Mendelsohn, Moholy-Nagy, and Giedion all approached science in awe of its inconceivable discoveries. Attacking the metaphysics of classicism, they cobbled together a new metaphysics out of the mysteries of nineteenth-century mathematics. Thus they gave the new an underlying structure just like the old. And thus they sustained a covert but strong strain of Christian Platonism beneath a denial of otherworldliness.

What I want to emphasize is that those aspects of nineteenth-century geometry that captured the imagination of modern architects could only be present in architecture *metaphorically*.[16] Architects could not use the fourth dimension or hyperbolic space in the same instrumental way in which they used triangles or projections, but they could allude to them, and that is what they did.

So this overtly contemporary, scientific subject matter, associated by proximity with the claims of functionalism, was given traditional, symbolic incorporation into architecture. The peculiar irony of the machine aesthetic, with architecture in pursuit of its own amputated tail, was dwelt on in the previous chapter. There were other peculiarities. Functionalists, borrowing the shapes of machines and things made with machines, sought to induce a smooth, mechanical functioning in buildings whose most conspicuous moving parts were human. It was a kind of totemism; the attributes of the mechanical beast were transferred to us by an act of symbolic appropriation. The same mentality affected the appropriation of new geometry, which became an inspirational talisman of considerable power.

Amongst these resurrections, residues, and atavisms something unprecedented had happened. If geometry can only be alluded to metaphorically, if it can be signified but not used, it follows that it cannot be the signifier in the way that, for example, the equilateral triangle was the signifier of the Holy Trinity in seventeenth-century art and architecture. Instead, *the geometry itself* has to be the thing symbolized or represented. It becomes the subject matter. This is indeed a momentous change of status. There had always been a sense in which geometry was architectural subject matter, but circumstances had never previously arisen where it was forced to be that alone. Van Doesburg's counter-compositions and Mendelsohn's Einstein Tower bore witness to its transfigured status, which entailed a form of cannibalism, as mystifying as the Eucharist; architecture consuming a transcendental version of its own constituents in an attempt to renew itself.

Of course, twentieth-century architecture, like that of preceding centuries, must make use of more familiar geometries in more direct fashion. There are thus three kinds of geometry in architecture. To compositional geometry and projective geometry we may now add signified geometry. At this moment the three appear stratified, each stratum representing, as it were, the sediment of an epoch of investigation: antique, earth-measure geometry trapped in the crystalline forms of composition; early modern projective geometry embedded in architectural drawing; the elusive new geometries not yet deposited but sometimes leaving traces of their existence on the older material. The picture is of a formation determined by the larger forces of history. But it only seems that way looking back after the events. Things could have turned out differently and, in any case, the arrangement as found is not immutable.

Conclusion

The Projective Cast

We seem to have come full circle. Back on home ground we can indulge our fascination with the remote and inconceivable. But there are reasons to believe that things are not entirely unchanged locally. Even Einstein's reassurance about the space of normal experience is not quite what it seems. His statement gives the impression that visual space is identical to Euclidean space, but he knew that visualization was a complex process dependent on more than the mere fact of sight. Investigations around the turn of the century had revealed remarkable differences between the nonuniform spaces of perception and the uniform metric space of everyday physics.

In the eighteenth century, Bishop Berkeley had described vision as a painting on the retina that could never divulge depth.[1] Distance was invisible. Only with the aid of touch could we infer the existence of the third dimension. Measurement and touch were intimately linked in this account of a collaborative sensorium in which division of labor was nevertheless absolute. Entirely consistent with the division of optic and haptic sensation was the distinction between projective geometry, concerned with the continuous transformation of images, and Euclidean geometry, concerned with the measurement of objects. Projective geometry was therefore identified as the geometry of vision; Euclidean geometry as the geometry of touch. This idea, presented at the beginning of this book as useful, is still widely accepted, but it too has its limits. Abiding by it, and founding an entire theory of artistic development on it, William Ivins described why he believed that space was created by hand: "Tactually, things exist in a series of heres in space, but where there are no things, space, even though 'empty,' continues to exist, because the exploring hand knows that it is in space even when it is in contact with nothing."[2] This same hand, putting one thing on another and feeling the equality of extent, judges congruence and measures distances. However, Ernst Mach, an Austrian physicist and philosopher, had already realized that hands by themselves would not survey a Euclidean space. Moreover, the soft, variously sensitive surfaces of the human body considered as a haptic organ stretched judgments of metric equivalence into anamorphoses. So it was clear to Mach that "haptic space, or the space of touch, has as little in common with the metric space as has the space of vision. Like the latter, it also is anisotropic and non-homogeneous."[3] Only when we introduce the ideal, infinitely rigid measuring rod will the space of touch correspond with Euclidean metric space. Henri Poincaré, a French mathematician and theorist of science, then gave grounds for rejecting Berkeley's depthless vision. He pointed out that no sensation is truly passive. Seeing is an *activity*, the normal performance of which is mobile in space, and which, even when the eyes are in a fixed position, requires two independent muscular adaptations, one to focus the lens of the eye, the other to coordinate the targeting of both eyes. These two movements are intrinsic to seeing, yet define depth.[4]

Visualization, in the sense invoked by Einstein, is more than just a picture therefore. It involves a balancing of sensations, motor activities, and concepts that will continue to corroborate the sharable, local truth of classical space as long as we maintain that particular balance. Einstein's defense of Euclidean space was judicious in 1922, when his theories were surrounded by overheated speculations. But his analogy gave an exceptionally clear demonstration of the interaction between three key varieties of geometry that apply in different spaces: non-Euclidean (Riemannian spherical), projective (stereographic perspectival), and Euclidean. Two dimensions are substituted for three and the activities of measuring and occupying are split apart; we measure on the globe and occupy the plane. Seen from the pole of the globe these two separate regions look identical. Standing outside, as observers of the model, we can appreciate how different they are and yet recognize how the translation between them works. Projection is once again the means of translation. Anamorphic vision is as important to Einstein's analogy as is the Euclidean metric.

By the time of Einstein's writing, an unexpected affinity was apparent between the nonhomogeneous spaces of perception and those of mathematics and physics, which may have been one reason why certain artists and architects were so enthusiastic about the new geometries: they were exotic but they brought things even closer to home. This alliance of the local and the remote, sensation and physics, was most vividly demonstrated in the manifestos and paintings of the futurists. As an explicit motive for art it receded afterward into the background, but the new geometries and relativistic physics helped keep sight of the idea that Euclidean space can only be made to correspond to the space of perception if a mental ruler is held over it. Tool and weapon, we feel its threat as we rely on its benefits. Modern architecture, the architecture that continues to epitomize the rule of taste, is a reinforcement of classical, metricated space that provides its own antidote by exaggerating material and visual ambiguity.

Reflection, luster, refraction, luminosity, darkness, color, softness, absorption, liquidity, atmospheric density, instability of shape: these and a host of other properties jeopardize perceptions of metric uniformity. It is perhaps ironic that the first challenge to Euclid came from mathematicians who were reputed to have stripped the world of all these properties, but it is not surprising that architects sought alliance and inspiration from those who had effected an escape from the tyrant of their own creation. Geometers had escaped from geometry; architects might escape from architecture. The routes led in opposite directions but seemed to converge in certain particulars.

Poincaré thought our definition of space should be decided on grounds of practicality. It is also an issue of sensibility. As the world fills up with reifications of Euclid, as points, lines, and planes crowd in on perception—the result of our own collective, constructive activity[5]—the urge to escape, widespread already, might increase. In consequence, we might respond to our own history by altering the subtle mental balance indicated by Poincaré, not as a private affair but as something that could be externalized. And it would be very interesting if this alteration turned out to be more than an act of vengeance against what we dislike, and more than a relieving delirium, but was worked out between consciousness and objects in the same way that classical space was.

In the meantime, plastic art is an arena in which trials, in both senses of the word, continue to be held. However, art need not be a portent of the future. Much of it is a holding operation. Throughout the twentieth century it has held out against, evaded, and played collaborator's tricks on the classical space that is supposed to be our past but is most decidedly our present. Projection, as always, has played an equivocal role in this, sometimes reinforcing, sometimes undermining classical presumptions.

In the sense that geometry is always ideal, never real, there can be no geometry in any architecture. The best that could be said is that buildings serve to bring the pure idea to mind. So arises the thought that since there is no such thing as a perfect materialized triangle, the essence not only of architectural geometry but of architecture *tout court* is not in the concrete object but in the cerebral design. We might complain that such a doctrine raises tremendous problems, without ever completely dislodging it from either architectural or mathematical thinking, where it lies embedded as a petrified Platonism. It is perhaps surprising, then, that with some modest revision this petrified relic proved capable of prompting the most momentous of modern thoughts: that humankind actively constructs its own perception of the world and, thereby, actively participates in the fabrication of the world.

Geometry was chosen as the key example of this process. In 1710, Giambattista Vico presented it thus to explain his *verum-factum* principle: "what is true is made." For Vico, the truth of geometry was assured because its elements were imaginary. We could be certain of them because we had invented everything about them. They could exist uncompromised by material incarnation. Nevertheless it was their mental constructibility, their imagined combination and testing within the confines of the mind, that enabled us to make use of their certainty outside the mind.[6] Since architecture is made with geometry, architecture should provide an evacuated sample of imaginary construction. Between geometry and architecture we have somehow hopped from inside the mind to outside. So when dealing with architectural geometry, we seem to be dealing with this route or doorway between mental and real.

In *The Ethics of Geometry*, David Lachterman proposes geometric constructibility as the thematic paradigm of modern thought.[7] He traces the shift from contemplation of ideal forms in Greek geometry to the manipulation of ideal forms in Descartes's geometry, then turns his attention to philosophers who used mathematical constructibility to exemplify the nature of knowledge, perception, representation, ethics, aesthetics, and the human condition in general, citing Kant, Fichte, Nietzsche, contemporary analytic philosophers, and postmodern critics. But a bunch of pure, mental forms, however acrobatic their combinations, could not hold our attention indefinitely unless they could be transported into external objects, for which vehicles are needed. Projection has provided a vehicle of this kind. Geometric projection transports images; because we also talk of projection as a psychological colonization of the real, as when we use the word in the Freudian sense, we can combine the two meanings and imagine that projection transports the properties of the unreal triangle in the mind out into things—an idea that antedates psychoanalysis.

I cannot think of a better illustration than that given by Howard Burns: Masolino's painting of *The Foundation of Santa Maria Maggiore*,[8] where Pope Liberius is shown tracing the outline of the building's plan revealing itself in a miraculously confined snowfall plummeting from God. This was parallel projection from on high. To describe events this way is to describe them in a classical/Christian domain occupied by classical/Christian subjects. The scenery has changed dramatically but the action has not. According to some accounts our fabricated modern world is comprised of exchangeable signs. It has no objective external location because we carry it with us, dress our own thinking with it, and distribute evidence of it with tokens that always refer to something else. In certain versions, reality dwindles to the size of a mirror held up to a fraction of human consciousness. In others, the mind and the objective exterior both contract, leaving an extensive web of coded representations in their stead. This is certainly the intellectual milieu that we occupy. There was once a shared world brought into being by God; it has for a long time been possible to envisage a shared world remodeled by constructive labor; it is increasingly easy to envisage a shared world modeled by constructive representation in which there are vehicles and there are stages en route but there are no longer any final destinations such as the self, or truth.

Architecture, which remodels nature on the pattern of geometry's fabricated truths, is just one step outside the circle of signs. If we insist that architecture is a code, divesting it of all its obdurate, thinglike qualities at a stroke, it is easily drawn back into the circle and we are alleviated of further responsibility to explain its relation to other events that do not show up so well in the endless chains of representation. It was often claimed that God created the world from thought, like a prescient architect building a house. In a modern revision of the simile, humankind may be envisaged as the architect-demiurges who make the perceptions and signs, as well as the artifacts, that constitute the world. Meanwhile, objects, subjects, and nature disappear behind the veil of signs. That is why, from now onward, we must seek an alternative to this vision of the world as a project, which can so easily turn into a pointless glorification

192 *Masolino*, The Foundation of Santa Maria Maggiore by Pope Liberius, *Museo Nazionale di Capodimonte, Naples.*

of an imagination victorious yet oblivious and inane. The imagination is not an emission. It is not exclusively legislative either. We need a rejoinder to the claims that mental construction is unlimited; otherwise, as Hamlet put it, we end up as kings of infinite space bounded in a nutshell. The unlimited imagination, unfed, presides over a very limited reserve.

It would be no answer, however, to restore the contemplative, mimetic ape of nature in place of our modern Leviathan/Narcissus, resolving the difficulties posed by self-creation by withdrawing from them. It is not a question of denying constructibility, but of recognizing its limits and showing what it confronts at the limits. It may also be useful to trace the conditions of its emergence. Both imagination and representation play a large part in the story of its development, and both of these have traditionally been related by, and sustained in, a projective space, although it is not always easy to recognize it as such. It could even be said that projection was the universal ether of constructibility.

The Arrested Image

Our ideas about our own thinking and perception are dominated by vision, and our ideas about our own vision are defined by tacit reference to pictures and projections. Neither pictures nor projections are fundamental to the visual experience—we see before we know anything of such abstractions—but they have become fundamental to our sharing of sight. Together with numbers and words, pictures help objectify what each private consciousness could never otherwise convey. The magic in pictures is often explained as due either to their transmission of feeling or to their mimetic properties, but it is more likely that their inexhaustible mystery arises from the fact that they externalize an aspect of perception, or that they appear to externalize it, as if one were seeing the thought itself, which does not happen with words or numbers in the same way. That is presumably what Michel Foucault meant when he referred to images as "quasi-perceptions."[9] That is also why David Hume described both memory and imagination as fading pictures of things seen.[10] I look at a painting by Vermeer and see a patch of grayish pigment that is indubitably a woman's head, although it has none of the characteristic signs that make pictures of Dutch heads easily identifiable, such as lips, eyes, or pinkness. I see it as a head despite the absence of these verifiers and conclude, recklessly and irresistibly, that in doing so I must surely be seeing it as Vermeer did, perhaps on the screen of a camera obscura, perhaps looking straight at a woman, and I am exalted at the thought that I see through his eyes as well as mine.

The traditional means of making pictures from observation ensures the repetition of this aftereffect. The astonishing penetration of another being that it entails has been so well assimilated that it slips from attention into a reservoir of commonsense assumptions about what is common about sense.

The mystery deepens when we find that there is no reason to suppose that what we see, when we open our eyes and look, are pictures. According to John Hyman, who argues trenchantly against the picture theory of vision, Descartes was the first to realize that the image on the retina was an interception of information en route to perception, not the perception itself, and therefore tells us little about the nature of that perception.[11] But if the retinal image has the same relation to vision as a bruise to pain, why all the confusion? The picture theory has been difficult to see through because pictures can stand for a large part of vision very convincingly. They stand for precisely the part that is picturelike. Even when a scene is dead still, in order to see it as if it were a picture, we need to frame a portion as when looking through a window, or we must almost close our eyes, or deliberately pull them out of focus. We are obliged, that is, to manipulate our perception so that it collaborates with the idea. If, on the other hand, pictures were indistinguishable from vision, we would be more confused about the difference between pictures and nonpictures than we are, and there would be no conceivable explanation for why it is so much more difficult to draw a picture from life than from another picture.[12]

It would nevertheless be disingenuous to suggest that the picture theory was just a naive mistake that can be cleared up now. A good deal of our understanding has been built on it, and if we are finally able to define its limitations and excesses, that is only because the shared knowledge of pictures made a substantial theory of vision possible in the first place. And not only a theory of vision but theories of cognition, imagination, and creativity too.[13] Since pictures need some eyelike organ to see them, consciousness became eyelike, so that we could see ourselves seeing, and so that we could survey the phantoms brought to mind by the imagination. This inner eye belonged to an inner being that constituted our subjectivity. What joined eye to picture was light, or some projective analogue of light. Thus subjectivity required not only interior pictures but also interior projections.

All these ideas have been under review for some time. The perspectival model of consciousness was challenged by Nietzsche, who wrote: "If our 'ego' is for us the sole being, after the model of which we fashion and understand all being: very well! Then there would be very much room to doubt whether what we have here is not a perspective illusion."[14] Latterly Gebser, Lacan, and Merleau-Ponty have attacked the same target. Richard Rorty and Jacques Derrida have laid siege to the notion of subjective interiority, with its concomitant inner eye.[15] J. J. Gibson raised the question of whether seeing is seeing pictures from the standpoint of experimental psychology, while Hyman raised the same question as a philosopher. I have the privilege of taking their work for granted. So it is now possible to say that projection has leapt from the drawing board into the mind, where it has been installed as an element that aids the construction and maintenance of the thinking self as we know it. This is constructibility in reverse. We have fabricated our subjectivity from manufactured parts that were the manifest evidence of constructive consciousness in the first place: the elements of geometry.

The internalizing of projection was a historical phenomenon. The only way of telling how it helped or hindered insight, or what part it played in a more general economy of theories and practices, is through historical inquiry. It became clear to me while doing this book that we are landed not only with a picture theory of vision, but with a pervasive picture method of construction for manufactured objects as well. If the extent of pictorial influence must in part have been due to the fascination of seeing what someone else had evacuated from their consciousness, it must also have been sustained by the incredible mobilization that pictures permit. They can be circulated more easily than any other commodity. Only words and numbers move as fast. Pictures have this advantage over buildings. At the same time we can see, from the diagram introduced later in this chapter, that pictures are, in another sense, fixations within a metamorphic flux of possible experiences. A picture is something that has to be arrested before it can be mobilized. That is why pictures play so ambivalent a role in architecture.

I have tried to describe that role: architecture begins and ends in pictures, but I would urge resistance to the idea that pictures give us all we need. From dust to dust, from image to image, but there is something else in between. Indeed, so far as architectural drawing is concerned, I conclude that there is little interest in the geometry of the arrested image per se, because it never told anyone anything about geometry that they could not find out more easily elsewhere. A certain investment in the arrested image and its measurements should not be discouraged, however. In a situation where everything is mobile, where only one thing at a time can be held down and kept still, images are the easiest items to immobilize. We should nevertheless recognize that possession of these easy captives is not a sign of victory, but a sign of fallibility. The art of composing images retains its preeminence largely because architecture has to be taught. The question is, how much more is ever brought within the scope of the architect's vision of a project than what can be drawn? The history of the arrested image includes the history of composition, but the composure of what is composed tends to dissolve in the surrounding streams of projective space. It is to these surrounding streams that we should now turn our attention.

I hope to have demonstrated that the geometrically simple composition of a Renaissance centralized church cannot be appreciated for its simplicity unless we recognize the ingenious results obtained in the realm of perception by virtue of this simplicity. I have argued similarly elsewhere that the Barcelona Pavilion and Mies van der Rohe's subsequent works are optical but not graphic in character.[16] Mies's own orthographic design drawings necessarily fail to register the salient properties of the buildings they represent. And the same also goes for geometrically complex designs such as Delorme's *trompe* at Anet, or Guarino Guarini's spectacular Chapel of the Santissima Sindone at the cathedral of Turin. Guarini presented orthographic projection as the very principle of architecture in his uniquely organized treatise *Architettura civile*, insisting that it was the only way to represent buildings. Each section of the book is devoted to a type of drawing or geometry: first a general introduction that deals with basic geometry, then *Della ichnographia* (on plans), *Della ortographia elevata* (on elevations and sections), *Della ortographia gettata* (on stereotomy), and finally *Della geodesia* (on mensuration and surveying).[17] All his known drawings are orthographic. But, as in the Barcelona Pavilion, the effect of the Sindone is not accessible from its drawings. Guarini espoused the traditional doctrine that the eye deceives, while his architecture made the most of its deceit. He was sensitive to consistent optical transformations of size and likewise, with changes in apparent brightness, roughness, and remoteness, listing examples of such dependable misreadings in his chapter on elevations and sections.[18] Thus did he create the impression of illimitable height in the polished black marble interior of the Sindone. H. A. Meek explains the effect as due, in part, to Guarini's leaving the stonework in the upper reaches of the structure unpolished so that it looked further away yet more luminous.[19] The effect was also due, in part, to reductions of the relative heights of the diminishing rings of superimposed arches as they approached the apex of the construction, turning what still appears to be a steep conic framework into a far less elevated domical form—a perspectival illusion.

193 Chapel of the Santissima Sindone, Turin, Guarino Guarini, 1667–1690, section, from Architettura civile, *1737.*

194 Chapel of the Santissima Sindone, view into dome.

In each case, composition is involved, but composition that achieves effects far beyond the pale of the composed design. Design is action at a distance. Projection fills the gaps; but to arrange the emanations first from drawings to buildings, then from buildings to the experience of the perceiving and moving subject, in such a way as to create in these unstable voids what cannot be adequately portrayed in designs—that was where the art lay. To suggest anything else is to pander to the lamentable ancient dogma that ideas always suffer from embodiment because they belong in the mind, or the equally lamentable modern assumption that the inscription of a sign is more significant than what follows from it. In either event, it is to fall victim to the illusion of plans.

Where, though, is the imagination? If it is not in the mind's picture gallery does it cease to exist? The chapters in this book suggest that it continues to exist but not in its traditional location. Imagination is not held within the mind, but is potentially active in all the areas of transition from persons to objects or pictures. It operates, in other words, in the same zones as projection and its metaphors.

Space

Space is often said to be the essence or fundamental medium of architecture. But if it is unwise to think that vision is made of pictures, why is it any wiser to think of vision as spatial? Architecture and geometry have become tightly entwined owing to the procedure of design. This brought the powers of the imagination into the foreground because of the pictorializing of subjective consciousness. Architecture, geometry, and pictures all bring space along with them as a silent partner.

If all three are artifacts, is it not possible that the space accompanying them is also an artifact? Kant thought so. He took things one step further than Vico and one step further than Ficino. Not geometry but space (for which geometry was a longhand notation, one might say) was the key example of constructibility. This could be demonstrated by the traditional means of subtracting sensible properties until even form, Ficino's irreducible residue, was expunged:

> *If we remove from our empirical concept of a body, one by one, every feature in it which is [merely] empirical, the colour, the hardness or softness, the weight, even the impenetrability, there still remains the space which the body (now entirely vanished) occupied, and this cannot be removed.*[20]

195 Luigi Moretti, plaster cast of interior space of Guarino Guarini's Santa Maria della Divina Providenza, Lisbon, from Spazio, *1952.*

196 Santa Maria della Divina Providenza, Lisbon, Guarino Guarini, designed c. 1681, begun 1698, section, from Architettura civile, *1737.*

From this thought experiment, he concluded that space is the precondition of perception. Only after we have manufactured an internal "representation" of space that is independent of all experience will we be able to have experience of the external world. From this same introspection, Kant discerned that there was one, and only one, space. Our intuition of it is singular and does not admit of multiplication. Furthermore, he understood the single space to be confirmed by the postulates of Euclidean geometry.

This is what we now find hard to accept. If space is devised by the mind as a universal repository, why is it always the same space with the same characteristics? Is it not merely dogmatic to assert that every mind at every moment abides by the same self-imposed legislation? And did not the development of mathematical geometry after Kant, with its non-Euclidean spaces and *n*-dimensional spaces, render his ideas obsolete?

I have used the term projective space. We also talk of visual space, motor-tactile space, imaginary space, social space, painterly space, and architectural space. We sometimes talk about it as if it were plastic and pliable, sometimes as if rigid. Roger Scruton, in an effort to preserve the commonsensical aspect of Kant and demolish the pretensions of architectural theory, holds the proliferation of spaces to be the consequence of linguistic confusion. In each instance we really mean a particular part of, or partial apprehension of, one great space. The locution "architectural space" is thus no more than an abbreviation of "the portions of space inside buildings and the portions of space between buildings."

Scruton maintains that the concept of space can be altogether eliminated from critical writing about architecture simply by using the word shape instead,[21] but the substitution does not always work. And for the same reason, Luigi Moretti's solid plaster models of great architectural interior spaces do not always work either.[22] The word *space* is as much about intimation as it is about surveyable dimensions. An intimation cannot be made of plaster, nor can it be described as a shape, nor can it easily be located as a portion of the one universal space. Right now I am looking out of my window. It is dark outside and parts of my room—the illuminated drawing table, the open door near the light, and the light shade—hang suspended, apparently much enlarged, about halfway across the street, interlaced with a tree and pinioned by a lamp post. Do these familiar ghosts belong to the space of the room, and if so, how far does the room extend? Do they belong in the street, and if so, how many spaces are out there? Do they belong to both at once or do they belong to neither, because contained within the reflective surface of the window? I

THE PROJECTIVE CAST

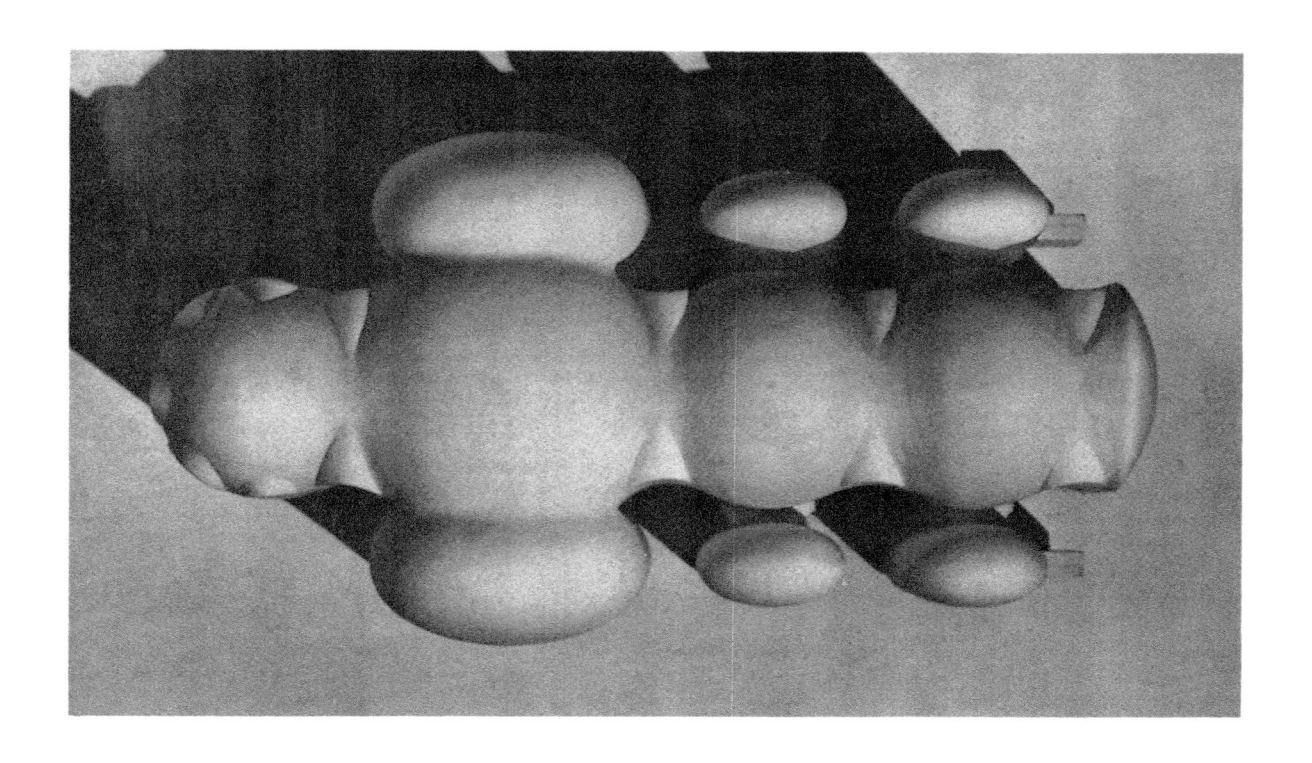

have no trouble grasping the situation in each of these four alternative states, none of which corresponds to Kant's one, and only one, space. All that is happening here is that I am assuming space is dependent on matter while Kant and Scruton assume it is not. This has nothing to do with theoretical physics. It is a question of how we choose to think of the experiences we have.

There are several ways of apprehending space and several ways of translating between the varied apprehensions. There are not an infinite number, but several related to each other and related to things outside their respective definitions. It used to be thought that a single, infinite, homogeneous, three-dimensional space corresponding to the axioms of Euclidean geometry had a special, ordaining status. Though still a force to be reckoned with, it is no longer the supreme authority. There is no question of having to decide which of the other spaces is the ultimately true, real, or right one in place of it, but questions do arise about their consistency and applicability, about the extent of their validity, and about the translations between them.

The two broadest routes from Euclidean space seem to lead in opposite directions, toward palpable experience and toward abstract mathematics. Of particular interest is the way in which projection has been used as a connecting thread between these extremes.

The Permeation of Projection

In the introduction I suggested that the geometry in architecture has always had a projective cast to it. I shall now try to say why in greater detail. The first thing to do is to define the different fields of projective transmission that concern architecture. A diagram is, I think, the most helpful way to do this. Projection operates in the intervals between things. It is always transitive. In the diagram, ten fields of projection are shown joining five types of target. Four of the targets are almost always thought of as pictures or picturelike. The one exception is the designed object, and that too can be endowed with picturelike qualities. I am attempting to portray the extent of projection and its metaphors, so the diagram treats varieties of real and imaginary spaces as if they were all the same. The part behind the dotted line cutting (2), (6), and (7) represents the observer—someone who is looking. As we shall see, the status of these lines as they pass across the border into consciousness is not at all clear.

The diagram is intended as a freely traversable guide to projective and quasi-projective transactions. All ten routes can be traveled in either di-

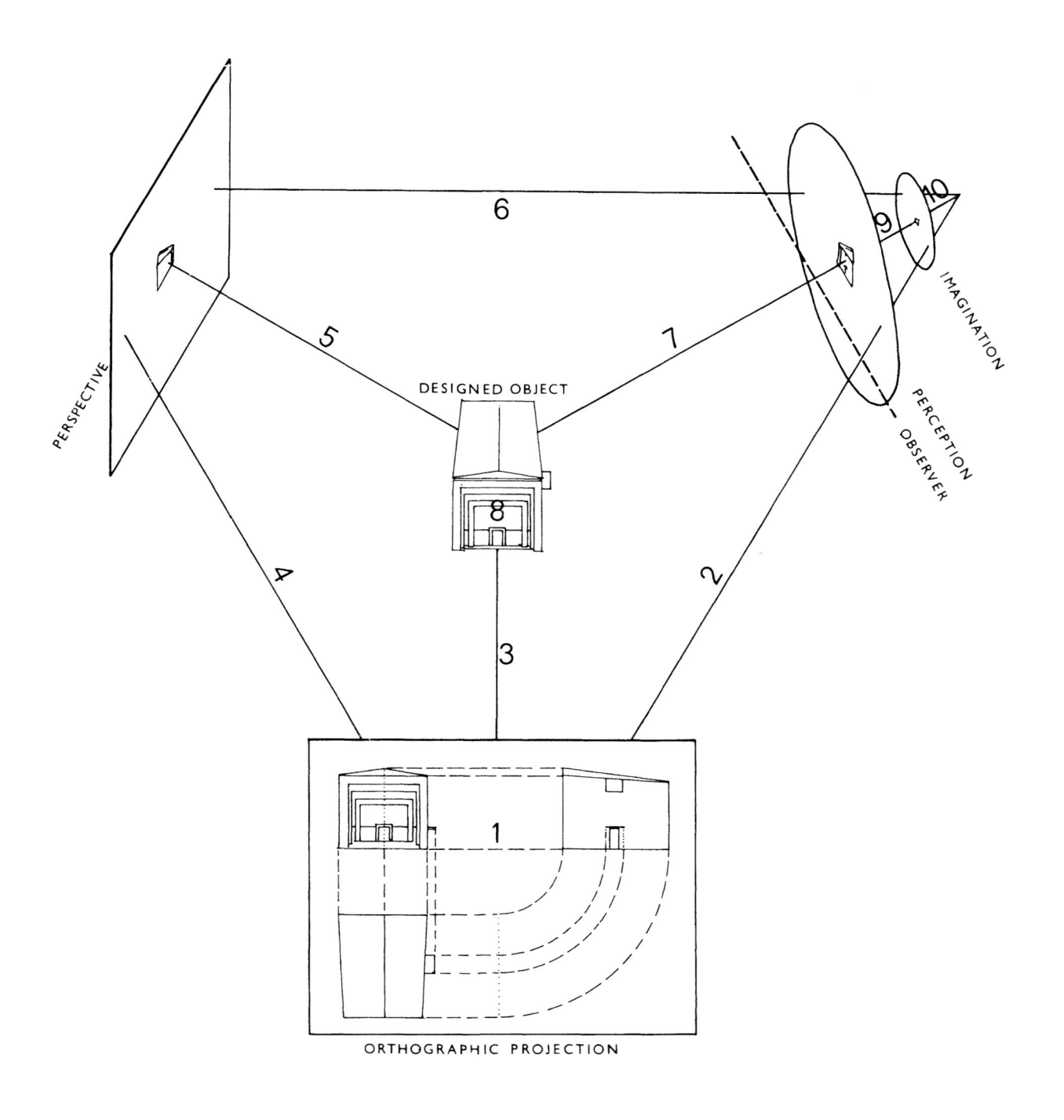

197 *Projection and its analogues:* The Arrested Image *(drawn by the author).*

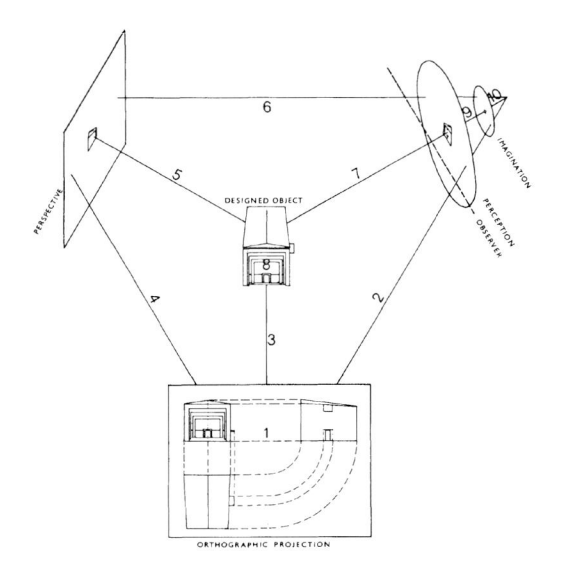

rection. There is no necessary starting point, and no necessary sequence. Four of the targets—orthographic projection, designed object, perspective, and observer—are symmetrically joined. Each is connected to all the others. The designed object is put in the center for convenience, although the diagram is best thought of as a tetrahedron, so that the center disappears. The fifth target, the odd one out, is imagination. The question arises as to why it has such a peculiar relation to all the rest.

There follows a brief itemization of the ten transitive spaces numbered in the diagram:

(1) Graphic projection (two-dimensional) correlating the several orthographic images of the object. The third dimension is imaginary within the drawing. This interaction between pictures was dealt with in chapters three and four.

(2) Perspectival space (three-dimensional) defining the optical route between a mobile, perceiving subject (observer or author) and an orthographic design. This route is fairly straightforward from the design drawing to the observer, but in the reverse direction it is far more complex. Designing is a performance during which vision maintains a constant interaction between manual movement and resulting inscriptions. But there are also "ideas" informing the performance. Somehow these are transferred from mind to page in a combination of visual and motor activity. In this direction, therefore, projective space is inextricably bound up with mobility and imagination. Sketching, which may be perspectival, orthographic, or indeterminate, can play a part here.

(3) Nonprojective space (three-dimensional) between the orthographic design and the designed object. The transmission of information between drawing and building is effected by scaled measurement and fabrication. No projection is involved; the result will nevertheless be projectively related to the design. This can solicit a pictorializing of the architectural object; see (8), below. The direction is reversed when measured drawings are taken from buildings.

(4) Graphic projection (two-dimensional) deriving perspective pictures directly from the orthographic design drawings, independent of the designed object's existence or realizability. The third dimension is imaginary within the picture. It is also possible to reverse direction along this line, as when plans and elevations were derived from Piero della Francesca's paintings and from G. B. Piranesi's perspective engravings.

(5) Perspectival space (three-dimensional) between the object and its derived pictures (photographs or drawings made on site). The viewpoint is undetermined and its choice involves what we call judgment, creativity, or imagination. The direction from object to picture can be reversed, although when that happens the route is usually diverted. Photographs have often been used to aid reconstruction and reproduction, but in such cases the journey from picture to object has mostly been via (4) and (3).

(6) Perspectival space (three-dimensional) between derived pictures and any perceiving subject. An increasing proportion of our information about architecture arrives along this path in journals, books, slides, and videos. Reversal would make the observer the manipulator/creator of the picture. This is the case when setting up a perspective or taking a photograph, as described in (4) and (5).

(7) Perspectival space (three-dimensional) between the object and any mobile observer (7a). This may seem the most dependable route. I hope to have shown in the chapter on the centralized churches that, with regard to design, it is the least predictable, the hardest to foresee, and therefore usually the most interesting. It is also more difficult to envisage how it is reversible (7b). The building is projected toward the observer, but what is projected toward the building? The building does not see us, unless we accept the mysterious tale told by Lacan, where a tin floating in the sea looks at him.[23] The answer has to be that the building presents itself to us as a field of action, as does a sheet of paper or a viewfinder in the context of designs and pictures. When we look at the building and imagine it otherwise, then decide to alter something about it, the building becomes as irresistibly impressed with us as we have been with it. Another instance would be a building erected without preliminary representation.

(8) Imaginary space produced within the object, so that its depth appears dilated, collapsed, or distorted. Such effects are sometimes obtained by transferring the implied depth of pictorial projection onto the modeled surfaces of buildings. This may involve a degree of collaboration on the part of the observer—another instance of (7b).

Thus far, I believe my diagram provides a reasonably good rough guide. But behind the dotted line, which represents any observer, it is not so dependable. My purpose here is to show how projection—or rather quasi-projection—breaches the boundary between world and self, the objective and the subjective. There are two further targets, the perception and the imagination belonging to the observer, and two further projective spaces behind them, (9) and (10). Imagination and visual perception are shown as pictures, because that is how they are normally described. They are not pictures, but the very fact that both are thought of in that way is very significant.

Notes

Introduction: Composition and Projection

1. Werner Oechslin notes that Daniele Barbaro, commenting on the fable of Aristippus finding geometric figures drawn in the sand, took the figures to be undeniable signs of higher thought. A footprint would have signified a man; a triangle signified a mind. See Oechslin, "Geometry and Line: The Vitruvian Science of Architectural Drawing," *Daidalos* (Berlin), 1, no. 1 (1981), 27–28.

2. Gustave Flaubert, *The Dictionary of Accepted Ideas*, trans. Jacques Barzun (Toronto, 1968), 43.

3. Joseph Conrad, *The Secret Agent* (Toronto, 1984), 32.

4. Sebastiano Serlio, *The Five Books of Architecture* (New York, 1982), fol. 1, recto.

5. "L'Architettura, sebbene dipenda dalla Matematica, nulla meno essa é un'Arte adulatrice." Guarino Guarini, *Architettura Civile* (Turin, 1737; reprinted Farnborough, 1964), Trat. I, iii, 3.

6. G. H. Hardy, *A Mathematician's Apology* (Cambridge, U.K., 1967), 84. See also P. J. Davis and R. Hersh, *The Mathematical Experience* (New York, 1980), 168–171; Ivar Ekeland, *Mathematics and the Unexpected* (Chicago, 1988), xii.

7. Jacques Hadamard, *The Psychology of Invention in the Mathematical Field* (Princeton, N.J., 1949).

8. Henri Poincaré, *Science and Hypothesis*, trans. W. J. G. (1905; reprinted New York, 1952), 49.

9. William M. Ivins, Jr., *Art and Geometry: A Study in Space Intuitions* (Cambridge, Mass., 1946), chap. 1.

10. H. Poincaré, *Science and Hypothesis*, 51–71.

Chapter One: Perturbed Circles

1. Jacob Burckhardt, *The Architecture of the Italian Renaissance*, trans. J. Palmes (London, 1985), 80.

2. Ibid., 95. See also Peter Murray's introduction to Heinrich Wölfflin, *Renaissance and Baroque*, trans. K. Simon (London, 1964).

3. Wölfflin, *Renaissance and Baroque*, 91ff.

4. Ibid., 38ff.

5. Ibid., 76–80.

6. The only mention of Wölfflin occurs in an appendix containing bibliographical notes on the theory of proportion: Rudolf Wittkower, *Architectural Principles in the Age of Humanism* (London, 1949), 163.

7. Ibid., 1. A very similar opinion is expressed by Sinding-Larsen in a more recent piece on the same subject: "The buildings alone do not normally provide a basis for interpretation of their possible meaning." See Staale Sinding-Larsen, "Some Functional and Iconographical Aspects of the Centralised Church in the Italian Renaissance," *Acta ad Archaeologiam et Artium Historiam Pertinentia* (Rome, 1965), vol. 2, 241.

8. Michael Ann Holly, *Panofsky and the Foundations of Art History* (Ithaca, 1984), 46–68; and Michael Podro, *The Critical Historians of Art* (New Haven, 1982), 98–116.

9. Wittkower, *Architectural Principles in the Age of Humanism*, 27–29, 117ff.

10. Stefano Ray, *Raffaello architetto* (Rome, 1974), 248ff.; Giuseppe Marchini, "Raphael the Architect," in *Complete Works of Raphael* (New York, 1969), 438–445; Simonetta Valtieri, "Sant'Eligio degli Orefici," in Frommel, Ray, and Tafuri, *Raffaello architetto* (Rome, 1984), 143–156.

11. Bramante's proposal for a circular colonnaded court round the Tempietto kiosk was published by Sebastiano Serlio, *Tutte l'opere d'architettura* (Venice, 1619), book 3, fol. 67.

12. The section of the Tempietto—the drawing that unlocks the secret—is very rarely published. Neither Serlio's half-section nor the section in the Codex Coner (Sir John Soane's Museum, London) show the undercroft, nor, apparently, did any other drawing until P. Letarouilly, *Edifices de Rome moderne* (Paris, 1874), vol. 1, pl. 103. See also Arnaldo Bruschi, *Bramante* (London, 1973), 132. The Letarouilly section is only illustrated in the complete Italian edition, *Bramante architetto* (Bari, 1969), 497.

13. Jean-Paul Richter, *The Notebooks of Leonardo da Vinci* (New York, 1970), vol. 2, 56–57; Ashmolean Museum, Oxford, Leonardo ms., vol. II, 8a. It is hard to tell what kind of oratory Leonardo had in mind for performance in such a place, since speech is directional. In his interesting account of the objections to centralized planning on acoustical grounds, Sinding-Larsen notes Serlio's counterargument that a centrally placed altar can be seen by everyone. Sinding-Larsen, "Some Functional and Iconographical Aspects of the Centralised Church," 215.

14. Claude Perrault, *A Treatise of the Five Orders*, trans. John James (London, 1708), 110.

15. William L. MacDonald, *The Pantheon* (Cambridge, Mass., 1976), 89; Wittkower, *Architectural Principles in the Age of Humanism*, 26.

16. Wolfgang Lotz, "Notes on the Centralised Churches of the Renaissance," in *Studies in Italian Renaissance Architecture* (Cambridge, Mass., 1977), 66.

17. See Richard Krautheimer, "Introduction to an Iconography of Medieval

18. Christine Smith, *The Baptistery at Pisa* (New York, 1978), 104, 114–116, 234–235; and G. B. Giovenale, *Il battistero lateranense* (Rome, 1929).

19. Charles E. Isabelle, *Parallèle des salles rondes de l'Italie*, 2d ed. (Paris, 1863). Of the 30 items illustrated in this admittedly highly selective work, only five are Renaissance buildings. See also Sinding-Larsen, "Some Functional and Iconographical Aspects of the Centralised Church," 242n.

    Architecture," *Journal of the Warburg & Courtauld Institutes*, 5 (1942), 21–33. Many examples of medieval baptisteries can be found in the published volumes of *Italica romanica* (Milan, 1979).

20. Leonardo has been cited as the inspiration behind many of the early sixteenth-century centralized churches. Leonardo was one of the few people with a grasp of the structural performance of arcuated vaults. In a little, annotated diagram he illustrates that the thrust from the base of the arch will be oblique, so that no support will be needed directly below the springing: Richter, *The Notebooks of Leonardo*, 89; Ms. H (1), 36a, Institut de France. Yet Leonardo's own centralized designs, like the medieval baptisteries, cautiously provide what he had judged to be redundant. On the other hand, works such as the St. Peter's project and Santa Maria delle Grazie by Bramante, or Santa Maria della Consolazione in Todi, which are said to be influenced by Leonardo, do not. See Ludwig H. Heydenreich, "Leonardo & Bramante," *Studien zur Architektur der Renaissance* (Munich, 1981); and Peter Murray, *The Architecture of the Italian Renaissance* (London, 1969), 112.

21. Rivoira, in his account of the evolution of the Renaissance dome, responded to those who held that Brunelleschi's Duomo cupola was an enlarged copy of the eleventh-century baptistery dome in front of the cathedral (to which it does bear a striking resemblance) by pointing out that Brunelleschi's dome had been constructed without timber supports filling the space beneath. I would add to this now invisible miracle, the still visible opening up of the lower part of the tiburio. G. T. Rivoira, *Roman Architecture* (Oxford, 1925), 284. An unlikely source of appreciation of this property was Giambattista Vico, *On the Study Methods of Our Time* (1709), trans. Elio Gianturco (Ithaca, 1990), 29. However, for emphasis on the continuities between medieval and Renaissance dome construction see Volker Hoffmann, "L'origine del sistema architettonico del Brunelleschi," and Giuseppe Marchini, "La cupola: medievale e no," in *Filippo Brunelleschi, La sua opera e il suo tempo* (Florence, 1980), vol. 2, 447ff., 915ff.

22. Wittkower, *Architectural Principles in the Age of Humanism*, 30. Again, a similar point is made by Sinding-Larsen, "Some Functional and Iconographical Aspects of the Centralised Church," 242.

23. It has been noticed by many historians that the reintroduction of these elements is easier to trace from Byzantine sources than from either Roman or medieval Italian ones, which makes it difficult to explain either in terms of the cultural continuity of established practices or the particular discontinuity we call the Renaissance.

24. Wittkower, *Architectural Principles in the Age of Humanism*, 10.

25. Ibid., 31.

26. My use of the word symbol needs an explanation, and perhaps an apology. To avoid unnecessary multiplication of terms within the text, I am using it very broadly, as did Wittkower, to cover any kind of sign. In C. S. Pierce's well-known classification, a symbol has an arbitrary relation to the thing signified, while signs that resemble what they signify are called icons. I have chosen not to use his terminology, preferring to highlight the difference between general resemblance and modeling. I hope that this does not increase confusion.

27. Wittkower, *Architectural Principles in the Age of Humanism*, 29.

28. Nikolaus Pevsner, *Outline of European Architecture*, 7th ed. (Harmondsworth, 1963), 182.

29. Frances Yates, *Giordano Bruno and the Hermetic Tradition* (London, 1964), 117–123; Paul Rorem, *Biblical and Liturgical Symbols within the Pseudo-Dionysian Synthesis* (Toronto, 1984), 69–73.

30. Rocco Montano, *Dante e il Rinascimento* (Naples, 1942), 141ff.; Vittorio Rossi, *Storia letteraria d'Italia: quattrocento* (Milan, 1953), 157–158.

31. Thomas Kuhn, *The Copernican Revolution* (Cambridge, Mass., 1957), 112–114.

32. Two dome frescos by Correggio in Parma (Christ and the Apostles in the cupola of San Giovanni Evangelista, 1518–1522, and the Duomo cupola fresco, c. 1524–1530) initiated this development by reviving the hieratic formation found in earlier dome decoration and subjecting it to perspective. John Shearman, "Correggio's Illusionism," in *La prospettiva rinascimentale*, ed. M. D. Emiliani (Florence, 1980), vol. 1, 281–295.

33. Martin Davis, *National Gallery: Italian School Catalogue* (London, 1951), 94–98; Rossi, *Storia letteraria d'Italia*, 177.

34. Matteo Palmieri, *Libro del poema chiamato Città di Vita* (Northampton, Mass., 1927), vol. 1, book 1, cap. VI, l. 51.

35. Dante, *Divina Commedia*, Inferno, canto 34, ll. 73–93. Also, Lamberto Donati, "Il Botticelli e le prime illustrazioni della Divina Commedia," in *Bibliofilia* (Florence), 62 (1960), 205–289; 63 (1961), 1–72.

36. Paul Oscar Kristeller, *The Philosophy of Marsilio Ficino*, trans. V. Conant (New York, 1943), 353–362.

37. Arthur O. Lovejoy, *The Great Chain of Being* (Cambridge, Mass., 1953), 102.

38. The issue of anthropocentrism is not simple. It is still widely held that it was characteristic of thinking prior to man's displacement from the center of the universe in the sixteenth and seventeenth centuries. Koyré, Kuhn, and Foucault all argue thus, though with reservations. Some seventeenth-century writers were certainly conscious of man having lost centrality, and certain Renaissance writers such as Pico della Mirandola and Paracelsus propounded it. In whatever scheme of thought this centrality was maintained, however, it could not have been cosmic and spatial. See Alexander Koyré, *From the Closed World to the Infinite Universe* (New York, 1957), 1–2; Kuhn, *The Copernican Revolution,* 7; Michel Foucault, *The Order of Things* (London, 1970), 22.

39. Pico della Mirandola, *On the Dignity of Man,* chap. 5, in P. O. Kristeller, ed., *Renaissance Philosophy* (Chicago, 1948), 226.

40. To see how the ambiguous registration of inner and outer rings was vastly enhanced by perspective, Lomazzo's Foppa Chapel dome fresco, which made such effective use of the perspective from below, might then be compared to the earlier fresco in the cupola of Santa Maria dei Miracoli, Saronno (1535), by Gaudenzio Ferrari, Lomazzo's teacher, which had not yet quite relinquished frontality to the plane of representation (in this case the surface of the dome), and then this might be compared to the entirely frontal but otherwise almost identically arranged fourteenth-century fresco in the cupola of the baptistery at Padua by Giusto de' Menabuoi, where the angelic faces look as if they have been pasted onto the dome. For Menabuoi see A. Smart, *The Dawn of Italian Painting* (London, 1978), 139–40. For Ferrari see Jean Villette, *L'Ange dans l'art d'Occident du 12ème au 16ème Siècle* (Paris, 1940).

41. Abel Letalle, *Les Fresques du Campo Santo de Pise* (Paris, n.d.), 103–104; Mario Bucci and Licia Bertolini, *Camposanto monumentale di Pisa; affreschi e sinopie* (Pisa, 1960), 104–105.

42. John Pope-Hennessy, *Giovanni di Paolo* (London, 1937), 20–21; see also S. K. Heninger, *The Cosmographical Glass: Renaissance Diagrams of the Universe* (San Marino, 1977), 16–20. For other examples, see esp. Charles de Bouelles, *Libellus de Nichilo* (Paris, 1510), fol. 63, and Hartman Schedel, *Liber Chronicarum* (Nuremberg, 1493); also, P. Brieger, M. Meiss, and C. Singleton, eds., *Illuminated Manuscripts of the Divine Comedy,* 2 vols. (Princeton, n.d.), vol. 2, 434–510. Another way of illustrating this dual centrality can be seen in the panel of the Fall of the Angels by the Sienese Master of the Rebel Angels (first half of fourteenth century) now in the Louvre: an enthroned God emanating light at the top; a spherical lump of blackness at the bottom. The angels tumble from one to the other, metamorphosing into demons as they fall.

43. David Lindberg, "The Genesis of Kepler's Theory of Light: Light Metaphysics from Plotinus to Kepler," *Osiris,* 2d series, 2 (1986), 14–17.

44. Dante, *Divina Commedia, Paradiso,* canto 28, ll. 41–42 and 49–51:
*ma nel mondo sensibile si puote
veder le volte tanto più divine,
quant'elle son dal centro più remote.*

45. Kenneth Clark, *The Drawings of Sandro Botticelli for Dante's Divine Comedy* (London, 1976), 200–206.

46. Jacques Derrida, *Writing and Difference,* trans. Alan Bass (Chicago, 1978), 27.

47. The sublimated character of solar imagery in the Middle Ages was made much of in H. Flanders Dunbar, *Symbolism in Medieval Thought* (New Haven, 1929), esp. chap. 3, 105–239, though he interprets its sublimated presence very differently.

48. Gaston Halsberghe, *The Cult of Sol Invictus* (Leiden, 1972), 167–175.

49. Walter Lowrie, *Art in the Early Church* (New York, 1947), 62.

50. For the medieval iconography of these see A. N. Didron, *Christian Iconography* (London, 1851), and George Fergusson, *Signs and Symbols in Christian Art* (New York, 1952), 56, 59, 267–268. For their early use see André Grabar, *Christian Iconography: A Study of Its Origins* (Princeton, 1968), 128–133. For Gerolamo da Vicenza, see Davis, *National Gallery: Italian School Catalogue,* 168–169. The Mariological connection with cosmic shape was strong in painting as well as architecture. For the latter see Sinding-Larsen, "Some Functional and Iconographical Aspects of the Centralised Church," 219–226; Lotz, "Notes on the Centralised Churches," 67–69.

51. Fergusson, *Signs and Symbols,* 59.

52. The same detail was used also by Giuliano da Sangallo at Santa Maria delle Carceri, Prato, 1485.

53. David Knowles, *The Evolution of Medieval Thought* (London, 1962), 49.

54. Yates, *Giordano Bruno,* 44–83.

55. Marsilio Ficino, *De lumine,* trans. Sylvain Matton, in Albin Michel, ed., *Lumière et Cosmos* (Paris, 1981), 57, 61.

56. Cited by Kuhn, *The Copernican Revolution,* 130.

57. Marsilio Ficino, *De sole,* chap. 3, trans. A. B. Fallico and H. Shapiro, in *Renaissance Philosophy* (New York, 1967), vol. 1.

58. "In medio vero omnium resided sol. Quis enim in hoc pulcherrimo templo lampadem hanc in alio vel meliori loco poneret, quam unde rotum simul possit illuminare?" Nicholas Copernicus, *Complete Works,* ed. E. Rosen (Cracow, 1978), vol. 2, 22; *De revolutionibus,* book 1, chap. 10; also Alexandre Koyré, *The Astronomical Revolution* (London, 1973), 65.

59. The first publication of Copernicus's heliocentric theory was by his disciple Johannes Rheticus in the *Narratio*

prima, 1539. Koyré suggests that knowledge of Copernicus's "secret," which he claimed to have kept into a fourth term of nine years (nine years being the period of secrecy recommended by the Pythagoreans), could not have been extensive prior to 1539. The years 1510 to 1514 are assumed to have been formative for Copernicus. Koyré, *The Astronomical Revolution*, 27–28.

60. Ficino, *De sole*, chaps. I, XII.

61. According to Seznec the gesture of the figure in the oculus is one of "benediction and command." According to Luisa Becherucci, Shearman observes a "gesture of welcome," but in fact Shearman does not see so unequivocal an expression. He calls it a "receiving figure" and also cites Fischel's brilliant triple simile that has the figure "like a cloud passing over the eye of the Pantheon." Jean Seznec, *The Survival of the Pagan Gods* (New York, 1953), 80; L. Becherucci, in *Complete Works of Raphael* (New York, 1969), 183; John Shearman, "The Chigi Chapel in S. Maria del Popolo," *Journal of the Warburg and Courtauld Institutes*, 24 (1961), 138–140.

62. Ludwig Pastor, *History of the Popes*, ed. F. I. Antrobus, 5th ed. (London, 1950), vol. 6, 560–579, gives a good account of the *Disputà* in relation to papal sentiment.

63. Sir Charles Holmes, *Raphael and the Modern Use of the Classical Tradition* (London, 1933), 91.

64. Pastor, *History of the Popes*, 560.

65. Oskar Fischel, *Raphael*, trans. B. Rackham (London, 1948), vol. 1, 84; J. A. Crowe and G. B. Cavalcaselle, *Raphael, His Life and His Works*, (London, 1885), vol. 2, 29–37.

66. Heinrich Wölfflin, *Classic Art*, trans. P. and L. Murray (London, 1952), 89.

67. Ibid., 88.

68. Giorgio Vasari, *Le Vite*, ed. C. L. Ragghianti (Milan and Rome, 1951), vol. 2, 138. "Ma molto piu arte ed ingegno [Raffaello] mostrò ne' santi [e] Dottori cristiani, i quali a sei, a tre, a due disputano per la storia; si vede nelle cere loro una certa curiositá et un affanno nel voler trovare il certo di quel che stanno in dubbio, facendone segno col disputar con le mani e col far certi atti con la persona, con attenzione degli orecchi, con lo increspare delle ciglia e con lo stupire in molte diverse maniere, certo variate e proprie."

69. B. Neunheuser, *L'Eucharistie*, trans. A. Liefooghe (Paris 1966), vol. 2, 110.

70. Jaroslav Pelikan, *The Christian Tradition*, vol. 4: *Reformation of Church and Dogma* (Chicago, 1984), 57–59; and Kenneth Scott Latourette, *A History of Christianity*, 2 vols. (New York, 1975), 360–361, 530–532, 712.

71. H. von Geymüller, *Raffaello Sanzio studiato come architetto* (Milan, 1884); Clemens Sommer, "A New Interpretation of Raphael's Disputà," *Gazette des Beaux-Arts*, 6th series, 27 (New York, 1945), 289–296.

72. George Hersey, *The Lost Meaning of Classical Architecture* (Cambridge, Mass., 1988), 115–116.

73. Pelikan, *Reformation of Church and Dogma*, 201.

74. A rough estimate from calculations of the volume of masonry obtained from available sections of the church.

75. See for instance Michael Baxandall, *Giotto and the Orators* (Oxford, 1971), for one aspect of this; and Ernst Gombrich, "Norm and Form: The Stylistic Categories of Art History and Their Origins in Renaissance Ideals," in *Norm and Form* (London, 1978), 81–98, for another.

76. See chapter 6 below.

77. Euclid's definition of points, lines, and planes as bodyless elements that were nevertheless capable of describing the properties of real bodies in space was the source of immense fascination inside and outside mathematics. One connotation of the tetractys ($1 + 2 + 3 + 4 = 10$) was its correspondence to the development from a point (1), to a line (defined by 2 points), to a plane (defined by 3 points not on a line), to a volume (defined by 4 points not in a plane). Euclid, beginning his work with their definition, calls a point "that which has no part" and a line "a breadthless length"; T. Heath, *The Thirteen Books of Euclid's Elements* (New York, 1956), book I, defs. 1 and 2.

78. Pico's spokesman for the Gods says to man: "We have set thee at the World's center that thou mayest from thence more easily observe whatever is in the world. We have made thee neither of heaven nor earth. . . . Thou shalt have the power to degenerate into the lower forms of life, which are brutish. Thou shalt have the power, out of thy soul's judgement, to be reborn into the higher forms which are divine." Pico della Mirandola, *On the Dignity of Man*, trans. E. L. Forbes, in *The Renaissance Philosophy of Man*, ed. E. Cassirer, P. O. Kristeller, and J. H. Randall, Jr. (Chicago, 1948), 225.

79. "Sforzati un poco a trarne la materia, se tu puoi: tu la puoi trarre col pensiero. Orsù trai a lo edificio la materia e lascia sospeso lo ordine: non ti resterà di corpo materiale." Ficino, *Commentary on the Symposium*, in E. Panofsky, *Idea: A Concept in Art Theory* (New York, 1968), 136–137.

80. "And we can in our Thought and Imagination contrive perfect Forms of Buildings entirely separate from Matter" (*Et lecibit integras formas praescribere animo et mente seclusa omni materia*); L. B. Alberti, *The Ten Books of Architecture*, trans. G. Leoni, ed. J. Rykwert (London, 1965), book 1, chap. 1.

81. This is shown in two drawings circa 1560–1565, one made by Baldassare Peruzzi's son Sallustio (Uffizi, 635r, v), the other an unattributed drawing (Berlin, Kunstbibliothek, 105, vol. D, Destailleur Codex).

82. Heath, *Euclid's Elements*, book I, def. 3.

83. Ibid., book XI, 14, 16. Euclid's definition of the sphere was more elaborate and cumbersome. Heath notes that this was because it followed from his earlier definition of the circle; vol. 3, 269.

84. The development of spherical geometry—i.e., geometry of properties intrinsic to the surface of the sphere—only occurred in the nineteenth century with the work of Riemann. Kline's observations on the retarded exploration of this geometry, considering the quantity of veneration heaped on the sphere from antiquity through to modern times, are germane here. See Morris Kline, *Mathematics and Western Culture* (Harmondsworth, 1972), 478–482.

There are three ways to describe a circle without reference to a center: (1) a line of constant curvature in a plane surface, such as the track that would be produced by a bicycle with handlebars fixed at a given angle to the frame; (2) the locus of the vertices of two triangles divided by a common base (a chord of the resulting circle) whose angles, opposite the base, are constant and add up to 180°; (3) the locus of the intersection of two lines of constant ratio from two fixed points (neither of which will be at the center unless the ratio is infinitely large).

85. N. Cusanus, *Of Learned Ignorance* (London, 1954), 46ff.; see also Maurice de Gandillac, "Le Rôle du soleil dans la pensée de Nicolas de Cues," in *Le Soleil à la Renaissance*, Colloque International (Brussels, 1965), 341–361. The same paradoxical formula for the infinite sphere was given by Thomas Bradwardine (1290–1349); see S. Y. Edgerton, Jr., *The Renaissance Rediscovery of Linear Perspective* (New York, 1976), 19.

86. Copernicus, *Complete Works*, vol. 2, 8, 9, 21; and Edward Rosen, *Three Copernican Treatises* (New York, 1971), 58–59.

87. Kuhn, *The Copernican Revolution*, 59–64, 66–72; and Koyré, *The Astronomical Revolution*, 55–66.

88. D. P. Walker, *Studies in Musical Science in the Late Renaissance* (London, 1978), 55, citing a letter from Kepler to Joachim Tankius in J. Kepler, *Gesammelte Werke*, ed. M. Caspar et al. (Munich, 1938), vol. 16, 158, letter 493; also H. F. Cohen, *Quantifying Music: The Science of Music at the First Stage of the Scientific Revolution, 1580–1650* (Dordrecht, 1984), 19.

89. Alberti, *The Ten Books of Architecture*, trans. Leoni, book IV, chap. 4; Kristeller, *Renaissance Philosophy*, 129; George Perrigo Conger, *Theories of Macrocosms and Microcosms in the History of Philosophy* (New York, 1922), 56.

90. Heninger, *The Cosmographical Glass*. Not one diagram in this well-illustrated compendium admits to anything but simple circular revolution. The full cosmic diagrams published with Copernicus's work are central and circular, although most of *De revolutionibus* was devoted to the elaboration of a geometry that would enable the determination, through observation, of eccentric and epicyclic aberrations. Armillary spheres, likewise, often simplified the motions involved in modeling the heavens. This is the case with the most monstrous of all armillaries, made by Santucci for Duke Ferdinand I in about 1590. Museo di Storia della Scienza, Florence, *Catalogo degli strumenti* (1954), 47–49. According to John of Scarobosco, original compiler of the key European text on astronomy from the thirteenth to the sixteenth century, there were three reasons for the sphericity of the heavens: (1) it is like the archetype of forms; (2) it is convenient, because the sphere is the largest and the most capacious of isoperimetric bodies; (3) it is necessary, for under rotation any other figure would create vacant spaces, which are against nature. *The Sphere of Scarobosco and Its Commentaries*, ed. L. Thorndike (Chicago, 1949), 120.

91. Arnold Hauser, *The Social History of Art*, 4 vols. (London, 1962), vol. 2, 82–84. For Hauser the relation between form and cultural expression was not quite direct, despite his manifest prejudice in favor of the connection. What was expressed in the formal constitution of a building or painting was the motive for its production, deceitful or otherwise. While this allows the straightforward, but in this context peculiar, observation that "a society based on the idea of authority and submission will, naturally, favour the manifestation of discipline and order in art" (this of the High Renaissance), it also allows for the more oblique view that classical art "stylises the whole pattern of its life in accordance with [a] fictitious scheme."

92. L. B. Alberti, *Della famiglia*, translated by Guido A. Guarino (Lewisburg, 1971), 214. For the prehistory of this metaphor see David Summers, *The Judgment of Sense* (Cambridge, U.K., 1987), 92.

93. Wittkower, *Architectural Principles in the Age of Humanism*, 30. Another aspect of this has been pointed out by Argan, who quotes Vasari and Bruni to suggest that, from outside, Brunelleschi's Duomo was the pivot of a politically defined landscape. Bruni uses the inverted image ("come . . . direbbe uno poeto") of the moon and stars encircling us to convey the proper relation of town and country to a castle. A further inversion is thus made apparent, for the mundane political centrality of the Duomo landmark turns, as one enters the comparative smallness of its interior, into a larger, cosmic, centrality. G. C. Argan, "Il significato della cupola," in *Brunelleschi: la sua opera e il suo tempo* (Florence, 1980), vol. 1, 12–15.

94. Although the word "ambiguity" would allow for any number of associations or interpretations arising from the same source, twoness usually prevails. In the visual arts, duality is encouraged through the definition of visual ambiguity as a dimensional matter, such that flat things may appear three-dimensional or three-dimensional things flat, but literary discussions of ambiguity also lean toward duality. Thus, Empson's seventh and most ambiguous category of ambiguity "occurs when the two meanings of the word, the two values of the ambiguity, are the two opposite meanings defined by the context, so that the total effect is to show a fundamental division in the writer's mind." See William Empson, *Seven Types of Ambiguity* (London, 1930), 244ff. Clearly, the ambiguities of centrality in

Renaissance architecture are not ambiguities of double meaning but of multiple. The result one might reasonably expect from such multiplication is decreased coherence and/or decreased intensity of expression. Thus Brunelleschi, Bramante, Raphael, et al. maintained both coherence and intensity when these would most easily have been lost.

95. This brings to mind the observation that art resolves contradictions unresolvable outside art. More familiar in literary studies, especially studies of the novel, it has been a way of exploring social themes. In this architectural version, however, it extends above and below the social into concepts and things, evading the more obviously human context. The wistful character of the centralized buildings under consideration lies in their revealing a beautiful coherence obtained by subtractions and disengagements that inevitably compromise fidelity. Their order is also fictional, but the fiction is not employed to restore social stability and harmony after the excitement is over, as in a conventional plot; it is used to produce a numinous diffusion of all relations of power and authority into the multiple centers of an apparently unified formation. So if the characteristic novel, with its vital, human subject matter, develops an irregular sequence of events into an orderly finale, the centralized churches, with their massive deadweight of cold geometric forms, develop a representation of ordered, graded sequence into an experience of vicarious release from the very hierarchy that was their principal content. See Frank Kermode, *The Sense of an Ending: Studies in the Theory of Fiction* (Oxford, 1967). The classic examples of its application to visual matters must be Claude Lévi-Strauss, "A Native Community and Its Lifestyle," in *Tristes Tropiques* (Harmondsworth, 1976), and "Do Dual Organisations Exist?," in *Structural Anthropology* (London, 1968), 132–163. Something not unlike this has been observed of the Renaissance by Manfredo Tafuri in "Discordant Harmony from Alberti to Zuccari," *Architectural Design*, 49, no. 5–6 (1979; AD Profile 21), 36–44.

96. Samuel Y. Edgerton, Jr., *The Renaissance Rediscovery of Linear Perspective* (New York, 1976), 115–118; Joan Gadol, *Leon Battista Alberti* (Chicago, 1969), 157–195.

97. Gadol, *Leon Battista Alberti*, 178.

98. Windsor Drawings, 12284. K. Clark and C. Pedretti, *The Italian Drawings at Windsor Castle: The Drawings of Leonardo da Vinci*, 2d ed. (London, 1968), vol. 1.

99. "Con questo adunque misureremo ogni sorte di edificio di che forma si sia, o tondo, o quadro, o con strani angoli e svolgimenti quanto si voglia." English translation by Anne Murray de Fort-Menares, "La Lettera a Leone X," M.Phil. dissertation, University of Cambridge, 1987. Original text in Vincenzo Golzio, *Raffaello nei documenti* (Vatican, 1936), 88.

100. Gadol, *Leon Battista Alberti*, 187–188.

Chapter Two: Persistent Breakage

1. Douglas Cooper, *The Cubist Epoch* (New York, 1970), 265.

2. Jean-François Lyotard, *The Postmodern Condition*, trans. Geoff Bennington and Brian Massumi (Minneapolis, 1984), 81.

3. Sigfried Giedion, *Space, Time and Architecture*, 1st ed. (Cambridge, Mass., 1941), 402–403.

4. Ibid., 401. See also Albert Einstein, *Relativity: The Special and the General Theory* (London, 1920), chap. 9.

5. Though not always. For example, Umberto Boccioni and Robert Delaunay argued over who introduced the term, which they both understood to mean something quite different. See Raffaele Carrieri, *Futurism*, trans. L. van Rensselaer White (Milan, 1965), 123–124.

6. See Beverly Fazio, ed., *The Machine Age* (New York, 1986), endpapers.

7. Giedion, *Space, Time and Architecture*, 569–580.

8. Gertrude Stein, *Picasso* (London, 1938), 49.

9. Eunice Lipton, *Picasso Criticism: 1901–1939* (New York, 1975), 249ff.

10. Gelett Burgess, "The Wild Men of Paris," *Architectural Record*, 27 (1910), 400–414. It is possible that Braque was prompted to this metaphor by Burgess, who had joked about young artists' ambitions to show more than can be seen in 1901. See Linda Dalrymple Henderson, *The Fourth Dimension and Non-Euclidean Geometry in Modern Art* (Princeton, 1983), 182–183.

11. Cooper, *The Cubist Epoch*, 33.

12. Daniel-Henry Kahnweiler, *The Rise of Cubism*, trans. H. Aronson (New York, 1949), 10.

13. Henderson, *The Fourth Dimension*, 57–58, 72. Jouffret was a graduate of the Ecole Polytechnique, Paris (see chap. eight).

14. It is pointed out, quite correctly, that such ideas may be countered by disclaimers from the same original sources. Gleizes and Metzinger, for example, were "frankly amused to think that many a novice may perhaps pay for his too literal comprehension of the Cubist's theory . . . by painfully juxtaposing the six faces of a cube or the two ears of a model seen in profile." Albert Gleizes and Jean Metzinger, *Cubism* (London, 1913), part IV, quoted in Herschell B. Chipp, *Theories of Modern Art* (Berkeley, 1968), 124. See also Mark Roskill, *The Interpretation of Cubism* (Philadelphia, 1985), 37.

15. Chipp, *Theories of Modern Art*, 197.

16. Lucretius, *On the Nature of Things*, trans. H. A. J. Munro (New York, n.d., 94; C. S. Peirce, "A Note on Percepts" (1901), in *Philosophical Writings of Peirce*, ed. J. Buchler (New York, 1955), 308–309.

17. Chipp, *Theories of Modern Art*, 232. Both Gray and Gamwell trace this notion to philosophical idealism. See Christopher Gray, *Cubist Aesthetic Theories* (Baltimore, 1953), chap. 2; Lynn Gamwell, *Cubist Criticism* (Ann Arbor, 1977), 3–4.

18. Franco Russoli and Fiorella Minervino, *L'opera completa di Picasso cubista* (Milan, 1972), cat. 528–548.

19. Guillaume Apollinaire, *Les Peintres cubistes* (Paris, 1913), 79–82; Nancy Troy, *Modernism and the Decorative Arts in France* (New Haven, 1991).

20. For example, P. Reyner Banham, *Theory and Design in the First Machine Age* (London, 1960), 203.

21. A thorough itemization of cubist pictorial techniques is found in Winthrop Judkins, *Fluctuant Representation in Synthetic Cubism* (New York, 1976). Followers of Gombrich may be inclined to think ambiguity the real subject matter of cubism, but whether we defer to Gombrich or whether we are persuaded by the breathless rhetoric of cubism's early commentators who claimed it divulged the power of perception to construct reality, we would have to concede that the attempts to make architecture look like cubist painting whilst suggesting the same things as cubist painting were doomed from the outset.

22. Ivan Margolius, *Cubism in Architecture and the Applied Arts* (Newton Abbot, 1979).

23. For Schwitters's *Merzbau*, see: Rosemarie Haag Bletter, "Kurt Schwitters' Unfinished Rooms," *Progressive Architecture*, 58, no. 9 (September 1977), 97–99; Dietmar Elger, *Der Merzbau: Werkmonographie* (Cologne, 1984); John Elderfield, "Phantasmagoria and Dream Grotto," chap. 7 of *Kurt Schwitters* (London, 1958), 144–171; Dorothea Dietrich, "The Merzbau; or The Cathedral of Erotic Misery," chap. 8 of *The Collages of Kurt Schwitters: Tradition and Innovation* (Cambridge and New York, 1993), 164–205.

24. Giovanni Lista, *Futurism,* trans. C. C. Clark (New York, 1986), 48–49.

25. S. O. Khan-Magomedov, *Pionier der sowjetischen Architektur* (Dresden, 1983); published in English as *Pioneers of Soviet Architecture* (New York, 1987).

26. For the cubist impact on writing see Wendy Steiner, *The Colors of Rhetoric: Problems in the Relation between Modern Literature and Painting* (Chicago, 1982).

27. Amédée Ozenfant and Charles-Edouard Jeanneret, *Après le cubisme* (Paris, 1918), 57–58.

28. Timothy Hilton, *Picasso* (London, 1975), 148.

29. Colin Rowe and Robert Slutzky, "Transparency: Literal and Phenomenal," in Rowe, *The Mathematics of the Ideal Villa and Other Essays* (Cambridge, Mass., 1982), 160–176.

30. Allan Doig, *Theo van Doesburg: Painting into Architecture, Theory into Practice,* (Cambridge, U.K., 1986), esp. 104–105.

31. Winfried Nerdinger, *Walter Gropius* (Berlin, 1985), 46–47; John Willett, *The New Sobriety 1917–1933* (London, 1978), 50.

32. José Ortega y Gasset, *The Dehumanization of Art and Other Essays,* trans. Helene Weyl (Princeton, 1968), 3–54.

33. Sigfried Giedion, *Space, Time and Architecture,* 2d ed. (Cambridge, Mass., 1949), 453–492.

34. Alvar Aalto, "The Humanizing of Architecture," in *Sketches,* ed. Göran Schildt, trans. Stuart Wrede (Cambridge, Mass., 1985), 77.

35. Ibid., 79.

36. Alvar Aalto, "Between Humanism and Materialism," in *Sketches,* 132.

37. Alvar Aalto, "Speech for the Hundred Year Jubilee of the Jyväskylä Lycée," in ibid., 165.

38. "The Trout and the Mountain Stream," in ibid., 136; "The Arts," in *Alvar Aalto,* ed. Karl Fleig (Zurich, 1971), 12; "Abstract Art and Architecture," in Aalto, *Synopsis* (Basel, 1970), 3–20.

39. Gerd Hatje, ed., *Encyclopaedia of Modern Architecture* (London, 1963), 30.

40. Nikolaus Pevsner, *Outline of European Architecture* (Harmondsworth, 1963), 425; Giedion, *Space, Time and Architecture.*

41. Vincent Scully, *Modern Architecture* (New York, 1961), "Fragmentation and Continuity," 10–32; Robert Venturi, *Complexity and Contradiction in Architecture* (New York, 1966), chap. 10 "The Obligation toward the Difficult Whole."

42. Christian Norberg-Schulz, *Meaning in Western Architecture* (London, 1975), 218–220; for a discussion of Rowe see below.

43. Lewis Mumford, *The Condition of Man* (London, 1944), 391ff.; Kenneth Frampton, "Towards a Critical Regionalism," in Hal Foster, ed., *The Anti-Aesthetic* (Port Townsend, Wash., 1983), 16–30.

44. Bruno Zevi, *The Modern Language of Architecture* (Seattle, 1978), 7–13.

45. Aalto, "Between Humanism and Materialism," 132.

46. Ibid., 132.

47. W. H. Auden, *The Dyer's Hand and Other Essays* (New York, 1962), 6–7.

48. W. H. Auden, "Horae Canonicae," in *Collected Shorter Poems 1927–1957* (New York, 1966), 333, cited in A. Arblaster and S. Lukes, *The Good Society* (London, 1971), 390–393.

49. Ernst Gombrich, *Meditations on a Hobby Horse* (London, 1963), 40–41.

50. Claude Lévi-Strauss, *The Savage Mind* (Chicago, 1966), 13–16.

51. Ibid., 30.

52. Colin Rowe and Fred Koetter, *Collage City* (Cambridge, Mass., 1978), frontispiece. The project was designed by David Griffin and Hans Kolhoff.

53. Michel de Certeau, *The Practice of Everyday Life*, trans. Steven Rendall (Berkeley, 1984), 40, 66.

54. Rowe and Koetter, *Collage City*, 117.

55. Alexis de Tocqueville, *Democracy in America*, trans. Henry Reeve (London, 1862), vol. 2, 389.

56. Rowe, *Mathematics of the Ideal Villa*, 134.

57. Michel Foucault, *Discipline and Punish*, trans. Alan Sheridan (Harmondsworth, 1977), 293ff.

58. David Harvey, *The Condition of Postmodernity* (Oxford, 1989), 271.

59. Stephen Kern, *The Culture of Time and Space* (Cambridge, Mass., 1983), 303ff. See also note 1 above.

60. Bernard Delthil, "Les Halles," *Architecture* (Paris), 1 (January 1979), 5–10; Michel Ragon, "Le Concours des Halles," *Architecture* (Paris), 13 (March 1980), 34–39.

61. Edward Fry, *Cubism* (London, 1966), figs. 14 and 15, p. 18.

62. Picasso moved away from architectural subject matter; the Bauhaus painter Lyonel Feininger made it his specialty. He claimed he was incapable of the picturesque, the insipid detail of which he despised, and yet his work pulled him in that direction by mere fact of the combination of cubism and architecture. Museum of Modern Art, *Lyonel Feininger* (New York, 1944), 8, citing a letter of 1919.

63. Thus Frederick Gutheim defends Aalto's work in the way that all such work must be defended: "It is not an exploitation of the picturesque or romantic." Gutheim, *Alvar Aalto* (New York, 1960), 31.

64. Humphry Repton, *Fragments on the Theory and Practice of Landscape Gardening* (London, 1816; reprinted New York, 1967); G. Carter, P. Goode, K. Laurie, *Humphry Repton, Landscape Gardener 1752–1818* (London, 1982), 77–78.

65. R. Evans, "Il mito dell'informalità," *Il progetto domestico: saggi*, XVII Triennale di Milano, ed. Georges Teyssot (Milan, 1986), 88–93; "The Developed Surface," *9H* (London), 8 (1989), 120–147.

66. For example: Eisenman, Tschumi, Libeskind, Koolhaas, Hadid, Morphosis. Exceptions would be Gehry and Coop Himmelblau.

67. *Skyline: The Architecture and Design Review* (March 1982), 21–23.

68. "Deconstruction and the Arts," symposium at the Tate Gallery, 11 May 1988, discussion between Norris, Bann, Benjamin et al.

69. Bernard Tschumi, "Parc de La Villette," *AD*, 58, no. 3/4 (1988), 33, and in the same issue, Eisenman interview with Charles Jencks, 50–51.

70. "The Architecture of Theories" is the title of an essay by Charles Sanders Peirce in 1891 that provides an interesting comparison to Derrida's view; in *Philosophical Writings of Peirce*, 315–323.

71. Jacques Derrida, "Point de folie—maintenant l'architecture," trans. Kate Linker, *La Case Vide: La Villette, folio VIII* (London, 1985), chap. 8.

72. B. Tschumi, "La Case Vide," *AA Files*, 12 (1986), 66.

73. See note 71.

74. B. Tschumi, *Manhattan Transcripts* (London, 1981).

75. Roland Barthes, *Sade, Fourier, Loyola*, trans. R. Miller (New York, 1976), 122ff.; Michel Foucault, *Madness and Civilization: A History of Insanity in the Age of Reason* (London and New York, 1967), 279–289; Jacques Derrida, *Writing and Difference*, trans. Alan Bass (London, 1978), chaps. 6 and 8, "La Parole soufflée," "The Theatre of Cruelty," both on Artaud.

76. Derrida, *La Case Vide*, "Point de folie," chap. 9.

77. Jacques Derrida, "Why Peter Eisenman Writes Such Good Books," trans. Sarah Whiting, in Marco Diani and Catherine Ingraham, eds., *Restructuring Architectural Theory* (Evanston, Ill., 1989), 104. Derrida was also concerned with the architecture for the Collège International de Philosophie in Paris, and alludes to the subject also in his writing on the Tower of Babel. The latter prefaced an essay on translation in which JHWH is portrayed as a deconstructive God. The biblical story maintains the authority of the word. When the Babylonians were cursed with the confusion of tongues, construction of the tower ceased. Architecture can progress no further than language allows. This arcane assumption is not subjected to interrogation by Derrida. See his "Des Tours de Babel," in Joseph F. Graham, ed., *Difference in Translation* (Ithaca, 1985), 165–207. For the Collège see "Architetture ove il desiderio può abitare," interview with Eva Meyer, Paris, 1986, in *Domus* (April 1986), 18–24.

78. The point was made in an untitled preview of the MoMA show *Deconstructivist Architecture* by Diane Ghirardo, discussing the history of the exhibition and deconstruction as a critical strategy, in *Architectural Review*, 183 (1988), 6/6. For an indication of writing's permeation into human activity according to Derrida in 1967, see *Of Grammatology*, trans. G. C. Spivak (Baltimore, 1976), 9.

79. This is so of all Derrida's writing on painting in *The Truth in Painting*. The one interesting but relatively minor exception is when he uses examples of framing to attack the limits set by Kant on what can be considered purely-aesthetic ("Parergon," 37–82). He does this in order to deconstruct Kant's system, not at its base but at its summit. It is not so much the architecture of Kant's system as the peculiar power and privilege allotted to aesthetic vision that Derrida wants to undermine. It is a kind of vision that could not signify anything and that was therefore outside the bounds of writing as conceived by Der-

79. (cont.) rida. See also Jacques Derrida, *Edmund Husserl's Origin of Geometry: An Introduction*, trans. J. P. Leavey, Jr. (Lincoln, Neb., 1989). In this long essay the subject of geometry is only touched upon insofar as it is already incorporated into Husserl's writing about it. It loses all definition outside of that context.

80. A similar conclusion is reached by Jeff Kipnis in "/Twisting the Separatrix/," *Assemblage* 14 (1991), 46.

81. Jacques Derrida, *The Truth in Painting*, trans. G. Bennington and I. McLeod (Chicago, 1987), "+ R (into the bargain)," 149-182, and "Cartouches," 183-253.

82. Marguerite Duras, "The Blue Blood of La Villette" (1957), in *Outside: Selected Writings*, trans. Arthur Goldhammer (London, 1987), 41-46.

83. Denis Hollier, *Against Architecture: The Writings of Georges Bataille*, trans. Betsy Wing (Cambridge, Mass., 1989), xii-xv.

84. Michel Foucault, *The Order of Things* (London, 1970), preface, xvii-xix; see also his "Other Spaces: The Principles of Heterotopia," *Lotus International*, 48/49 (1986), 9-17, and Dimitri Porphyrios, *Sources of Modern Eclecticism* (London, 1982), 1-12.

85. Gilles Deleuze and Félix Guattari, *A Thousand Plateaus: Capitalism and Schizophrenia*, trans. Brian Massumi (Minneapolis, 1987), chaps. 1, 3, 6, 9, 14.

86. Michel Serres, *Hermes*, ed. J. V. Harari and D. F. Bell (Baltimore, 1982), chaps. 3, 4, 5, 9.

87. Lyotard, *The Postmodern Condition*, 82; David Carroll, *Paraesthetics* (New York, 1987), 37-43; Sigmund Freud, *The Interpretation of Dreams*, vol. 4, trans. James Strachey (Harmondsworth, 1976), 446.

88. Fredric Jameson, "Postmodernism and Consumer Society," in Foster, ed., *The Anti-Aesthetic* 118-125; Victor Burgin, "Paranoiac Space," *New Formations*, 12 (1991), 1-23.

89. Félix Guattari, "Space and Corporeality," paper delivered at Columbia University School of Architecture, 7 April 1990, 8.

90. Peter Dews, *Logics of Disintegration* (London, 1987), 224-226. The phrase is taken from Adorno's *Negative Dialectics*.

91. Diani and Ingraham, eds., *Restructuring Architectural Theory*, 102; see also Kipnis, "/Twisting the Separatrix/," 31-61, for a detailed account of the collaboration.

92. Mark Wigley, introductory essay in *Deconstructivist Architecture* (New York, 1988), 11.

93. The story was related by the curator, Andrew MacNair, in *Gordon Matta-Clark: A Retrospective*, ed. Mary Jane Jacob (Chicago, 1985), 96. It was recounted by Sorkin in his review of the MoMA show (see note 95).

94. Ibid. My description is based on inference from photographs and drawings. I did not see the work.

95. Michael Sorkin, "Decon Job," *Village Voice* (5 July 1988), 81-83; James Wines, "The Slippery Floor," in A. Papadakis, C. Cooke, and A. Benjamin, eds., *Deconstruction: Omnibus Volume* (London, 1989), 135-138, first published in *Stroll* (June 1988), 15-23. The contrast between Matta-Clark and the MoMA exhibitors was prompted by the almost simultaneous opening of the Matta-Clark retrospective at the Brooklyn Museum. I am grateful to Margi Reeve for these references and other material relating to the MoMA exhibition.

96. Jürgen Habermas, "Modernity—An Incomplete Project," *New German Critique*, 22 (Winter 1981), reprinted in Foster, ed., *The Anti-Aesthetic*, 3-15.

97. Leo Steinberg, *Other Criteria: Confrontations with 20th Century Art* (London, 1972), "The Algerian Woman and Picasso at Large," 125-235.

98. Ibid., 234.

99. Although, it should be added, the claim that visual art extends beyond the reach of language is a commonplace of art criticism and aesthetics.

100. Jean-François Lyotard, *Discours, figure* (Paris, 1978), 27-28, 271-279, fig. p. 276. The pertinent chapter is translated in Lyotard, *Driftworks*, ed. R. McKeon (New York, 1984), 57-68.

101. For example: Mary-Alice Dixon Hinson declares the cubist landscape the best reference map to take the place of the academic plan, in "Vico's Dictionary," *Volume Zero* (New York), 1 (1985), 10; Wendy Steiner seeks to continue the cubist inspiration of literature in *The Colors of Rhetoric*, "A Cubist Historiography," 177-221; as does Gregory Ulmer in "The Object of Post Criticism," in Foster, ed., The *Anti Aesthetic*, 83-110.

102. Alvaro Siza, *Poetic Profession* (Milan, 1986), 9.

103. Edward Robbins, *Why Architects Draw* (Cambridge, Mass., 1994), 284-293. I am grateful to the author for letting me read the manuscript of this book before publication.

104. The competition was won by Jørn Utzon, the architect, who eventually resigned, his professional abilities having been called in question. The consultant engineers, Ove Arup, carried on to complete the project, considerably enhancing their professional credibility.

105. E. M. Farrelly, "Songs of Innocence and Experience: Co-op-Himmelblau," *Architectural Review*, 180 (1986), 17-24; "Prix and Swiczinsky in Conversation with Alvin Boyarsky," *AA Files*, 19 (1990), 70-73; Wolf Prix, "On the Edge," *AD, Deconstruction III* (London, 1990), 55-71.

106. Prix, "On the Edge," 67.

107. Peter Pfankuch, *Hans Scharoun; Bauten, Entwürfe, Texte* (Berlin, 1974), 8.

108. See also Kenneth Frampton, "Genesis of the Philharmonie," *Architectural Design*, 35 (1965), 112.

109. Harvey, *The Condition of Postmodernity*, 314–323.

110. Jürgen Joedicke, entry for Hugo Häring, in Gerd Hatje, ed., *Encyclopedia of Modern Architecture* (London, 1963). I also thank Peter Blundell-Jones for his translation of Häring's 1934 essay on proportion.

111. Giedion did not mention either Scharoun or Häring in any edition of *Space, Time and Architecture*, but he condemned expressionism in architecture as both transitory and unhealthy (1st ed., 1941, p. 393).

112. Mark Jarzombek, "Mies van der Rohe's New National Gallery and the Problem of Context," *Assemblage*, 2 (1987), 33–43; Alan Balfour, *Berlin: The Politics of Order, 1737–1989* (New York, 1990), 214–218. Wilfried Wang pointed out to me the relation between Scharoun's Tiergarten site and Hitler's plan. I am also indebted to René Lotz, Polytechnic of Central London, 1991.

113. Wilfried Wang, lecture, Harvard University Graduate School of Design, Fall 1991.

114. R. S. Lanier, "Acoustics in the Round at the Berlin Philharmonie," *Architectural Forum*, 120 (May 1964), 85–105; Peter Blundell-Jones, *Hans Scharoun* (London, 1978), 36–37. Blundell-Jones has said that the acoustics consultant Lothar Cremer proclaimed Scharoun the best architect he had worked with because he would always respond to suggestion and alter his designs.

115. Julius Posener, "The Philharmonic Concert Hall, Berlin," *Architectural Review*, 135 (1964), 207.

116. It should be noted however, that far too much has been made of this unexceptional characteristic of Piranesi's architectural fantasies.

117. Posener, "The Philharmonic Concert Hall."

118. Posener, ibid.: "Man is to be the centre, vaulting round himself, skylike, his desire" (1919); "The orchestra and conductor stand spatially and optically in the very middle" (of the Philharmonie, 1963).

119. Cremer designed the acoustics of the Philharmonie for the performance of nineteenth-century romantic music. That is why the reverberation time is longer than usual in modern auditoria. See Lanier, "Acoustics in the Round," 102.

120. Foucault, *The Order of Things*, 387.

121. *AD, Deconstruction III*, 33.

122. Peter Eisenman, *House X* (New York, 1982), 24, 34.

123. Anthony Vidler, "The Building in Pain: The Body and Architecture in Post-Modern Culture," *AA Files*, 19 (1990), 3–10.

124. Arthur I. Miller, "Visualization Lost and Regained," in Judith Wechsler, ed., *On Aesthetics in Science* (Cambridge, Mass., 1978), 77; Steiner, *The Colors of Rhetoric*, 195–196; and Victor Erlich, *Russian Formalism: History—Doctrine*, 3d ed. (New Haven, 1981), 65.

125. Rowe and Koetter, *Collage City*, endpaper, title page, 13, 19, 119.

126. Wigley, *Deconstructivist Architecture*, 10.

127. Vidler, "The Building in Pain."

128. Jacques Lacan, "The Mirror Stage" (1949), in *Ecrits: A Selection*, trans. Alan Sheridan (London, 1977), 5.

129. Ortega y Gasset, *The Dehumanization of Art*, 3–54; Nicholas Berdyaev, *The Fate of Man in the Modern World* trans. D. A. Lowrie (London, 1935), chap. 2, "Dehumanization"; Karsten Harries, *The Broken Frame: 3 Lectures*, (Washington, D.C., 1989).

130. Hans Sedlmayr, *Art in Crisis: The Lost Centre* (London, 1957), 153.

131. Ibid., 239.

132. Ibid., 231–232.

133. Boethius, *The Consolation of Philosophy*, trans. Richard Green (Indianapolis, 1962). Alberti seemed to regard architecture this way at times. See Christine Smith, *Architecture in the Culture of Early Humanism: Ethics, Aesthetics, and Eloquence, 1400–1470* (New York, 1992), 12–16.

134. Ian Hacking, *The Taming of Chance* (Cambridge, U.K., 1990), 2.

135. Charles Jencks, letter, *Architectural Review*, 158 (1975), 322.

Chapter Three: Seeing through Paper

1. Republished in Wolfgang Lotz, *Studies in Italian Renaissance Architecture* (Cambridge, Mass., 1977), 1–65.

2. Karl Marx, in *The German Ideology*, picked on Raphael—presumably because he represented genius—as the example of an artist whose work was anyway "determined by the technical advances in art made before him, by the organization of society and the division of labour in his locality, and, finally, by the division of labour in all the countries with which his locality had intercourse." David McLellan, ed., *Karl Marx: Selected Writings* (Oxford, 1977), 189. Admitting the value of his refreshing iconoclasm, Marx's assessment of art as based entirely on existing demand and methods of production denies the artist any possibility of creating demand or making technical advances in the means of production, both of which Raphael did (and both of which were recognized by Marx in other circumstances). Marx immobilized the artist to give a brilliant animated picture of the artist's situation. I want to give back some movement to the figure of the artist in the broad context brought to life by Marx.

3. A. M. de Fort-Menares, "La lettera a Leone X," M.Phil. dissertation, Department of Architecture, Cambridge University, 1987, appendix (translation from letter text in V. Golzio, *Raffaello: nei documenti, nelle testimonianze dei contemporanei e nella letteratura del suo secolo* (Vatican City, 1936), 16.

4. Lotz, *Studies in Italian Renaissance Architecture*, 32.

5. Cited by Erwin Panofsky, *Idea: A Concept in Art Theory*, trans. J. J. S. Peake (New York, 1968), 49.

6. Sir Thomas L. Heath, ed. and trans., *Euclid: The Elements* (New York, 1956), Book 1, def. 4, pp. 153, 165–169.

7. John Dee, *The Mathematical Praeface to the Elements of Geometrie of Euclid of Megara* (London, 1570), aij(v). Dee's words are suggestive, but his Megethologia was an odd assortment of contemporary topics including measures of movement and time as well as space.

8. That projectors, and therefore projection, could nevertheless be devised with no reference to light was made apparent in the proto-projective layout for a sphinx in the Neues Museum, Berlin. Papyrus fragments show part of a plan outline (from the top), and front elevation, ready to be transferred to a rectangular block of stone. Perpendicular lines of chiseling, not perpendicular rays of light, would be the equivalent of projectors in the act of realization. These fragments were discussed by Erwin Panofsky in another context. See *Meaning in the Visual Arts* (New York, 1955), 58–61.

9. To appreciate the similarity, think of a perspective view of a system of orthographic projection (as in fig. 50). The parallel projectors would converge on a vanishing point (the infinitely distant eye, or light source), and the orthographic system would be indistinguishable from a perspective system.

10. Lotz, *Studies in Italian Renaissance Architecture*, 32.

11. Leone Battista Alberti, *On Painting*, trans. John R. Spencer (New Haven, 1966), 48.

12. The architecture of *The School of Athens*, often compared to Bramante's St. Peter's project, is more easily comparable to the layout of the complete scheme for the Villa Madama. The broad, well-lit foreground of the fresco is equivalent to the loggia of the villa. Then, in the painting and the building, a niched recess leads into a three-bay, barrel-vaulted passage, to a bright, circular space (in the fresco it is unclear whether this is domed or open), through another barrel-vaulted passage, via a rectangular open court to an arched opening in a wall.

13. Fort-Menares, "La lettera a Leone X," appendix, p. 17.

14. White and Panofsky (in his later writing on perspective) have established that many of the geometric schemata that we see behind perspective painting were being used prior to the invention of linear perspective, while the astounding realism of Flemish art was accomplished without its aid (which, incidentally, is not the same as saying that it was accomplished without the aid of geometry). They have shown that linear perspective was as much the effect as it was the cause of realism and spatial illusion. Because of their work it is easy for us now to recognize that the evolution of perspective geometry interlaces with intuitive developments of spatial illusion in painting. See John White, *The Birth and Rebirth of Pictorial Space* (London, 1957); Erwin Panofsky, "I primi lumi" (1952), in *Renaissance and Renascences in Western Art* (New York, 1965), 114–161.

15. Fort-Menares, "La lettera a Leone X," 15.

16. Vitruvius, *The Ten Books on Architecture*, trans. M. H. Morgan (New York, 1914), 14.

17. Fort-Menares, "La lettera a Leone X," 14.

18. Roberto Weiss, *The Renaissance Discovery of Classical Antiquity* (Oxford, 1969), 118.

19. Jacques Guillerme and Hélène Vérin, "The Archaeology of Section," *Perspecta*, no. 25 (1989), 226. I am obliged to Stanford Anderson for bringing the role of the section to my attention.

20. The collusion between drafting board and building has been nowhere so persistent as here, where the orthogonal framework of the drawings is similar to that of the buildings drawn. This was not inevitable. Albrecht Dürer, first to publish architectural plan, elevation, and section together in 1528, had already devised a projective method for drawing conic sections (ellipse, parabola, and hyperbola) that used plan and elevation to obtain sections of a cone. See R. Evans, "Projection," in E. Blau and E. Kaufman, eds., *Architecture and Its Image* (Montreal, 1989), 22–23; and Albrecht Dürer, *Underweysung der Messung* (Nuremberg, 1525), 33–37. Guillerme and Vérin describe the evolution of three mutually perpendicular types of drawing in the designing of ship's hulls too ("Archaeology of the Section," 240–253). While there are some common features (vertical, bilateral symmetry of hulls; axiality of cones) these nonrectilinear forms made use of the three drawings in a different way. The conic sections were oblique cuts showing simple plane profiles; the multiple cross sections of a hull were also structural profiles related by smooth curves in plan. In neither case were they pictorial sections of the sort that made so dramatic an impact on architectural representation. It may be that more complex forms could only be dealt with by this method if the potentially pictorial character of the projective section were suppressed (see below).

21. Fort-Menares, "La lettera a Leone X," 15.

22. This can be demonstrated by reference to a rare exception. Henri Labrouste's 1828 reconstruction of the Temple of Hera at Paestum flew in the face of received opinion. He interpreted the temple as a protomodern enterprise from an ingenious breakaway culture that was radically altering its Greek prototypes, a culture that was graduating from a religious to a secular stage of historical development. That was why the Poseidonians, as he called them, had been so bold as to break the spell of classical order. Labrouste thought he was reconstructing a stoa, not a temple. Processional axis and

hierarchical sequence were obliterated by a central row of columns that made the interior into a continuous circuit rather than an approach to a terminus. Labrouste forced attention on this by taking a conventional long section through the apex of the truss that cut straight through the columns, exaggerating the cancellation. It looks very strange. See Neil Levine, "The Romantic Idea of Architectural Legibility: H. Labrouste and the Neo-Grec," in Arthur Drexler, ed., *The Architecture of the École des Beaux Arts* (New York, 1977), 377–391.

23. Peter Blundell-Jones, "From the Neoclassical Axis to Aperspectival Space," *Architectural Review*, 183 (1988), 18–27, in particular note 14.

24. Uwe Brandes and Wilfried Wang have recently studied the drawings for the Philharmonie, but have not as yet published their observations on the subject.

25. The process is not unlike that employed to design hulls and fuselages. The crucial difference is that in these the sections correspond to structural ribs, which can be fabricated directly from the drawing, whereas in the Philharmonie they do not.

26. I intend, sometime, to publish the results of this fruitless labor. Only Leo Steinberg seems to suggest that the shape of the dome may not be as drawn in the accepted definitive Albertina (Vienna) drawing 173, though, in his extensive treatment of San Carlo's form, he never spells it out. Leo Steinberg, *Borromini's San Carlo alle Quattro Fontane* (New York, 1977), 77–93.

Chapter Four: Piero's Heads

1. For a discussion of this see W. J. T. Mitchell, *Iconology: Image, Text, Ideology* (Chicago, 1986), esp. 37–52.

2. David Summers, *The Judgment of Sense* (Cambridge, U.K., 1987), chap. 2.

3. Maurice Merleau-Ponty, *The Visible and the Invisible*, ed. C. Lefort, trans. A. Lingis (Evanston, Ill., 1968), 212.

4. Jacques Lacan, "Anamorphosis," in *The Four Fundamental Concepts of Psycho-Analysis*, ed. Jacques-Alain Miller, trans. Alan Sheridan (New York, 1977), 79–90.

5. Martin Jay, "The Empire of the Gaze: Foucault and the Denigration of Vision in 20th Century French Thought," in D. C. Hoy, ed., *Foucault: A Critical Reader* (Oxford, 1986).

6. Michel Foucault, *Discipline and Punish*, trans. Alan Sheridan (London, 1977), part 3, chap. 3.

7. Lacan, *The Four Fundamental Concepts*, 86. See also Hal Foster's introduction to *Vision and Visuality* (Seattle, 1988), x–xiv.

8. Erwin Panofsky, "Die Perspektive als 'symbolische Form,'" in *Vorträge der Bibliothek Warburg, 1924–5* (Leipzig and Berlin, 1927), reprinted as *Perspective as Symbolic Form*, trans. Christopher Wood (Cambridge, Mass., 1991). The background to this essay is discussed in M. A. Holly, *Panofsky and the Foundations of Art History* (Ithaca, 1984), 131–157.

9. Michael Kubovy, *The Psychology of Perspective and Renaissance Art* (Cambridge, U.K., 1986). Kubovy is particularly interested in what he calls the robustness of perspective eccentrically viewed and makes an excellent argument that Renaissance artists quickly understood and exploited this property. See also Richard Rosinski and James Farber, "Compensation for Viewing Point in the Perception of Pictorial Space," in M. Hagen, ed., *Alberti's Window: The Projective Model of Pictorial Information* (New York, 1980).

10. Nelson Goodman, *Languages of Art* (Indianapolis, 1976), 15–19; Kubovy, *The Psychology of Perspective*, 122–126.

11. Albrecht Dürer, *Underweysung der Messung* (Nuremberg, 1525), frontispiece, and pp. 183, 185.

12. A broader theory of optical perspective might take account of the gravitational deflection of light and/or the atmospheric modulation of light rays.

For a version of the latter see Walter Tape, "The Topology of Mirages," *Scientific American* (June 1985), 120–129.

13. See also Ernst Gombrich, "Standards of Truth," in W. J. T. Mitchell, ed., *The Language of Images* (Chicago, 1974), 195–197; John White, *The Birth and Rebirth of Pictorial Space* (London, 1957), chaps. 14, 15; Martin Kemp, "Perspective Rectified: Some Alternative Systems in the 19th Century," *AA Files*, 15 (1987), 30–34.

14. Euclid, *Liber de visu* (Ann Arbor, 1984). Euclid's second assumption was as follows: "The figure contained under the visual rays is a cone having its vertex in the eye but its base at the limits of the object viewed" (185).

15. Samuel Edgerton, *The Renaissance Rediscovery of Linear Perspective* (New York, 1975), 25.

16. Thomas Frangenberg, "The Image and the Moving Eye: Jean Pelerin to Guidobaldo del Monte," *Journal of the Warburg and Courtauld Institutes*, 49 (1986), 150.

17. A picture of a perspective with the observer included could be a perfectly legitimate orthographic projection of someone looking at a perspective picture set parallel to the plane of projection. The confusion arises when one thinks of the figure existing in the same space as the perspective scene, as is indicated by du Cerceau, who puts the observer's feet *in* the perspective. The picture is then impossible.

18. Leone Battista Alberti, *On Painting*, trans. J. R. Spencer (New Haven, 1966), 56.

19. Norman Bryson, "The Gaze in the Expanded Field," in H. Foster, ed., *Vision and Visuality*, 89–91.

20. Luigi Vagnetti, *Prospettiva: studi e documenti di architettura* (Florence), nos. 9–10 (March 1979), 41; and Kirsti Andersen, "The Problems of Scaling in Perspective Constructions," in *Proceedings of the Institute of the History of Science* (December 1985), 33.

21. Hubert Damisch, *L'Origine de la perspective* (Paris, 1987), 153, 312–315; Michel Foucault, *The Order of Things* (London, 1974), 3–16.

22. Vitruvius, *The Ten Books on Architecture*, trans. M. H. Morgan (Cambridge, Mass., 1914), book VII, preface, chap. 2.

23. Hubert Damisch, *Théorie du nuage. Pour une histoire de la peinture* (Paris, 1972).

24. Francis Ames-Lewis, *Drawing in Early Renaissance Italy* (New Haven, 1981).

25. Susanne Lang, "Brunelleschi's Panels," in Marisa Dalai Emiliani, ed., *La prospettiva rinascimentale: codificazione e trasgressione* (Florence, 1980), 70; Damisch, *L'Origine de la perspective*, chap. 12.

26. The fact that both paintings have identical figure compositions, except for a minor change in the posture of Mary Magdalen, means that one of the two must have been conceived with the figures prior to the architecture and landscape background, which alters quite drastically. According to the architecture, the horizon should be at a very low level in both versions, as it is shown in the Uffizi painting. But in the Dublin version the horizon is hoisted to the eye level of the virgin, which makes sense with regard to the figures but not with regard to the architecture. See Carlo Castellaneta, *L'opera completa del Perugino* (Milan, 1969), 93–94.

27. Jane Roberts, in *Leonardo Da Vinci* (London, 1989), 26.

28. Giorgio Vasari, *Le Vite*, ed. C. L. Ragghianti (Milan and Rome, 1951), preface to Part Three.

29. Alberti, *On Painting*, 48.

30. Lucretius, *On the Nature of Things*, book IV; cited by Summers, *The Judgment of Sense*, 45.

31. Manetti's description of Brunelleschi's panels, which he claims to have held in his own hands, stress the verisimilitude of the results. Manetti shows that Brunelleschi made every effort to increase the exactitude of correspondence. Thus the clouds would be seen to move behind the baptistery of San Giovanni because the image was painted on a mirror that reflected the sky. Antonio Manetti, *The Life of Brunelleschi*, ed. H. Saalman (University Park, Pa., 1970), 44–46.

32. M. Brion-Guerry, *Jean Pélerin Viator, sa place dans l'histoire de la perspective* (Paris, 1962), 217–227, gives a French translation of the 1505 Latin text.

33. The relation between eye and picture is an aftereffect of Pélerin's construction, not an ingredient as in Alberti's. Pélerin took advantage of the fact that by choosing, quite arbitrarily, a horizon line, a central vanishing point, a diagonal vanishing point, and the width of one tile, sufficient was determined to define a consistent measuring grid in perspective. Things can then be plotted into this foreshortened net, which will necessarily correspond to equal subdivisions seen from some real viewing point. Even the most unlikely-looking choices are correct. The combination of these elements also provides an exhaustive description of all possible viewpoints. For an appreciative account see William M. Ivins, *On the Rationalization of Sight* (New York, 1938).

34. Lise Bek, *Towards Paradise on Earth: Modern Space Conception in Architecture a Creation of Renaissance Humanism* (Odense, 1980), part II, 95–163.

35. Leone Battista Alberti, *The Ten Books on Architecture*, ed. Joseph Rykwert (London, 1965), book V, chap. 1, and book IV, chap. 5.

36. Bek, *Towards Paradise on Earth*, 123.

37. In writing this and the following sections I consulted the notes and materials for a book on Piero planned by the late Roger Jones. I wish to record my debt to him, and my thanks to Mary Wall for her generosity in making them available.

38. Giorgio Vasari, *The Lives of the Artists*, trans. George Bull (Harmondsworth, 1965), 251.

39. Bernard Berenson, *Italian Painters of the Renaissance* (Oxford, 1930), 135.

40. Kenneth Clark, *Piero della Francesca* (London, 1969), 75.

41. Luca Pacioli, *De divina proportione*, facsimile of 1498 manuscript (Milan, 1956); Margaret Daly Davis, *Piero della Francesca's Mathematical Treatises* (Ravenna, 1977), 11.

42. M. D. Davis, "Carpaccio and the Perspective of Regular Bodies," in Dalai Emiliani, ed., *La prospettiva rinascimentale*, 196.

43. Roberto Longhi, cited in Pierluigi de Vecchi, *The Complete Paintings of Piero della Francesca* (London, 1970), 11.

44. For example: Lionello Venturi, *Piero della Francesca*, trans. J. Emmons (Geneva, 1954), 14 ("In any number of his paintings we find that each figure and object aspires to the state of a geometrical form"); Millard Meiss, *The Great Age of Fresco* (London, 1970), 139 ("Piero has given these volumes [women] an exceptionally high degree of stereometric abstraction.... Late in life, some 20 years after painting the fresco [at Arezzo] he wrote a treatise in which he explored the Platonic concept that the whole complex of appearances is reducible to five geometric forms which have a divine perfection. Here in his painting he affirmed a belief in a rational world given shape and structure by geometry and number").

45. Davis, *Piero della Francesca's Mathematical Treatises*, chap. 3.

46. Berenson, *Italian Painters of the Renaissance*, 134–138.

47. Michael Baxandall, *Piero della Francesca* (London, 1966), 6.

48. Creighton Gilbert, "On Subject and Not Subject in Italian Renaissance Pictures," *Art Bulletin*, 34 (1952), 202–216.

49. The following are taken from a summary of the literature by Roger Jones: C. Gilbert, "Piero della Francesca's Flagellation: The Figures in the Fore-

ground," *Art Bulletin*, 53 (1971), 45–51; Thalia Gouma Peterson, "Piero della Francesca's Flagellation: An Historical Interpretation," *Storia dell'Arte*, 28 (1976), 217–233; Ludovico Borgo, "New Questions for Piero's Flagellation," *Burlington Magazine*, 121 (1979), 547–553; J. Hoffman, "A Reading from Jewish History," *Zeitschrift für Kunstgeschichte* (1981), 340–357. See also J. Pope-Hennessy, "Whose Flagellation?," *Apollo* (September 1986), 162–165.

50. For instance, Carlo Ginsburg, *The Enigma of Piero*, trans. M. Ryle and K. Soper (London, 1985), who is particularly dependent on Lavin's insight while being particularly unforgiving of her faults.

51. Peter Murray, introduction to de Vecchi, *Complete Paintings of Piero della Francesca*, 6.

52. For recent attempts to interpret the figures in the foreground see note 49. For attempts to divine the geometry in the perspective architecture of the painting, see: R. Wittkower and B. A. R. Carter, "The Perspective of Piero della Francesca's 'Flagellation,'" *Journal of the Warburg and Courtauld Institutes*, 16 (1953), 292–302; Fernando Casalini, "Corrispondenza fra teoria e pratica nell'opera di Piero della Francesca," in *L'Arte*, n.s. 1 (1968) 62–95; Warman Welliver, "The Symbolic Architecture of Domenico Veneziano and Piero della Francesca," *Art Quarterly*, 36 (1973), 1–30; Ruth B. Warner Brocone, "An Analysis of the Perspective in Piero della Francesca's *Flagellation*," in *The Oakland Review* (Rochester, Mich.), 1 (1968), 37–50; Corrado Verga, "L'architettura nella 'Flagellazione' di Urbino," *Critica d'Arte* (1976–1977), 7–19, 31–44, 52–59; Eugenio Battisti, *Piero della Francesca* (Milan, 1971), vol. 1, 326–330.

53. Vasari, *The Lives of the Artists*, 196.

54. Alberti, *On Painting*, 72.

55. Ibid.

56. Joseph Meder, *Die Handzeichnung* (Vienna, 1919); Lawrence Wright, *Perspective in Perspective* (London, 1983), 100–101.

57. Peter J. Booker, in his seminal *A History of Engineering Drawing* (London, 1963), makes a distinction between the secondary geometry of perspective, dependent on the convergence of parallels, and the primary geometry of perspective, dependent on the mapping of sight lines from eye to object. The primary geometry, is what we find in Piero's Other Method. Booker, after dispensing with the claim that the method from primary geometry was invented by Gaspard Monge at the end of the eighteenth century, traces its first appearance to Samuel Marolois's *Perspective* (Amsterdam, 1629[sic]), 31–34. Booker seems to have been unaware of this aspect of Piero's treatise, as also was René Taton; see his *L'Histoire de la géometrie descriptive* (Paris, 1954).

58. M. Boskovits, "Quello che i dipintori oggi dicono prospettiva, parte 2," in *Acta Historiae Artium Academicae Scientarium Hungaricae*, 9, nos. 1–2 (1963), 144, 147.

59. Piero della Francesca, *De prospectiva pingendi*, ed. G. Nicco-Fasola (Florence, 1984), 64.

60. He does so as follows: he begins by establishing that perspective diminutions of any line parallel to the picture plane will be in a ratio expressed as the distance between the eye and the picture divided by the distance between the eye and the line. Then he says: "If over a straight line divided in many parts, another line equidistant from it is drawn, and from the division of the first line, other lines are drawn to meet in a point, they will divide the equidistant line in proportion with the given line [*Sopra a la recta linea data in più parti devisa, se un'altra linea equidistante a quella se mena et da la divisioni de la prima se tira linee che terminino ad un puncto, devidaranno la equidistante in una proportione che è la linea data*]." He then gives a simple proof of this uncontentious theorem, showing—without drawing attention to it—that his earlier definition of proportional diminution is identical to the convergence of orthogonals into a single point. Lawrence Wright observed that Piero never made use of oblique vanishing points in his drawings. He only used convergence to foreshorten lines parallel to the picture plane. Piero della Francesca, *De prospectiva pingendi*, 70; also Wright, *Perspective in Perspective*, 78.

61. "Ma perchè hora in questo terzo intendo tractare de le degradationi de corpi compresi da diverse superficie et diversamente posti, però avendo a tractare de corpi più deficili, pigliarò altra via et altro modo nelle loro degradationi, che non ò facto nelle dimostrationi passate; ma nello effecto sirà una cosa medesima, e quello che fa l'uno fa l'altro. Ma per due cagioni mutarò l'ordine passato; l'uno è perchè sirà più facile nel dimostrare et nello intendere; l'altro si è per la gran multitudine de linee, che in essi corpi bisognaria de fare seguendo il modo primo, sì che l'occhio et l'intellecto abagliaria in esse linee, senza le quali tali corpi non se possono in perfetione degradare, nè senza gran deficultà." Piero della Francesca, *De prospectiva pingendi*, 129.

62. Lutes and violas turned up in several later perspective treatises expounding the Other Method, for example Lorenzo Sirigatti, *La pratica de prospettiva* (Venice, 1596); Salomon de Caus, *La Perspective avec la raison des ombres et miroirs* (London, 1612); Pietro Accolti, *Lo inganno degl'occhi* (Florence, 1625).

63. See Frederic G. Higbee's introduction to selected passages from *De prospectiva pingendi*, in Elizabeth G. Holt, *A Documentary History of Art*, vol. 2 (New York, 1957), 254–256; also Franco Ghione, "Breve introduzione sul contenuto matematico del 'De prospectiva pingendi' di Piero della Francesca," in Nicco-Fasola's edition of *De prospectiva pingendi*, xxix.

64. In this respect Piero's Other Method is the opposite of Pélerin's tiers point method; Pélerin's is exclusively perspective projection, Piero's exclusively orthographic projection.

65. "Il cubo dato posante sopra ad uno suo angulo, et che nisuno suo lato sia equi-

distante al termine posto, proportionalmente degradare." Piero della Francesca, *De prospectiva pingendi*, 145.

66. Booker, *A History of Engineering Drawing*, 138–140. Booker notes that Dürer understood this method of projective rotation. It is also found in Hans Lencker, *Perspectiva* (Nuremberg, 1571), ix, xvi, xviii. Dürer may either have learnt it from Piero indirectly or devised his own version of it. The former seems more likely since there are many things in Dürer's drawings that resemble Piero's studies.

67. Piero's techniques of mapping and moving objects in perspective space are similar to those now made familiar in computer graphics. The computer takes the labor out of the reiterative procedures for mapping points. This does not mean we are about to enter a new era of Francescan sensibility. The relation between Piero and computer graphics is probably much the same as the relation between Vermeer and photography. Vermeer may have made use of the *camera obscura*. We get reminders of this in some photographs, but photographs do not look like Vermeer paintings.

68. ". . . profili et contorni proportionalmente posti nei luoghi loro." Piero della Francesca, *De prospectiva pingendi*, 63.

69. Leone Battista Alberti, *On Painting and On Sculpture*, Latin texts ed. with trans. by Cecil Grayson (London, 1972), 129; also Joan Gadol, *Leon Battista Alberti: Universal Man of the Early Renaissance* (Chicago, 1969), 76–81.

70. Alberti, *On Painting and On Sculpture*, 133.

71. Piero della Francesca, *De prospectiva pingendi*, figs. lxviii, lxix.

72. For examples of *mazzocchi* and similar figures by the Other Method, see: Daniele Barbaro, *La pratica della perspettiva* (Venice, 1569), 118–128; Peter Halt, *Perspectivische Reiss Kunst* (1625), plates 158–164; Wenzel Jamnizer, *Perspectiva corporum regularium* (Nuremberg, 1568; reprint, ed. A. Flocon, 1964), I.i–I iii. Jamnizer did not show his construction, but from his brief introductory statement one may infer that the Other Method was likely. See also Davis, "Carpaccio and the Perspective of Regular Bodies."

73. Sigfried Giedion, *Space, Time and Architecture* (Cambridge, Mass., 1941), 57.

74. Thomas Martone, "Piero e la prospettiva dell'intelletto," in *Piero: teorico dell'arte*, ed. O. Calabrese (Rome, 1984), 183. As to outlining, the frescos at Arezzo are remarkable for having no painted outline. Close inspection reveals that the contours were carefully marked by pricking into the damp plaster. Similar pricked outlines have come to light during restoration of the Nativity. For the frescos, see: Russell Cowles, "The Frescos of Piero della Francesca," *Journal of the American Institute of Architects*, 4 (1916), 274; also Bruce Cole, *The Renaissance Artist at Work* (London, 1893), 76–95. For the Nativity see National Gallery, London, "Dossier on Piero's Nativity," conservation record, vol. 6. H. Ruhemann discovered the prick holes in 1950.

75. Decio Gioseffi, "Introduzione all'arte, introduzione a Piero," *Arte in Friuli, Arte a Trieste*, no. 4 (1980), 9–25, fig. 5 on p. 144. Gioseffi mixed perspective and orthographic views of Piero's head, overlooking the difference. This is easy to do when the angle of vision subtended by a head is 12°, as in the treatise. When it is $1\frac{1}{2}°$, as in the fresco, the difference between perspective and orthographic projection is infinitesimal. That is why Jean Cousin later proposed that painters could use orthographic projection instead of perspective for human figures. See Jean Cousin, *Livre de pourtraiture* (Paris, 1618).

76. See also note 14 of chapter 3.

77. See Brocone, "An Analysis of the Perspective," 37.

78. The Brera barrel vault is similar to the last example in the third book of *De prospectiva Pingendi*, figs. lxxiii–lxxvii.

79. Battisti, *Piero della Francesca*, 440, thinks it his last work; C. Gilbert, *Change in Piero della Francesca* (Locust Valley, N.Y., [1968]), xiv, 115, thinks it late (c. 1475).

80. In 1942 G. Nicco-Fasola pointed out that the bench and bed in *The Dream of Constantine* were slightly out of alignment with the picture plane, which he regards as a compositional ploy. Its slight measurable disparity is, however, beyond the range of normal discernment; see Nicco-Fasola's introduction to Piero della Francesca, *De prospectiva pingendi*, 53–54.

81. Paul Barolsky, "Piero's Native Wit," source unknown.

82. Prolongation of the bottom and top edges indicates a vanishing point some way above the horizon line. This could be an accident (again, it is not visually discernible without the superimposed construction). This means that the roof tilts upward as it recedes to the right; its edge is not quite horizontal.

83. Nicco-Fasola, however, regards it as deliberately made very slightly out of line. See note 80.

84. Caroline Feudale, "The Iconography of the Madonna del Parto," *Marsyas*, 7 (1954–1957), 8–24.

85. The *Battle between Constantine and Maxentius* is very badly damaged, and the *Battle between Heraclius and Chosroes* is generally thought to be the work of assistants and pupils.

86. Ernst Gombrich, "Leonardo's Method for Working out Compositions" (1952), in *Norm and Form* (London, 1966), 58–63.

87. Vasari, *The Lives of the Artists*, 95–104.

88. The crossing of the two vanishing points is not a mannerism in *Subsiding of the Flood*, but results from the two long floating arks, between which the view is cast, not being quite parallel to each other.

89. Ashmolean Museum, Oxford, *Paolo Uccello's Hunt in the Forest* (Oxford, 1981).

90. John Pope-Hennessy, *Paolo Uccello: The Rout of San Romano* (London, 1952), 5. A revised order for the panels is given by Umberto Baldini, "Restauri di dipinti fiorentini," *Bollettino d'Arte* (July/September 1954).

91. In the London panel, 9 of the 13 lances converge approximately on one spot (or patch!) near the center of the picture; in the Paris panel no three converge on one point.

92. Vasari, *The Lives of the Artists*, 96.

93. See note 72, and Piero della Francesca, *De prospectiva pingendi*, 138–145, figs. xlix-li.

94. The asymmetry of ring patterns projected at an angle to the picture plane was illustrated by Piero, *De prospectiva pingendi*, figs. lvi-lviii. In an essay, Marianne Shin (Graduate School of Design, Harvard, 1990) showed that several partially hidden horses in the panels are made of visible parts that do not add up properly.

95. John Pope-Hennessy, *The Complete Work of Paolo Uccello* (London, 1950), 11.

96. Vasari, *The Lives of the Artists*, 96.

97. Richard Krautheimer, *Lorenzo Ghiberti* (Princeton, 1956), 236–240.

98. Manetti, *The Life of Brunelleschi*, 52–53.

99. Vasari, *The Lives of the Artists*, 136.

100. Damisch, *L'Origine de la perspective*, 113–121; Edgerton, *The Renaissance Rediscovery of Linear Perspective*, 124–150.

101. How many and how long depends on the visual angle subtended in the perspective. As Martin Kemp points out, there is no way to decide this from the evidence we have. But even with the widest possible angle (approximately 90°) they would not be prominent, less so in fact than shown by Kemp, since the two lateral ranges of building along the piazza were not built parallel, though drawn that way in Kemp's diagram. See Lang, "Brunelleschi's Panels"; also Martin Kemp, "Science, Non-science and Nonsense: The Interpretation of Brunelleschi's Perspective," *Art History*, 1, no. 2 (June 1978), 134–146, esp. fig. 3.

102. Here scholars have found themselves in a quandary. The plausible suggestion that Piero was responsible for much of the ducal palace at Urbino, either directly or as advisor to Laurana, backed up by stylistic evidence, has not produced one scrap of supporting evidence from documentary sources. See Mario Salmi, *Piero della Francesca e il Palazzo Ducale di Urbino* (Florence, 1945); Pasquale Rotundi, *The Ducal Palace of Urbino* (London, 1969), 9–10.

Chapter Five: Drawn Stone

1. I am indebted to Giorgio Cucci for information on seventeenth-century use. In the eighteenth century, La Rue used the word *stéréotomie* not to describe the *traits* but as title for a supplementary essay dealing at a more abstract and generalized level with the geometry involved. Likewise Frézier divided his work into a preliminary volume on *stéréotomie* in general, and two further volumes on *timotechnie* which applied *stéréotomie* to the particular circumstances of stonecutting. See Jean Baptiste de La Rue, *Traité de la coupe des pierres* (Paris, 1728), 167–183; and A. F. Frézier, *La Théorie et la pratique de la coupe des pierres*, 3 vols. (Strasbourg, 1737–1739).

2. See for example François Derand, *L'Architecture des voûtes* (Paris, 1643), i, ii, and 97–153, esp. 137–139; La Rue, *Traité de la coupe des pierres,* plate xlvi, pp. 93–95; Frézier, *La Théorie et la pratique de la coupe des pierres*, vol. 2, 265; Jean Rondelet, *Traité théorique et pratique de l'art de bâtir* (Paris, 1838), vol. 4, plates 30–58.

3. This is dealt with by Wolfgang Herrmann in *Laugier and Eighteenth Century French Theory* (London, 1962), 74–78, 237–242; and Marc Grignon in "Pozzo, Blondel, and the Structure of the Supplement," *Assemblage*, no. 2 (1987), 97–109.

4. Jean-Marie Pérouse de Montclos, *L'Architecture à la française* (Paris, 1982). The greater part of this remarkable work is concerned with the development and persistence of stereotomic masonry in French architecture.

5. See Herrmann, *Laugier and Eighteenth Century French Theory*; Pérouse de Montclos, *L'Architecture à la française*, 183, 222, 224; W. Simonin, *Traité Elémentaire de la coupe des pierres* (Paris, 1792), 3.

6. Philibert Delorme, *L'Architecture* (1981), 91r. The first edition of Delorme's *Premier tome de l'architecture* was published in Paris in 1567. This was reprinted together with the earlier *Nouvelles inventions pour bien bastir* (1561) in 1568. The 1648 integrated edition has been reprinted by Pierre Mardaga as *Architecture de Philibert de l'Orme* (Brussels, 1981); the above and all subsequent references are to the Mardaga edition.

7. Ibid., Bekaert's introduction, xxii.

8. Ibid., 90r, v.

9. Anthony Blunt, *Philibert de l'Orme* (London, 1958), 19–20.

10. It was Delorme, however, who emphasized this more than any of the later authors. See Delorme, *L'Architecture*, 65r–67r; Derand, *L'Architecture des voûtes*, preface, avis VIII.

11. ". . . ie fus redigé en grande perplexité, car ie ne pouvois trouver le dit cabinet sans gaster le logis & les chambres, qui estoient faites suyvant les vieux fondemens & autres murs, que l'on avoit commencez premier que i'y fusse. Or qu'advint-il?" Delorme, *L'Architecture*, 88r.

12. Ibid.

13. Most of the palace was demolished after the French Revolution. Blunt identified a drawing, signed Barbier and dated 1698, in the Bibliothèque Nationale, Paris, that shows the garden facade in elevation. Blunt, *Philibert de l'Orme*, 30, note 1, and plate 5.

14. It is by no means certain that this bending and distortion of the balustrade motif took place. The Barbier drawing (see note 13) shows the *trompe* just cutting into the base molding, which anyway continues horizontally round this part of the cabinet as it does around the rest, while in similar *trompes*, such as that beneath the organ loft at La Flèche (Jacques Nadreau?, 1636), the *trompe* is set decisively beneath the balustrading.

15. Delorme, *L'Architecture*, 166r.

16. As built, these seven wedges were no doubt subdivided into blocks of a more convenient size. In all later examples the wedges do not taper into sharp chisel ends, but are truncated and replaced at the point of convergence with a single stone, called a *trompillon*. Nothing of the kind is indicated by Delorme, but he probably omitted it for the sake of simplicity and clarity in the demonstration.

17. Edward Dobson, *Masonry and Stonecutting*, 12th ed. (London, 1903), 85ff.; and George Barham, *Masonry* (London, 1914), 15–29.

18. Delorme, *L'Architecture*, 22v–24v. See Massimo Scolari, "L'idea di modello," *Eidos* (Asolo), 2 (1988), 22, for a proposal that Delorme did use models this way.

19. John Shearman, *Mannerism* (Harmondsworth, 1967), 81–91.

20. The biblical story does not correspond with the fresco. In the former, she did not come of her own accord; she was taken. David had Bathsheba, wife of Uriah, brought to him by his servants, laid with her, and made her pregnant (*2 Samuel* 11:2–5).

21. David Summers, *Michelangelo and the Language of Art* (Princeton, 1981), 82–84.

22. The words used by Vasari to describe the Palazzo Sacchetti frescos. See Giorgio Vasari, *Lives of the Artists*, vol. 2, trans. George Bull (Harmondsworth, 1987), 297.

23. The date 1720–1739 is given by Joan Evans. Hautecoeur had it as 1770. See Joan Evans, *Monastic Architecture in France from the Renaissance to the Revolution* (Cambridge, U.K., 1964), 49; Louis Hautecoeur, *Histoire d'architecture classique en France* (Paris, 1950–1953), vol. 3, 383. Neither gives specific evidence, but Evans's supporting evidence of other, similar contemporary works is convincing.

24. A tabulation of examples is given in Pérouse de Montclos, *L'Architecture à la française*, 198.

25. Michel Gallet, Monique Mosser, and Daniel Rabreau, *Charles de Wailly: peintre architecte dans l'Europe des Lumières* (Paris, 1979), cat. 176–184.

26. Ibid., cat. 184, lists plans and elevations in the Archives Nationale, Paris.

27. Lomazzo (*Trattato*, 1584) discussed the serpentine line under the heading of proportion. He presented it as a trick played by the painter against the true but unedifying proportions of the flat surface on which he paints. See Elizabeth G. Holt, *A Documentary History of Art* (New York, 1958), vol. 2, 78; see also Anthony Blunt, *Artistic Theory in Italy 1450–1600* (Oxford, 1956), 144–145.

28. Vasari, *Lives of the Artists*, 290.

29. ". . . comme si on dépouilloit la figure proposée de la robe dont elle est enveloppée." Frézier, *La Théorie et la pratique de la coupe des pierres*, vol. 1, 12; also cited by Pérouse de Montclos, *L'Architecture à la française*, 90.

30. James Hall, *Dictionary of Subjects and Symbols in Art* (London, 1974), 171.

31. There are at least two other paintings by Goltzius with this same diagonal square format. One of these, the *Man of Sorrows*, 1616 (Marienkirche, Ültzen, Hanover), is very similar to *Job in Distress*. O. Hirschmann, *Hendrick Goltzius als Maler: 1600–1617* (The Hague, 1916), cat. 26, p. 81.

32. See also note 27.

33. ". . . que premier ils ne tirent sur une ligne droicte, une autre perpendiculaire, ou traict d'équierre (comme l'appellent les ouvriers) soit simplement, ou dedans la circonference d'un cercle. Ils y peuvent semblablement, proceder par deux lignes parallelles, pourveu que tousjours au bout d'icelles, ou bien au milieu, on en tire une perpendiculaire. On peut aussi tirer la ligne perpendiculaire sur le bout de la ligne droicte. . . ." Delorme, *L'Architecture*, 31v.

34. "Paravanture pour figurer que la vie & le salut devoit advenir aux hommes, par la mort d'un seul mediateur Iesus Christ, qui seroit attaché au bois, portant figure de Croix, quie est la premiere que Dieu son pere a figuré au monde. Mais nous laisserons tels propos aux Theologiens, & reprendrons nos lignes & traicts de Geometrie, en tant que l'Architecte s'en peut ayder." Ibid., 32v.

35. Joseph Rykwert, "On the Oral Transmission of Architectural Theory," *AA Files*, no. 6 (Spring 1984), 15–27.

36. Delorme, *L'Architecture*, 62r.

37. It should be said that he did refer to projection twice, if only indirectly. However, neither of these references had to do with architectural drawing as such. The first suggested (inconclusively) knowledge of central projection for making architectural views; the second described a method for making sundials that involved parallel projection. See Vitruvius, *The Ten Books on Architecture*, trans. Morris Hicky Morgan (Cambridge, Mass., 1914), 14, 270–272.

38. Douglas Knoop and G. P. Jones, *A Short History of Freemasonry to 1730* (Manchester, 1940), 30–31.

39. Another, later ancient geometer, Claudius Ptolemy, could conceivably

have been a source of the trait, but that, for various reasons, would rule out a Gothic origin for the technique, and anyway his name is never mentioned in connection with it. I would tentatively agree with Pérouse de Montclos, who concludes that the renewed study of Euclidean geometry in the French universities in the sixteenth century was unconnected to the contemporary development of the *trait;* and with Shelby, who regards the practical geometrical techniques of Gothic masons as autonomous from mathematical geometry. See Pérouse de Montclos, *L'Architecture à la française,* 184; and Lon R. Shelby, "The Geometrical Knowledge of Mediaeval Master Masons," *Speculum,* 47 (July 1972), 395–421. But while there is little to suggest direct links, we should remember that the art of stonecutting was only one of numerous branches of practical projective geometry flourishing in this period as never before. We are not warranted to consider it as ever and always hermetic. Questions as to its provenance and inspiration remain open. For example, the method of construction to calculate sun angles shown in Ptolemy's *Analemma,* translated by Commandino and published in 1562, is very similar indeed to the methods presented by Delorme and Jousse for stonecutting. See Thomas Heath, *A History of Greek Mathematics* (New York, 1981), vol. 2, 286–292.

40. Abraham Bosse, *Manière universelle de Mr. Desargues pour pratiquer la perspective* (Paris, 1648). These few added sheets had nothing to do with the "universal" perspective technique for painters that was described in the main part of the book. Translation in J. V. Field and J. Gray, *The Geometrical Works of Girard Desargues* (New York, 1987).

41. Abraham Bosse, *La Pratique du trait* (Paris, 1643), 29. This has been noticed by Field and Gray, *Geometrical Works of Desargues,* 14. They do not include *La Pratique du trait,* because "it makes a sophisticated use of three-dimensional geometry, but seems to be unrelated to Desargues' work on conics."

42. The earliest known drawing demonstrative of modern stereotomy is Jean Chéreau's plan and elevation of the vault of Saint-Jean, Joigny. This was not a working *trait,* but part of a manuscript (1567–1574) on stonecutting. See Pérouse de Montclos, *L'Architecture à la française,* 95, 97.

43. For de la Hire see Field and Gray, *Geometrical Works of Desargues,* chap. 3.

44. Nicolas Le Camus de Mézières, *Guide de ceux qui veulent bâtir* (Paris, 1786), vol. 1, 111.

45. M. Postan and E. E. Rich, eds., *The Cambridge Economic History of Europe* (Cambridge, U.K., 1952), vol. 11, 511; H. M. Colvin, ed., *History of the King's Works* (London, 1963), vol. 1, 183.

46. The very title of Jousse's book—"The Secret of Architecture"—was of course suggestive of some kind of espionage.

47. As stated for example in Alberto Pérez-Gómez, *Architecture and the Crisis of Modern Science* (Cambridge, Mass., 1983).

48. Mathurin Jousse, *Le Secret d'architecture* (La Flèche, 1642).

49. A well-observed instance of a related transfer from one aesthetic location to another is to be found in Werner Müller, "Guarini e la Stereotomia," in *Guarino Guarini e l'internazionalità del barocco* (Turin, 1970), vol. 1, 532–556. Müller traces the diagrammatic figure for the *trait* of a vaulted spiral stair from its French source, through to its publication in Turin in Guarini's treatise, to its inventive redeployment as the plan for a church in the hands of Benedetto Alfieri (started 1756).

50. George Hersey, *The Lost Meaning of Classical Architecture* (Cambridge, Mass., 1988), chaps. 4–6.

51. " L'homme n'est pas fait pour supporter un fardeau immuable." A. F. Frézier, Dissertation sur les ordres d'architecture (Strasbourg, 1738), 26; see also his "Remarques sur quelques livres nouveaux," *Mercure de France* (July 1754), 14–15.

52. Frézier, *La Théorie et la pratique de la coupe des pierres,* Discours préliminaire.

53. Herrmann, *Laugier and Eighteenth Century French Theory,* 92–95, 245–246.

54. Robin Middleton and David Watkin, *Neoclassical and 19th Century Architecture* (New York, 1977), 12–28.

55. Delagardette's introduction to W. Simonin, *Traité élémentaire de la coupe des pierres, ou Art du trait* (Paris, 1792), 4.

56. G. W. F. Hegel, *The Philosophy of History,* trans. J. Sibree (New York, 1956), 27.

57. Denis Diderot and Jean d'Alembert, *Encyclopédie* (Paris, 1754), vol. 4, 347; see also Pérouse de Montclos, *L'Architecture à la française,* 85, citing Charles Perrault as the source of the idea.

58. Marc-Antoine Laugier, *An Essay on Architecture,* trans. W. and A. Herrmann (Los Angeles, 1977), 32–33; Wolfgang Herrmann, *Laugier and Eighteenth Century French Theory* (London, 1962), 237b; Diderot and d'Alembert, *Encyclopédie* (Neufchâtel, 1765), vol. 9, 808.

59. "De même qu'il y a des bornes à la solidité, il y en a aussi à la délicatesse et à la legereté d'un édifice; car quand même il seroit solide par la consistance de ses materiaux, et par l'artifice de leur liaison, lorsque l'idée que nous avons naturellement de la proportion qui doit être entre le suport et la charge, nous fait paroitre un suport trop foible ou trop étroit, nous n'en pouvons aprouver la construction, notre esprit se révolte contre ce qui paroit hazardé, nous voulons non seulement une solidité réelle, mais encore aparente, qui ne donne pas occasion au Spectateur de craindre que l'édifice culbute. Nous admirons un homme qui danse sur la corde, mais dans le fond on le condamne de s'exposer mal à propos, et on sent de la peine à le voir." Frézier, *Dissertation,* 16.

60. Ibid., 14ff.

61. Marcello Fagiolo, ed., *Natura e artificio* (Rome, 1979), esp. Gabriele Morolli, "A

quegli idei selvestri," 55ff.; and Amedeo Belluzzi, "L'Opera Rustica nell'architettura italiana del primo Cinquecento," 98ff.

62. Delorme, *L'Architecture*. The design has no explanation. It was placed with some other unrealized proposals at the end of book 8 of the *Livre*. Another of these added plates shows a pavilion held aloft on an open spiral tower built of rusticated stone; the heaviness of rustication combines incongruously with the lightness of stereotomy.

63. Frézier, *Dissertation*, 65.

64. Laugier, *Essay*, 14: "The Column must be strictly perpendicular, because, being intended to support the whole load, perfect verticality gives it its greatest strength."

65. Ibid., 77.

66. Ibid., 15, 21. Laugier's recommendation of Derand's text is doubly peculiar because Derand did not deal with these flat ceiling vaults, whereas Frézier, whom Laugier passed over, did. Frézier, clearly irritated, criticized Laugier's odd choice in his review of the *Essai*; see A. F. Frézier, "Remarques," *Mercure de France* (July 1754), 35.

67. Reginald Blomfield, *A History of French Architecture* (London, 1911), vol. 2, 68–70. Blomfield held Lemercier as much responsible for the importation of the Roman Jesuit type of church to France as Martellange. Both architects had altered the Roman original by adding galleries between aisle and clerestory.

68. These cartouche-like carvings bear a striking resemblance to the ornamental title vignettes for the *traits* in Derand's *Architecture des voûtes*.

69. "... on porte, sitôt qu'on y a mis le pied, la vûe en haut, comme au lieu d'où les yeux espèrent tirer plus de contentement, par la considération des figures, et des rares et agréables diversitez qui se voyent aux traits." Derand, *Architecture des voûtes*, viii.

70. La Flèche was the Jesuit seminary where Descartes was educated. Derand also trained there and later taught mathematics there. See Pierre Moisy, *Les Eglises des Jésuites de l'Ancienne Assistance de France* (Rome, 1958), 131–141.

71. François Blondel, *Resolution des quatre principaux problemes d'architecture* (Amsterdam, 1677), 32ff; *Cours d'architecture* (Paris, 1698), vol. 2, 425–434. Unsightly *jarrets* were mentioned in most treatises on stonecutting from Delorme onward. Projectively derived curves were preferable to ovals made from circular arcs of different radii because they did away with such discontinuities. Blondel set about the task of describing an inventory of projective curves—the conic sections—without recourse to projection. In this form they could then be applied directly to architecture.

72. Anthony Blunt, *Art and Architecture in France, 1500–1700* (Harmondsworth, 1982), 176.

73. See note 23.

74. The vaults resting on freestanding columns in Sainte-Geneviève had to be carefully arranged so that the normally asymmetric forces from aisle and nave canceled each other out, leaving no lateral thrust to push the vertical columns over.

75. Mark Deming describes the plan as a conventional market hall in a town square turned inside-out. The Halle au Blé is a town square in a market hall. See Le Camus de Mézières, *Guide*, 148; and M. Deming, *La Halle au Blé de Paris 1762–1813* (Brussels, 1984), 55–63.

76. Deming, *La Halle au Blé*, 88–89. Deming points out the similarity between this canvas and another, later painting by Machy of a Gothic church under demolition.

77. Allan Braham, *The Architecture of the French Enlightenment* (London, 1980), 109.

78. Robert Willis, "On the Construction of the Vaults of the Middle Ages," *RIBA Transactions*, 1st series, vol. 1, pt. 2 (1842). The paper from which the article was drawn was delivered in 1839.

79. Pérouse de Montclos, *L'Architecture à la française*, 181–183.

80. "La révolution ogivale, c'est l'adoption d'intersection segmentaire, coplanaire, nervurée et de surfaces gauches; la contre-révolution, c'est le retour à des surfaces réglés et à des pénétrations elliptiques et gauches." Ibid., 183.

81. Frézier, *La Theorie et la pratique de la coupe des pierres*, vol. 3, 171.

82. Delorme, *L'Architecture*, 107r.

83. For French lierne vaults see Roland Sanfaçon, *L'Architecture flamboyante en France* (Québec, 1971).

84. See for example Paul Frankl, *Gothic Architecture*, trans. Dieter Pevsner (Harmondsworth, 1962), 115, 146–149. Bond pointed out our thoroughly uncharacteristic modesty in this matter, putting it to rights thus: "Truly, if vault construction is the be-all and end-all of Gothic architecture, it is not with any foreign country, but with England, that the artistic supremacy in mediaeval architecture rests." Francis Bond, *Gothic Architecture in England* (London, 1905), 349n.

85. Karl Heinz Clasen, *Deutsche Gewolbe der Spätgotik* (Berlin, 1958); Goetz Fehr, *Benedikt Ried* (Munich, 1961).

86. Between 1842 and 1869 Willis published successive volumes on the description of major cathedrals and other medieval buildings. His voluminous papers are now deposited in the Cambridge University Library.

87. William Halfpenny, *The Art of Sound Building* (London, 1725), section II, 14–23.

88. Pevsner singled out Willis's contribution. Appreciative more of his indefatigable labor and encyclopedism, he nevertheless acknowledged the essay on vaults to be unsurpassed. Nikolaus Pevsner, *Some Architectural Writers of the*

Nineteenth Century (Oxford, 1972), 52–61. Blouet and Viollet-le-Duc both took illustrations as well as explanations from Willis. Abel Blouet, *Traité de l'art de bâtir: supplément* (Paris, 1843), 12–22; Eugène-Emmanuel Viollet-le-Duc, *Dictionnaire raisonné de l'architecture française* (Paris, 1868), vol. 9, 533–537.

89. *Dictionary of National Biography*, vol. 21, 492–494.

90. Willis, "On the Construction of the Vaults of the Middle Ages," 3.

91. Ibid., 10–13.

92. Ibid., "Additional Remarks," 61–69.

93. See for examples: François Bucher, *Architector: The Lodge Books and Sketch Books of Mediaeval Architects*, vol. 1 (New York, 1979); and F. Bucher, "Design in Gothic Architecture: A Preliminary Assessment," *Journal of the Society of Architectural Historians*, 27 (March 1968), 49–71.

94. I wish to thank Canon David Welander of Gloucester Cathedral for his generous help and for acting as guide within the vaults.

Willis seems to have envisaged the masons at Gloucester "adjusting" the net of segmental ribs to conform as closely as they might to an imaginary cylindrical surface. It is not clear whether he thought the surface was, at this stage, defined by projection.

95. See John Harvey, *The Perpendicular Style* (London, 1978), 30.

96. The toy vaults in the canopy of the Despenser tomb, Tewkesbury Abbey, 1359, are particularly interesting, as each miniature bay of four fans is monolithic and the surface is left smooth and ribless.

97. In the Dean's Chantry (1441–1449), only 7 feet wide, the complex of mural and pendant fans seems to prefigure later developments at the Oxford Divinity Schools and the Henry VII Chapel, Westminster Abbey. Yet, for all its apparent complexity, it was easy to design and cut. It was made from rectangular blocks from which whole pendants or half pendants were carved. The jointing is both simple and crude.

98. In the Gloucester cloisters the jointed rings of stone in the fans are arranged to correspond to three shallow bands of ogival carving: the pattern of joints was therefore adapted to the pattern of decorative ribs, but was not identical to it.

99. The size of this flat keystone in the Gloucester cloisters was, however, reduced by the carefully planned extension of the stones round the rim of the fans into the horizontal plane.

100. Willis, "On the Construction of the Vaults of the Middle Ages," 43ff. Willis praised the Peterborough fan vaults as the most perfectly cut, but the Westminster, Cambridge, and Bath fans are all of masterly workmanship with very fine joints.

101. Harvey maintains this attribution to be the most likely, although the evidence is only circumstantial. Recently it has been suggested on stylistic grounds that the designer was Robert Janyns, another of the King's Masons. However, as regards the formation and construction of vaults, the circumstantial evidence still points toward the Vertues. See John Harvey, *English Mediaeval Architects* (Gloucester, 1984), entries for R. and W. Vertue; Christopher Wilson et al., *Westminster Abbey* (London, 1986), 70–71.

102. Sacheverell Sitwell, *Gothic Europe* (London, 1969), 61.

103. P. Kidson, P. Murray, and P. Thompson, *A History of English Architecture* (Harmondsworth, 1965), 138.

104. Harvey, *English Mediaeval Architects*, 309; Banister Fletcher, *A History of Architecture*, 17th ed. (London, 1961), 427; Bond, *Gothic Architecture in England*, 349.

105. Harvey, *The Perpendicular Style*, 13–14.

106. Harvey, *English Mediaeval Architects*, 309.

107. John Henry Parker, *An Introduction to the Study of Gothic Architecture* (Oxford and London, 1849), 191, and John Ruskin, *The Seven Lamps of Architecture* (Oxford and London, 1849), 191.

108. Harvey, *English Medieval Architects*, 309.

109. W. R. Lethaby, *Westminster Abbey and Craftsmen* (London, 1906), 230.

110. Willis, "On the Construction of the Vaults of the Middle Ages," 52. Leedy also takes this view; see Walter C. Leedy, *Fan Vaulting: A Study of Form, Technology, and Meaning* (London, 1980), 27.

111. I am very grateful to Peter Foster, not only for this fascinating suggestion, but for imparting a great deal of other information regarding the chapel. Foster says one might reasonably assume that the stones have recently been subject to movement because of modern traffic vibration, but points out that this explanation does not account for the shaving of the ribs.

112. George Herbert West, *Gothic Architecture in England and France*, 2d ed. (London, 1927), 85–86. West's dedication was to Viollet-le-Duc and A. de Baudot.

113. Leedy has suggested that the vaults are capable of ambiguous behavior; they may act as frames or surface structures. Leedy, *Fan Vaulting*, 24.

114. West was also of the opinion that it was unfair to condemn the vaults as false construction just because the ribs were "merely decorative." He realized that this loss of status encouraged emphasis of the surface, which then replaced the rib as structure. West, *Gothic Architecture in England and France*.

115. Standing on it induces a mixture of intellectual and physical elation as you look through one of the small apertures in its surface down to the chapel floor 60 feet below, while measuring with your finger the unbelievable thinness of its stone panels—at one point less than 3 inches. I very much appreciate the indulgence of Harry Tooze, the former Clerk of Works, in this matter.

Observation confirms Willis's drawing to have been quite remarkably accurate. Only one small difference is worth noting. Willis had observed the persistence of surfaces of operation on the upper side of the Cambridge and Peterborough fans. He thought these to have been present originally in the Henry VII Chapel vault too. In the drawing there are, in acknowledgment of this idea, flat horizontal areas around the pendant bosses on the longitudinal ridge. In fact this area is not flat, but in continuous modulation. The absence of horizontal plane surfaces does not mean that Willis was wrong in assuming that the stones were cut from orthogonally oriented blocks; it does mean that the English masons had already performed as complete an undressing of the original rectangular box as would the French practitioners of the *trait*. Measurement of a set of similar stones on the lateral ridge of the vault (where any adaptation due to cumulative excess or defect would presumably have been made) suggests a maximum difference of 5 percent in linear dimension, and a norm around 1 percent. Circumstances made exact measurements difficult to obtain, however.

116. There were about 70 drawings of Tudor tracery and vaulting, a number of them showing the fans of the Henry VII Chapel. This emerges from an acrimonious correspondence between the architect John Carter disputing their authenticity and the antiquary John Britton proclaiming it. It was Carter, however, who had raised the issue of working drawings. Britton did not think they were working drawings but "memoranda." *The Gentleman's Magazine*, 77, no. 2 (1807), 1187–1190; 78, no. 1 (1808), 286, 296, 399.

117. H. M. Colvin, ed., *History of the King's Works*, vol. 3 (London, 1975), 211 and plate 15.

118. British Library, Cotton Mss., Augustus II.1. The drawing was plotted into a feint-ruled, two-sided axonometric construction of guide lines. The detail deploys perspectival recession in parts and frontal orthographic representation in other parts. Several conventions, including the consistent tinting of all recessed molding, are held to, but inside the clear schematic structure chaos reigns.

119. There has been much debate on the role of drawing in Gothic architecture in general. Shelby inclines to the view that templates were more important than drawings. By contrast, Branner holds that the introduction of more or less orthographic elevations prompted the development of French Gothic into its *rayonnant* phase. Bucher also believes Gothic was made with drawings. He claims also that he can show how Gothic vaults were determined by stereotomy, although he has not published the promised demonstration from the 2,200 items of Gothic architectural drawing that he has been able to identify. It should be noted that the great majority of these examples are from the late fifteenth and sixteenth centuries. Lon R. Shelby, "Mediaeval Mason's Templates," *Journal of the Society of Architectural Historians*, 30 (March 1971), 140–154; François Bucher, "Design in Gothic Architecture," *Journal of the Society of Architectural Historians*, 27 (March 1968), 49–71; Robert Branner, "Villard de Honnecourt, Rheims and the Origin of Gothic Architectural Drawing," *Gazette des Beaux Arts*, 6th series (Paris/New York), 61 (March 1963), 129–146.

Chapter Six: The Trouble with Numbers

1. François Blondel, *Cours d'architecture enseigné dans l'Académie Royale*, 2d ed. (Paris, 1698), vol. 2, 759; Bernardo Vittone, *Istruzioni elementari per indirizzo di giovani allo studio dell'architettura civile* (Lugano, 1760), vol. 1, 368.

2. "... il y a une telle analogie entre les Proportions de la Musique et celles de l'Architecture, que ce qui choque l'orielle en celle-là, blesse la veüe en celle-cy, et qu'un bâtiment ne peut être parfait, s'il n'est dans les mêmes Règles que celles de la Composition ou mélanges des accords de la Musique." René Ouvrard, *Architecture harmonique ou l'application de la doctrine des proportions de la musique à l'architecture*, reprinted in part in Françoise Fichet, *La Théorie architecturale à l'âge classique* (Brussels, 1979), 177.

3. Rudolf Wittkower, *Architectural Principles in the Age of Humanism* (London, 1949), 142–154, "The Breakaway from the Laws of Harmonic Proportion in Architecture"; and Le Corbusier, *The Modulor I and II*, trans. P. de Francia and A. Bostock (Cambridge, Mass., 1982), 15–17.

4. By "ourselves" I mean architectural critics and historians. The idea is particularly virulent among this group.

5. Regarding the rose see F. Alberto Gallo, *Music of the Middle Ages II*, trans. K. Eales (Cambridge, U.K., 1985), 94. For the circumstances of performance see Gustave Reese, *Music in the Renaissance* (London, 1954), 79. For the position of the choir see Frank A. D'Accone, "The performance of Sacred Music in Italy in Josquin's Time," in E. E. Lowinsky, ed., *Josquin des Prez: Proceedings of the International Josquin Festival Conference* (New York, 1971), 614.

6. Charles W. Warren, "Brunelleschi's Dome and Dufay's Motet," *Musical Quarterly*, 59 (1973), 92–105.

7. Rudolf Wittkower, "Brunelleschi and Proportion in Perspective," *Journal of the Warburg and Courtauld Institutes*, 16 (1953). Also in *Idea and Image* (London, 1978), 125–136.

8. Opinion differs as to the degree to which fourteenth- and fifteenth-century isorhythm and mensural diminution were discernible. At one extreme, Brian Trowell regards isorhythm as the most accessible aspect of Dunstable's works, for "they are built on mathematical and proportional principles that the ear can discern and delight in." He goes on to discuss many more recondite structures buried in the motets. More usually such rhythmic effects are regarded as largely subliminal. Richard Crocker, describing the mensural acceleration of the repeated tenor in Dufay's mass *Se la face ay pale*, writes: "Thus even if the cantus firmus is imperceptible, the textural plan derived from it is clearly apparent to the listener." "There is a good chance the cantus firmus will be imperceptible . . . [since] the countertenor . . . goes out of its way to obscure the formal plan of the cantus firmus." See Brian Trowell, "Proportion in the Music of Dunstable," *Proceedings of the Royal Musical Association*, 105 (1978/9), 100; and Richard L. Crocker, *A History of Musical Style* (New York, 1966), 149. Certainly, after many attempts, I am still unable to recognize it, but I am not a musician. See also note 26.

9. Gallo, *Music of the Middle Ages II*, 39, quoting the fourteenth-century writer Johannes Boen, who had said that these procedures were more easily seen than heard. Gallo adds that this "confirms that written notation had become an essential part of composing. Only on paper could the composer calculate exact correspondences, and only on paper could the skill with which the piece had been composed be appreciated." For fifteenth-century color notation see M. Bent, "Notation 3," in S. Sadie, ed., *New Grove Dictionary of Music and Musicians*, vol. 13 (London, 1980), 369.

10. Willi Apel, *The Notation of Polyphonic Music 900–1600*, 5th ed. (Cambridge, Mass., 1953), 403.

11. Johannes Tinctoris, *Proportions in Music (Proportionale Musices)*, trans. A. Seay (Colorado Springs, 1979), 7. Another important work on this aspect of music was Prosdocimus de Beldamandis, *A Treatise on the Practice of Mensural Music in the Italian Manner* (1408), trans. J. A. Huff, American Institute of Musicology Study no. 29 (1972).

12. For examples see Trowell, "Proportion in the Music of Dunstable," 139.

13. The ratios of nesting squares rotated 45 degrees that Warren refers to are alternately rational and irrational. More closely approximate values would be 6:4.24:3:2.12.

14. Warren, "Brunelleschi's Dome and Dufay's Motet."

15. I came across the following criticism of Warren's conclusions while browsing in a bookshop. It started me off on the trail that led to this chapter. Unfortunately I have not been able to trace it since.

16. Samuel E. Brown, Jr., "New Evidence of Isomelic Design in Dufay's Isorhythmic Motets," *Journal of the American Musicological Society*, 10 (1957), 7–13.

17. Warren, "Brunelleschi's Dome and Dufay's Motet," 102. The document in question is a notarial record of the Florentine Woollen Guild of 1420.

18. Charles Hamm, "Dufay," *New Grove*, vol. 5, 676.

19. Johannes Tinctoris, *Opera Theoretica*, ed. A. Seay, vol. IIa (Stuttgart, 1978), 11: "ad honorem tuæ proportionatissimæ capellæ."

20. Girolamo Mei, *Letters on Ancient and Modern Music*, ed. and trans. C. V. Palisca, American Institute of Musicology Study no. 3 (1960), 71.

21. Manfred Bukofzer, "John Dunstable and the Music of His Time," *Proceedings of the Royal Musical Association*, 65 (1938/9), 31.

22. Margaret Bent, *Dunstable* (London, 1981), 1–3.

23. Vitruvius, *The Ten Books on Architecture*, trans. M. H. Morgan (Cambridge, Mass., 1914), book 1, chap. 1, cap. 16.

24. Trowell, "Proportion in the Music of Dunstable," 102; Bent, *Dunstable*, 64–65.

25. Brown, "New Evidence of Isomelic Design."

26. Warren, "Brunelleschi's Dome and Dufay's Motet," 104; and Howard Mayer Brown, "Guillaume Dufay and the Early Renaissance," *Early Music*, 2 (October 1974), 225–226. "In a sense," says Brown, "Dufay continued in his cantus-firmus masses the isorhythmic motets, with the possible difference that the mass tenor may have been intended to have an audible effect on the listener's perception of the form." Also, H. M. Brown, *Music in the Renaissance* (Englewood Cliffs, 1976), 30–52.

27. I am indebted to James Bradburne for bringing to my attention the peculiar difficulties entailed in claiming that something was indiscernible to people whose sensibilities were not necessarily identical to ours. I take the view that as I cannot rely entirely on my own judgment I must rely on contemporary statements directly or indirectly supporting my contention. These are few and far between, but do exist.

28. The cryptic patterning of the isorhythmic motet reached its climax by the mid-fifteenth century. Dunstable and Dufay were its most eminent expositors, although others developed more elaborate arrangements. Afterward isorhythm was on the wane, but the concern to give large, architectonic proportion to music had not entirely disappeared. A case in point is Josquin's *Missa Di dadi* (Dice Mass), circa 1470, so called because the notation of a surviving copy showed the faces of a die to indicate the different mensurations of the cantus firmus in the different sections of the mass (6:1, 5:1, 4:1, 2:1). Concern with proportion in mensural polyphony continued through the sixteenth and even into the seventeenth century. Ernest H. Sanders, "Isorhythm," *New Grove*, vol. 9, 351–353. See also "Proportion," *New Grove*, vol. 15, 306–307.

29. Panofsky notes the difference between Plotinus's concept of ideal forms as unitary and simple and the Renaissance concept of proportion which treated them always as compound and complex. See Erwin Panofsky, *Idea: A Concept in Art Theory*, trans. J. J. S. Peake (New York, 1968), 29. The point I am making is that the Plotinian definition would be destructive of art as we understand it.

30. Colin Rowe, *The Mathematics of the Ideal Villa and Other Essays* (Cambridge, Mass., 1982), 1–27.

31. Roger Scruton, *The Aesthetics of Architecture* (London, 1979), 63–65.

32. Heinrich Wölfflin, "Zur Lehre von den Proportionen," in *Kleine Schriften*, ed. Joseph Gantner (Basel, 1946), 48–50; Wittkower, *Architectural Principles*, 41–47.

33. Leone Battista Alberti, *Ten Books on Architecture*, trans. G. Leoni, ed. Joseph Rykwert (London, 1965), book VI, chap. 2, 113.

34. Ibid., book IX, chap. 5, 194.

35. For a selective review see P. H. Scholfield, *The Theory of Proportion in Architecture* (Cambridge, U.K., 1958), 98–125.

36. Cesare Cesariano, *Vitruvius* (Como, 1521; reprint, New York, 1986), xv recto and verso; Sebastiano Serlio, *The Five Books of Architecture* (London, 1611; reprint, New York, 1982), fol. 12v, 13r; Philibert Delorme, *Architecture* (Rouen, 1648; reprint, Brussels, 1981), 225–236.

37. Leo Steinberg, "The Line of Fate in Michelangelo's Painting," in W. J. T. Mitchell, ed., *The Language of Images* (Chicago, 1980), 107–108.

38. Galileo Galilei, *Dialogues Concerning Two New Sciences*, trans. H. Crew and A. de Salvio (New York, 1954), 4.

39. Thus Charles Bouleau, discussing the "internal construction" of paintings, says: "The framework of a painting or carving, like that of the human body or that of a building is discrete; sometimes, indeed it makes one forget its existence; but it cannot be absent." C. Bouleau, *The Painters' Secret Geometry* (London, 1963), 9.

40. L. B. Alberti, *I dieci libri de l'architettura*, trans. from Latin by Cosimo Bartoli (Venice, 1546), book IX, chap. 5, 202.

41. Wittkower (see note 7 above) took up a theme that had been presented by G. C. Argan in a more general and bombastic essay, "The Architecture of Brunelleschi and the Origins of Perspective Theory in the 15th Century," *Journal of the Warburg and Courtauld Institutes*, 9 (1946), 96–121.

42. Jean Paul Richter, ed., *The Notebooks of Leonardo Da Vinci* (New York, 1970), vol. 1, 60.

43. Wittkower, *Idea and Image*, 132.

44. Alberti, *Ten Books on Architecture*, 22.

45. Albrecht Dürer, *Vier Bücher van menschlicher Proportion* (Nuremberg, 1528).

46. See note 77. The origin of this disdain for vision, so much more pronounced in architectural literature than in writings on other visual arts, was with Vitruvius. See Vitruvius, *Ten Books on Architecture*, 84–86. See also David Summers, *The Judgement of Sense* (Cambridge, U.K., 1987), chap. 2, "The Fallacies of Sight," 42–49.

47. "La prospettiva subalterna, discendante e sigliuola della geometrica, conchiudesi essere scienza delle linee visibili." Paolo Giovanni Lomazzo, *A Tracte Containing the Artes of Curious Paintinge* (London, 1598), book 5, chap. 3, 188.

48. The words are Bernhard Schneider's; see "Perspective Refers to the Viewer, Axonometry to the Object," *Daidalos*, 1, no. 1 (1981), 81.

49. Perspective constructions were sometimes fitted into a Procrustean bed of circles, squares, and equilateral triangles as if these were essential to the construction. The method described by Pomponius Gauricus in 1504 is of this kind; see Robert Klein, *Form and Meaning* (Princeton, 1979), 116. Another version is found in a sketch by Peruzzi; see Heinrich Wurm, *Baldassare Peruzzi, Architekturzeichnungen* (Tübingen, 1984), 270.

50. Michael Baxandall, *Painting and Experience in Fifteenth Century Italy* (Oxford, 1972), 96–97.

51. L. B. Alberti, *On Painting*, trans. John R. Spencer (New Haven, 1956), 56. Baxandall was not alone in assuming perspective diminution to be related to geometric series. Panofsky made this mistake and then corrected it. The same misunderstanding is found in El Lissitzky's "A and Pangeometry" and in Le Corbusier's Modulor, where he identified perspective diminution with the golden section. See Erwin Panofsky, *Renaissance and Renascences in Western Art* (London, 1970), 127, n. 1; Sophie Lissitzky-Küppers, *El Lissitzky* (London, 1968), 349; Le Corbusier, *The Modulor I and II*, 79–80.

52. Desargues's architectural work is currently being studied by Giorgio Ciucci.

53. David Eugene Smith, *A Source Book in Mathematics* (New York, 1929), 311–314. Also Julian Lowell Coolidge, *A History of the Conic Sections and Quadric Surfaces* (New York, 1968), 28–29.

54. Books VII–IX of Euclid's *Elements* dealt with proportions between numbers, restricting the treatment quite deliberately to commensurables, which was particularly telling because in the two preceding books he had dealt with proportions as continuously variable sets of magnitudes. See Sir Thomas Heath, *The Elements* (London, 1913), vol. 2, 112–113.

55. Dürer, *Vier Bücher van menschlicher Proportion*, 2v.

56. Albrecht Dürer, *Underweysung der Messung* (Nuremberg, 1532), 6, 7, 12, 13, 95.

57. H. F. Cohen, *Quantifying Music* (Dordrecht, 1984), 3ff.; Bent, *Dunstable*, 31–35; Brown, *Music in the Renaissance*, 3, 30.

58. J. Murray Barbour, *Tuning and Temperament: A Historical Survey* (East Lansing, 1953), 15–24.

59. Cohen, *Quantifying Music*, 36–39.

60. Barbour, *Tuning and Temperament*, 3.

61. Ibid., 8.

62. Ibid., 9.

63. As we have seen, René Ouvrard, a musician, could still maintain as his principal thesis in 1679 that the beauty of architecture derived from musical harmonies. This in itself suggests that the situation was complex, involving a distinction between perfect theoretical form and imperfect practical form, as well as between informing and informed subjects. Certainly, the architectural commentators, although their ideas seem

arcane, were not merely out of touch with musical theory. See Blondel, *Cours d'architecture*, vol. 2, 756–761; also Joseph Rykwert, *The First Moderns* (Cambridge, Mass., 1980), 13.

64. Alberti, *Ten Books on Architecture*, 197–199. Antonio Manetti, circa 1480, said that Brunelleschi was the first to introduce musical proportion into architecture, but there is some doubt about the credibility of this claim. See Wittkower, *Architectural Principles*, 117.

65. I would therefore take issue with Samuel Edgerton's characterization of Renaissance perspective as quantifiable; it was potentially so but not actually so. Samuel Y. Edgerton, Jr., "The Renaissance Artist as Quantifier," in Margaret A. Hagen, ed., *The Perception of Pictures*, vol. 1: *Alberti's Window: The Projective Model of Pictorial Information* (New York, 1980), 179–212.

66. The invention of linear perspective was the first modern prelude to the history of projective geometry. Projective geometry is now recognized as logically prior to and therefore fundamental to Euclidean geometry. Bruce Meserve, *Fundamental Concepts in Geometry*, 2d ed. (New York, 1983), 181–182, 310.

67. Cohen, *Quantifying Music*, 51–58.

68. This assumption underlies William Ivins's *Art and Geometry* (Cambridge, Mass., 1946), esp. chap. 10.

69. David Summers, *Michelangelo and the Language of Art* (Princeton, 1981), 358–379.

70. Barbour, *Tuning and Temperament*, 5–13.

71. The same could be said of some modern treatments of the history of perspective. Searches for its origins in optics are of great interest, but the pertinence of optics to the advancement of pictorial practice has yet to be demonstrated. For the optical origins see Samuel Y. Edgerton, Jr., *The Renaissance Rediscovery of Linear Perspective* (New York, 1975), chaps. 5 and 6; for reference to optics in Renaissance writings on perspective see David C. Lindberg, *Theories of Vision from Al-Kindi to Kepler* (Chicago, 1976), 147–177.

72. Alberti and Gauricus were both concerned to show that perspective related to *istoria* and composition. Gauricus's essay on perspective indicates that he was nearly ignorant of the principles of perspective construction. As a humanist, his interest was in its deployment for rhetorical purposes. See Klein, *Form and Meaning*.

73. Daniele Barbaro, *La pratica della perspettiva* (Venice, 1569), proemio.

74. Cohen, *Quantifying Music*, 29; D. P. Walker, "Musical Humanism in the Sixteenth and Early Seventeenth Centuries," *Musical Review*, 2–3 (1941–1942) (5 parts).

75. Rykwert, *The First Moderns*, 33–47. Wolfgang Herrmann notes that while Perrault was openly critical of ancient science he did not criticize ancient art in the same way; W. Herrmann, *The Theory of Claude Perrault* (London, 1973), 48.

76. Claude Perrault, *A Treatise of the Five Orders of Columns in Architecture*, trans. John James (London, 1708), 131.

77. Ibid., iii–v.

78. Claude Perrault, "De la musique des anciens," *Essais de Physique*, vol. 2 (Paris, 1680), 337–402. See also Albert Cohen, *Music in the French Royal Academy of Science* (Princeton, 1981), 9–10.

79. Perrault, *A Treatise of the Five Orders*, 107–113.

Chapter Seven: Comic Lines

1. Trans. Simon Watson Taylor, in R. Shattuck and S. W. Taylor, eds., *Selected Works of Alfred Jarry* (London, 1965).

2. Le Corbusier, *The Chapel at Ronchamp*, trans. Jacqueline Cullen (London, 1957), 120. The French edition is *Ronchamp: oeuvre de Notre-Dame du Haut*, (Stuttgart, 1957).

3. Le Corbusier, *Modulor I and II*, trans. Peter de Francia and Anna Bostock (Cambridge, Mass., 1982), p. 254 of Modulor II (the two components are separately paginated in this edition, and are henceforth cited as *Modulor I* and *Modulor II*). The first French editions were published in 1950 and 1955 respectively.

4. This aphorism may have been Le Corbusier's rather than Einstein's. See Stanislaus von Moos, *Le Corbusier: Elements of a Synthesis* (Cambridge, Mass., 1979), 367, n. 101.

5. Information from conversation with Jerzy Soltan.

6. Jean Petit, *Textes et dessins pour Ronchamp* (1967), 8.

7. Jerzy Soltan says the doubling was Le Corbusier's idea. Though he is reticent about his own contribution, it must have been Soltan who introduced the Fibonacci series. I am very grateful to Soltan for the wealth of information he supplied me in conversation and letter.

8. When Le Corbusier found out from the mathematician René Taton that his original diagram involved a mismeasurement of $^6/_{1000}$ (in fact $^{12}/_{1000}$; Taton was making approximate, long-hand calculations), he concluded that these $^6/_{1000}$ were of "infinitely precious importance." *Modulor I*, 229–235.

9. *Modulor I*, 15.

10. Le Corbusier, *The Four Routes*, trans. Dorothy Todd (London, 1947), 139.

11. Richard A. Moore, "Alchemical and Mythical Themes in the Poem of the Right Angle, 1947–65," *Oppositions*, 19/20 (1980), 111–138.

12. Sigfried Giedion, "Le Corbusier and the Contemporary Means of Architectural Expression," in Le Corbusier: Architecture, Painting, Sculpture, Tapestries (Liverpool, 1958–1959), 10. Conversation with Guillermo Jullian de la Fuente, December 1989.

13. Charles Jencks, *Le Corbusier and the Tragic View of Architecture*, rev. ed. (Harmondsworth, 1987), 7.

14. Jean Labatut, "Le Corbusier's Notre Dame du Haut at Ronchamp," *Architectural Record*, 118, no. 4 (October 1955), 169–172.

15. James Stirling, "Ronchamp: Le Corbusier's Chapel and the Crisis of Rationalism," *Architectural Review*, 119 (March 1956), 155–161; Nikolaus Pevsner, *An Outline of European Architecture*, rev. ed. (Harmondsworth, 1968), 429.

16. ". . . un refus des formes trop rigoureuses et du système orthogonal qui primait depuis décennies." Danièle Pauly, *Ronchamp: lecture d'une architecture* (Paris, 1980), 152.

17. Ibid., 151.

18. See Le Corbusier, *Ronchamp: oeuvre de Notre-Dame du Haut*; Jean Petit, ed., *Le Livre de Ronchamp* (Paris, n.d.); and Le Corbusier, *Oeuvre complète*, vol. 5 (Zurich, 1953), 72–78.

19. Petit, *Le Livre de Ronchamp*, 119.

20. The major changes in the general layout were subtractions of elements outside the chapel such as the crescent embankment to the east and the carillon frame to the north.

21. *Modulor II*, 279–280.

22. *Modulor I*, 37.

23. Hanning realized his error later on and informed Le Corbusier, who seemed to think it was Hanning rather than his instruction that was at fault (*Modulor I*, 41–42). The error, it has to be said, is very basic, and it is also very large. The difference between a correct version and the original version is easily seen (fig. 150).

24. *Modulor II*, 20, 45.

25. *Modulor I*, 26.

26. P. H. Scholfield, *The Theory of Proportion in Architecture* (Cambridge, U.K., 1958), 102–104.

27. Fondation Le Corbusier, *The Le Corbusier Archive*, ed. H. Allen Brooks (New York, 1982–1984), vol. 13, FLC 31191.

28. Le Corbusier, *Aircraft* (London, 1935), 6.

29. Le Corbusier, *The Chapel at Ronchamp*, 88.

30. Ibid., 126.

31. Fondation Le Corbusier, *Le Corbusier / Savina: Dessins et Sculptures* (Paris, 1984). In 1950 Le Corbusier wrote to Savina telling him of the Ronchamp commission (". . . sensationelle. sans décor mais quelle forme . . ."). Then he wrote that this generation of artists is too pretentious to make an impact on buildings, and then commands Savina: ". . . marchez, allez de l'avant." Correspondence, 1 July 1950.

32. Petit, *Le Livre de Ronchamp*, 17.

33. Le Corbusier, *Journey to the East*, trans. Ivan Žaknić and Nicole Pertuiset (Cambridge, Mass., 1987), 15–16.

34. Martin Heidegger, "The Origin of the Work of Art," trans. Albert Hofstadter, in *Poetry, Language, Thought* (New York, 1971).

35. Le Corbusier, *Journey to the East*, 18.

36. Le Corbusier, *The Decorative Art of Today*, trans. James I. Dunnett (London, 1987), 126, 34–35.

37. Ibid., 106.

38. Stanislaus von Moos, "Corbusier as Painter," *Oppositions*, 19/20 (Winter/Spring 1980), 89–107; Mary McLeod, "Le Corbusier and Algiers," ibid., 55–85; Christopher Green, "The Architect as Artist," in *Le Corbusier: Architect of the Century* (London, 1987), 124–126.

39. Stephen Gardiner, *Le Corbusier* (New York, 1974), 115.

40. Bruno Salvatore Messina, *Le Corbusier: Eros e Logos* (Naples, 1987), 15.

41. Conversation with Guillermo Jullian de la Fuente, December 1989. I wish to record my thanks to Jullian for providing so much that was so useful to me in our several meetings.

42. Gardiner, *Le Corbusier*, 96.

43. *Modulor II*, 52.

44. Jerzy Soltan, "Working with Le Corbusier," in *Le Corbusier: The Garland Essays*, ed. H. Allen Brooks (New York, 1987), 10; see also André Wogensky, introduction to the *Le Corbusier Archive*, vol. 16, xvi.

45. *Modulor I*, 235–238.

46. *Modulor I*, 221–225.

47. *Modulor I*, 224.

48. Jencks, *Le Corbusier*, and Messina, *Le Corbusier*.

49. Richard Allen Moore, *Le Corbusier and the "Méchanique Spirituelle"* (Ann Arbor, 1982), which includes an English translation of the *Poème*, appendix E, cap. E2.

50. Diana Agrest has noticed a similar tendency in Filarete's fifteenth-century account of architecture. She calls it "transexuality." D. Agrest, "Architecture from Without: Body, Logic, Sex," *Assemblage*, no. 7 (1988), 29–41.

51. These are the easiest and the most obvious things to measure. They offer themselves up to the observer, but one must suppose that Le Corbusier was not thinking of them when he issued his challenge.

52. Fondation Le Corbusier, *Le Corbusier Archive*, vol. 20, FLC 7.167.

53. For the early sketch: Fondation Le Corbusier, *Le Corbusier Sketchbooks* (New York, 1981), vol. 2, sketchbook E.18 (February 1951), fig. 326. For dimensioned drawings: *Le Corbusier Archive*, vol. 20, FLC 7.207/8.

54. *Modulor I*, 92–99.

55. For a relatively recent example see H. E. Huntley, *The Divine Proportion: A Study in Mathematical Beauty* (New York, 1970). Chapter 2 deals with the pentagram.

56. Le Corbusier nevertheless took some heed of Soltan's observation in 1945 that they were treating the Modulor as a problem in plane geometry, when it should be treated as a line of measurements. "After that we were out of the doldrums," wrote Le Corbusier; *Modulor I*, 48–49.

57. *Modulor I*, 79; *Modulor II*, 44–48.

58. Moore, *Le Corbusier and the "Méchanique Spirituelle,"* 336.

59. *Casabella*, 207 (September/October 1955), 25.

60. Le Corbusier, *The Chapel at Ronchamp*, 123.

61. The easiest way to do this is to compare the drawings of the Garches facades with *traces régulateurs* (published by Matila Ghyka, *Le Nombre d'or*, 1931) with photos of the same. Can the rectangles defined by the diagonal traces be seen as similar just by looking at the photo?

62. "Pour que le mécanisme des rapports soit efficace, les quantités que ceux-ci génèrent seront saisissables, lisibles." Le Corbusier, *Une maison—un palais* (Paris, 1928; reprint, Paris, 1989), 3.

63. Angel Guido, *La Machinolatrie de Le Corbusier* (Argentina, 1930), 48. Tim Benton has noted that in the 1920s Le Corbusier usually applied *traces régulateurs* toward the end of the design process rather than at the beginning; Benton, *The Villas of Le Corbusier 1920–1930* (New Haven, 1987), 27.

64. This was conveyed to me by Caroline Constant via Peter Carl. See Alberto Sartoris, *Encyclopédie de l'architecture nouvelle* (Milan, 1948), vol. 1, 30. Le Corbusier, whose own work was prominently featured, provided the introduction.

65. Peter Weed, "Ronchamp: Le Corbusier's Chapel and the Crisis Caused by Reversal," essay, Harvard Graduate School of Design, 1989.

66. Le Corbusier, *Towards a New Architecture*, trans. F. Etchells (London, 1946), 167–170.

67. Petit, ed., *Le Livre de Ronchamp*, 119.

68. This was noticed by Vincent Scully, *Modern Architecture* (New York, 1961), 46.

69. Le Corbusier, *Ronchamp: oeuvre de Notre-Dame du Haut*, 103.

70. What we call the open plan (which was not so open) had already offered revision to this architectural metaphor of subjectivity, but did not assault it.

71. Nouritza Matossian, *Xenakis* (London, 1986), 34–35 (an invaluable source).

72. *Modulor II*, 326.

73. Matossian, *Xenakis*, 66.

74. Sergio Ferro, Chérif Kebbal, Philippe Potié, and Cyrille Simonnet, *Le Corbusier: Le Couvent de La Tourette* (Marseilles, 1987), 90–92.

75. Xenakis the engineer-musician was wrestling with the difference between division and subtraction, as had the Renaissance musical theorists. They privileged whole numbers at the expense of equal temperament; he privileged equal temperament and still maintained a hold on rationality, although the Fibonacci series drifted toward irrationality as surely as equal temperament itself. Matossian, *Xenakis*, 60–64; *Modulor II*, 327.

76. Conversation with Guillermo Jullian de la Fuente, December 1989.

77. Iannis Xenakis, "Le Corbusier's Electronic Poem," in *Gravesaner Blätter*, 9 (1957), 51.

78. Ibid., 53.

79. Although it seems that neither building construction nor descriptive geometry had been Xenakis's strong subjects at the Athens Polytechnic. Matossian, *Xenakis*, 28.

80. Ibid., 58.

81. "The Philips Pavilion at the 1958 Brussels World Fair," *Philips Technical Review*, 20, no. 1 (1958–1959), 3; Xenakis, "Le Corbusier's Electronic Poem," 52; Institut Français d'Architecture, *Le Corbusier: l'atelier 35 rue de Sèvres*, Bulletin d'Informations Architecturales, no. 114 suppl. (Summer 1987), 16.

82. Institut Français d'Architecture, *Le Corbusier: l'atelier*, and Matossian, *Xenakis*, 42.

83. Xenakis, interview with Matossian in 1977, in Matossian, *Xenakis*, 113.

84. Fondation Le Corbusier, *Le Corbusier Archive*, vol. 20, 34a.

85. Mircea Eliade, *The Forge and the Crucible*, trans. S. Corrin (Chicago, 1978), 7.

86. Stirling, "Ronchamp."

87. First, at the top left Olek drew the simplest long section profile using 3 straight lines and an arc, adding horizontal and vertical section lines at convenient intervals. Then at top right he showed the corresponding cross sections. Again he used only arcs and straight lines to define the profiles, adjusting the position of the arcs to produce a flattened, splayed crest and asymmetrical sides. The free-looking contours in the drawing below are the resultant sections along the horizontal cuts. They were completely defined by the other two drawings.

88. Institut Français d'Architecture, *Le Corbusier: l'atelier*, 15.

89. "Le Corbusier revoit encore les lignes essentielles de la silhouette générale." Petit, *Le Livre de Ronchamp*, 119.

90. ". . . tous les grands volumes sont définis à partir de directrices courbes et de génératrices qui sont rectilignes." Ibid.

91. "Nous avions fait le maximum de l'étude en géométrie descriptive; mais aux extrémités de la coque la géométrie ne permettait pas de trouver les raccordements et d'exprimer les doubles courbures. J'ai travaillé très étroitement avec l'entreprise. Une ingénieur très qualifié a traduit nos plans en ferraillage, en structure, très fidèlement. Au bureau, il y a toujours eu des ingénieurs qui calculaient ce qui était nécessaire. A Chandigarh comme à Ronchamp, nous donnions des dimensionnements qui correspondaient presque exactement à ce qui s'avérait juste par des calculs plus poussés." Institut Français d'Architecture, *Le Corbusier: l'atelier*, 15.

92. "Les éléments géométriques plus typiques sont la toiture-coque, composée de deux conoïdes renversés et parallèles (ce sont les deux dalles en béton de la coque), et le mur sud composé de deux surfaces réglées opposées qui partent obliquement à la Grande Porte, se redressent en parcourant le plan du mur et arrivent à former deux verticales à l'angle sud-est de l'édifice." Petit, *Le Livre de Ronchamp*, 119.

93. M. and H. U. Gasser, "Chapel of Notre Dame de Ronchamp," *Architectural Design*, 25, no. 7 (July 1955), 215.

94. Guarino Guarini, *Euclides adauctus* (Turin, 1671); *Architettura civile* (Turin, 1737), vol. 1, 231–234; vol. 2, trat. IV, plate IX.

95. Werner Müller, "The Authenticity of Guarini's Stereotomy," *Journal of the Society of Architectural Historians*, 27 (March 1968), 202–208.

96. In particular, the consistent parallelism of the generators on a (horizontal) plane of projection. The Ronchamp roof is labeled as a conoid in an office drawing (FLC 7.120, 9 September 1952), but its continuous surface divides into three strips: a conoid between two hyperbolic paraboloidal areas. One of the latter (toward the southern edge) is almost flat, the other (on the north) more distinctly warped.

97. Jerzy Soltan, "Working with Le Corbusier," 10.

98. Le Corbusier, *Ronchamp* (Stuttgart, 1975), 22, photograph by Franz Hubmann; Le Corbusier, *Oeuvre complète*, vol. 6 (Zurich, 1958), 26.

99. H. Allen Brooks, introduction to Brooks, ed., *Le Corbusier: The Garland Essays*.

100. Pointed out to me by Paul Shepheard.

101. John Farmer, "Battered Bunkers," *Architectural Review*, 181 (January 1987), 60–65.

102. Le Corbusier, *Aircraft*, fig. 18.

103. Fondation Le Corbusier, *Le Corbusier Archive*, vol. 22, 5–7, unsigned, dated 25 January 1952.

104. Ferro et al., *Le Corbusier: Le Couvent de la Tourette*, 110, 111, 117. The canons de lumière were the work of Xenakis.

105. Guillermo Jullian de la Fuente, *Atelier rue de Sèvres 35* (Lexington, 1975), 26–27.

106. Anthony Eardley, *Le Corbusier's Firminy Church* (New York, 1981), 4–23.

107. Jullian de la Fuente, *Atelier rue de Sèvres 35*, 46.

108. Fondation Le Corbusier, *Le Corbusier Archive*, vol. 16, FLC 25.174 and FLC 25.183 for the pilotis; FLC 25.261 for the rooftop nursery roof shell, showing a ruled surface and signed by Olek.

109. Von Moos, *Le Corbusier: Elements of a Synthesis*, 136.

110. Fondation Le Corbusier, *Le Corbusier Archive*, vol. 14, FLC 24290-2.

111. Matossian tells us that just prior to the Philips Pavilion Xenakis was working on the elliptical roof truss over the Chandigarh Assembly Hall tower (*Xenakis*, 111–112). The tower was a hyperboloid of revolution. An imposing model at the rue de Sèvres showed its skeleton of crisscrossing linear generators.

112. Fondation Le Corbusier, *Le Corbusier Sketchbooks*, vol. 2, fig. 275.

113. As can be seen from FLC 24298, *Le Corbusier Archive*, vol. 14.

114. "Depuis longtemps, j'ai pensé qu'en certains lieux que j'ai qualifiés 'd'acoustiques' (parce qu'ils sont les foyers régissant des espaces) les grandes formes faites des surfaces gauches d'une géométrie intelligente pourraient habiter nos grandes bâtisses de béton, de fer, ou de verre. J'ai cherché l'homme qui, comme l'ancien constructeur des nefs joindrait les charpentes et des planches pour constituer des coffrages dans lesquels le béton de statues inattendues serait coulé. Devant les bâtisses, à leur flanc ou sur leur front, les formes en appelleraient à l'espace. La sculpture en appelant à l'espace . . . il me semble que c'est être dans la ligne même de ces destinées plastiques. Architecture et sculpture: jeu savant, correct et magnifique, des formes sous la lumière." In Antoine Pevsner, *La Première exposition des oeuvres de Pevsner* (Paris, 1947).

115. Although this is not the way he ended up constructing the ruled surfaces at Ronchamp.

116. Naum Gabo, "Art and Science," in *Naum Gabo* (Cambridge, Mass., 1957), 180–181.

117. William Hogarth, *The Analysis of Beauty* (London, 1753), 7.

118. Ibid., 9.

119. Fred Angerer, *Surface Structures in Building* (London, 1961), 107.

120. Mario G. Salvadori, "Thin Shells," in Robert E. Fischer, ed., *Architectural Engineering: New Structures* (New York, 1964), 38–42.

121. Felix Candela, "Understanding the Hyperbolic Paraboloid: Part 2," *Architectural Record* (August 1958), 205. Candela, an advocate of intuition in earlier years, was at pains to point out that such intuition of structural form only developed from a clear understanding of the limits of calculability.

122. An early model of Firminy shows the square base and the tilted oval top held

apart by two metal rods. The shell surface is implied by tying strings between base and top. The string generators are spaced equally along the four sides of the base and equally around the perimeter of the oval, which seems a good idea but is very difficult to draw since the generators are neither parallel in any plane nor are they convergent. Several experiments later, with the Dirlik model of December 1963, a much simpler distribution producing a surface with greater visual complexity had been hit upon. Equal spacing is maintained, but the threads from the four sides of the square converge on the four axial points on the perimeter of the oval (making four plane triangles), while the threads from the four segments of the oval collect on the four corners of the square (making four elliptical cone segments). It was easier to draw, easier to build, and more interesting to look at. For the models see Eardley, *Le Corbusier's Firminy Church*, 44–58.

123. The formwork rested on a substructure of straight trusses that did follow the generator lines. The formwork was bent to shape over this simple frame, clearly shown in a construction photo taken by Maisonnier; see Le Corbusier, *Ronchamp: oeuvre de Notre-Dame du Haut*, 91.

124. Ibid., 120–121.

125. "Modulor reduit a 4m.52 = 2 × 2m.26. Le défie; je défie le visiteur de découvrir cela de lui-même. Si ce n'avait pas été tendu comme les cordes de l'arc, le jeu des proportions n'eût pas été joué!" Pauly, *Ronchamp: lecture d'une architecture*, 48. The dimension of 4.52 meters refers to the floor-to-ceiling height where the concave roof shell dips to its lowest along the west wall.

126. *Modulor II*, 153.

127. Le Corbusier, *Ronchamp: oeuvre de Notre-Dame du Haut*, 87.

128. Moore, *Le Corbusier and the "Méchanique Spirituelle,"* appendix E, cap. E4.

129. Ibid., chap. 15.

130. Le Corbusier, *When the Cathedrals Were White*, trans. F. E. Hyslop (London, 1947), 6.

131. *Modulor II*, 320.

132. Le Corbusier, *When the Cathedrals Were White*, 6n.

133. Le Corbusier, *The Chapel at Ronchamp*, 25.

134. "Il faut toujours dire ce que l'on voit, surtout il faut toujours, ce qui est plus difficile, voir ce que l'on voit." Le Corbusier, *Le Corbusier Talks with Students* (New York, 1963).

135. Rudolf Wittkower, "Le Corbusier's Modulor," in Columbia University School of Architecture, *Four Great Makers of Modern Architecture* (New York, 1963), 196.

136. Especially Jencks, *Le Corbusier and the Tragic View of Architecture*, 178.

137. Conversation with Guillermo Jullian de la Fuente, December 1989.

138. Stanislaus von Moos, *Le Corbusier l'architect et son mythe* (Frauenfeld, Switzerland, 1968), 245.

139. Source unknown: idiomatic punning phrase used in school to describe sums about the volume of water in containers.

140. Le Corbusier, *Ronchamp: oeuvre de Notre-Dame du Haut*, 128.

141. Basil Spence, "The Modern Church," *RIBA Journal*, 3rd series, 63 (July 1956), 372.

142. Sean Russell observed that Aldo Rossi's floating Theatre of the World borrowed the properties of whatever it stood next to. I found this idea very illuminating when considering Ronchamp.

143. Henri Bergson, "Laughter," in *Comedy*, ed. Wylie Sypher (New York, 1956), 80, 85.

144. Le Corbusier, *Précisions* (Paris, 1930), 37; translated as *Precisions on the Present State of Architecture and City Planning* (Cambridge, Mass., 1991).

145. Le Corbusier, *Ronchamp: oeuvre de Notre-Dame du Haut*, 20, 112. The photograph was taken by Franz Hubmann.

146. *Modulor II*, 196–200.

147. François Rabelais, *Gargantua and Pantagruel* (Harmondsworth, 1955), introduction by J. M. Cohen, 26.

148. Le Corbusier thought some of his buildings oscillated in and out of accord. Soltan recalls that Le Corbusier explained Garches as alternating *points vibrants* and *points morts* as one approached. Conversation with Jerzy Soltan, December 1992.

149. After writing this with some hesitation I learned that Hélène Lipstadt is looking into Rousseau's influence on Le Corbusier.

Chapter Eight: Forms Lost and Found Again

1. Le Corbusier, *The Four Routes*, trans. Dorothy Todd (London, 1947), 138–139, 141. What upset Le Corbusier was that when they taught architecture they taught the styles, which led to the kind of travesty he witnessed at Marathon, where a stupendous dam was "desecrated" with a period parapet.

2. Terry Shinn, *L'École polytechnique: 1794–1914* (Paris, 1980), 13.

3. Tableau indicatif de l'organisation de l'École central des travaux publics (Paris, 1795), and A. Fourcy, *Histoire de l'École polytechnique* (Paris, 1828), 378–379; Janis Langins, *La République avait besoin de savants* (Paris, 1987), appendix B1.

4. *Tableau, indicatif de l'organisation de l'École*.

5. See, for example, Allan Braham, *The Architecture of the French Enlightenment* (London, 1980), 255; Alberto Pérez-Gómez, *Architecture and the Crisis of Modern Science* (Cambridge, Mass., 1983), 280–282.

6. Werner Szambien, "Durand and the Continuity of Tradition," in R. Middleton, ed., *The Beaux Arts and 19th Century French Architecture* (London, 1982), 29.

7. Pérez-Gómez, *Architecture and the Crisis of Modern Science*, 309–311.

8. Even as a perspectivist, before his professorship, Durand showed a consistent preference for frontal views. See Testard, Sergent, and Durand, *Vues pittoresques de Paris* (Paris, 1795). See also Jacques Guillerme, "La tirannia dell'idealizzazione," *Casabella*, 520–521 (1986), 78–82, for an appreciation of the difference between Monge and Durand.

9. Descartes (book II of the *Geometry*), like Monge, foresaw the principal advantage of algebraic geometry in its capacity to describe more complex curves produced by compound mechanical motions.

10. The methodical variations shown in Durand's plates for the *Précis* are encyclopedic, not rational. They seem to be rule-regulated, but when the rule produces a result that is unacceptable within the established norms of architectural composition, the result is omitted from the tabulation.

11. Robin Evans, "Architectural Projection," in E. Blau and E. Kaufman, eds., *Architecture and Its Image* (Montreal, 1989), 27–30.

12. Bernard Huet, introduction to Werner Szambien, *J. N. L. Durand: 1760–1834* (Paris, 1984), 7.

13. Peter Booker, *A History of Engineering Drawing* (London, 1963), 107.

14. René Taton, *L'Histoire de la géométrie descriptive* (Paris, 1954), 19.

15. Amédée Frézier, *La Théorie et la pratique de la coupe des pierres* (Strasbourg, 1737–1738), vol. 1, 42–43; vol. 2, 7–12.

16. René Taton, article on Gaspard Monge, *Dictionary of Scientific Biography*, ed. C. C. Gillispie (New York, 1970–1976), vol. 9, 475.

17. M. Hachette and G. Monge, *Géométrie descriptive avec supplement* (Paris, 1811), supplement, 1ff. Their definition is as follows: "La surfáce la plus générale qu'on puisse engendrer par une droite mobile dans l'espace est déterminée, lorsqu'on assujettit cette droite à s'appuyer constamment sur trois courbes fixes, dont la forme et la position sont données." ("The most general surface that can be engendered from a line moving through space is determined when this line passes continuously through three fixed curves, the form and position of which are given.")

18. See Julian Lowell Coolidge, *A History of the Conic Sections and Quadric Surfaces* (Oxford, 1945), chaps. 11 and 12. According to Coolidge, in the work done by Monge and Hachette on quadrics (which include ruled surfaces) it is difficult to distinguish what should be attributed to which author. Hachette also collaborated with Monge on the algebraic development of quadrics in *Applications de l'analyse à la géométrie*. Coolidge, 169–177.

19. See for example Thomas Bradley, *Elements of Geometrical Drawing* (London, 1861), part II, plates 37–38, p. 45; and Joseph Woolley, *The Elements of Descriptive Geometry . . . and Its Application to Ship Building* (London, 1850), part I, 142–193. Practical applications were envisaged by the originators of the analytical treatment of ruled surfaces. Monge and Hachette wrote that: "à cause de l'usage fréquent de ces surfaces dans les arts graphiques, il est indispensable d'en bien connaître, sinon les propriétés, du moins la forme et la génération, et on peut acquérir ces connaissances par des simples considérations de géométrie." ("Because of the frequent use of these surfaces in the graphic arts, it is necessary to understand, if not their properties, at least their form and generation, and one might acquire this knowledge by the simple consideration of geometry.") *Géométrie descriptive*, 7. Charles Dupin, who developed the understanding of the ruled surfaces inspired by Monge, his teacher, titled the work in which they were expounded *Developments of Geometry with Applications to the Stability of Vessels, the Cutting and Filling of Earthworks, Optics, etc., and to Connect the Descriptive and Analytic Geometries of Monge* (Paris, 1813).

20. Joaquin Gomís and J. Prats Vallés, eds., *Gaudí*, with a preface by Le Corbusier (Barcelona, 1958).

21. Salvador Dalí, preface to Robert Descharnes and Clovis Prévost, *Gaudí the Visionary*, trans. F. Hill (London, 1971).

22. Cesar Martinell y Brunet, *Gaudí: His Life, His Theories, His Work*, trans. Judith Rohrer (Barcelona, 1975), 158.

23. Ibid., 128.

24. Le Corbusier, *Sketchbooks* (New York and Cambridge, Mass., 1981), vol. 1, C.11, fig. 700.

Chapter Nine: Rumors at the Extremities

1. Evert van Straaten, *Theo van Doesburg; Schilder en Architect* (The Hague, 1988), 114–137; Nancy Troy, *The De Stijl Environment* (Cambridge, Mass., 1983), 106–110; Allan Doig, *Theo van Doesburg: Painting into Architecture, Theory into Practice* (Cambridge, U.K., 1986), 149–152.

2. Yve-Alain Bois, "Metamorphosis of Axonometry," *Daidalos*, 1 (1981), 40–58.

3. Reyner Banham, *Theory and Design in the First Machine Age* (London, 1960), 191.

4. Manuel Corrada, "On Some Vistas Disclosed by Mathematics to the Russian Avant-Garde: Geometry, El Lissitzky and Gabo," *Leonardo*, 25 (1992), 377–378.

5. Charles Howard Hinton, *The Fourth Dimension* (New York, 1904), frontispiece. Richard Difford describes Hinton's color coding and its relation to van Doesburg's work in his dissertation, Polytechnic of Central London, 1992.

6. Linda Dalrymple Henderson, *The Fourth Dimension* (Princeton, 1983), 332.

7. Robin Evans, "Architectural Projection," in E. Blau and E. Kaufman, eds., *Architecture and Its Image* (Montreal, 1989), 34. This same hypercube illustration was certainly the inspiration for Peter Eisenman's complex of parallelogram frames in the Carnegie Mellon Research Institute project; see *Casabella*, 586–587 (1992), 102.

8. This does not mean that no such exports can be made, but when they are it is usually because an object designed in orthographic projection is drawn as if in oblique (axonometric) projection. Things made this way substitute parallelograms in place of rectangles. For examples see Robin Evans, "Not to Be Used for Wrapping Purposes," *AA Files*, 10 (1985), 70–71. Also the Eisenman project cited in note 7. Le Corbusier made use of this property in the upper stories of the Salvation Army hostel in Paris and the church at Firminy; see Le Fondation Le Corbusier, *The Corbusier Archive*, ed. H. Allen Brooks (New York, 1982–1984), vol. 31, 299, 302, 306, 360.

9. Albert Einstein, *Sidelights on Relativity*, trans. G. B. Jeffery and W. Perrett (New York, 1922), 50.

10. Ibid., 56.

11. Werner Heisenberg, *Physics and Philosophy* (New York, 1958), 122.

12. Erich Mendelsohn, *Letters of an Architect*, ed. Oskar Beyer, trans. G. Strachan (London, 1967), 167.

13. Ibid.

14. Erich Mendelsohn, "Dynamics and Function" (1923), in Ulrich Conrads, *Programs and Manifestoes on 20th-Century Architecture*, trans. Michael Bullock (Cambridge, Mass., 1971), 72.

15. Henderson, *The Fourth Dimension*.

16. Manuel Corrada, a mathematician studying modern art, arrives at a similar conclusion. He describes three roles for mathematics in art, one of which is as a metaphor of progress. Corrada, "On Some Visions Disclosed by Mathematics."

Conclusion: The Projective Cast

1. George Berkeley, *An Essay Towards a New Theory of Vision*, 4th ed. (Dublin, 1732), secs. 2, 50.

2. William M. Ivins, *Art and Geometry: A Study in Space Intuitions* (Cambridge, Mass., 1946), 5.

3. Ernst Mach, *Space and Geometry in the Light of Physiological, Psychological and Physical Enquiry*, trans. T. J. McCormack (La Salle, Illinois, 1906), 10.

4. Henri Poincaré, *Science and Hypothesis* (New York, 1952), 51–59.

5. This is the so-called "carpentered" world that surrounds us. Patrick Heelan describes it as follows: "The fact that we can and do in fact see in a Euclidean way indicates that some technological assistance has been provided. This in fact we have provided for ourselves by making and dispersing widely in all inhabited space engineered shapes; these together with the sensory organs and neurophysiological system linked functionally by a (virtually) instantaneous medium of communication constitute the visual Euclidean frame of reference." Heelan, *Space-Perception and the Philosophy of Science* (Berkeley, 1983), 277.

6. Giambattista Vico, *On the Most Ancient Wisdom of the Italians*, trans. L. M. Palmer (Ithaca, 1988), 50–52.

7. David R. Lachterman, *The Ethics of Geometry: A Genealogy of Modernity* (New York, 1989).

8. Howard Burns, Symposium on Architectural Representation, Harvard University Graduate School of Design, 1987.

9. Michel Foucault, "Dream, Imagination and Existence," trans. F. Williams, *Review of Existential Psychology and Psychiatry*, 19, no. 1 (1984–1985), 41.

10. David Hume, *A Treatise of Human Nature* (London, 1739), book 1, part 1, sec. 3.

11. John Hyman, *The Imitation of Nature* (Oxford, 1989).

12. See Daniel C. Dennett, *Consciousness Explained* (Boston, 1991), 53–55.

13. See Alan R. White, *The Language of the Imagination* (Oxford, 1990); Michael Tye, *The Imagery Debate* (Cambridge, Mass., 1991).

14. Friedrich Nietzsche, *The Will to Power*, ed. W. Kaufmann, trans. W. Kaufmann and R. S. Hollingdale (New York, 1968), cap. 518. See also caps. 481–482, 489–490, 499–500, 506, 565.

15. Richard Rorty, *Philosophy and the Mirror of Nature* (Princeton, 1979); Rodolphe Gasché, *The Tain of the Mirror: Derrida and the Philosophy of Reflection* (Cambridge, Mass., 1986).

16. Robin Evans, "Mies van der Rohe's Paradoxical Symmetries," *AA Files*, 19 (1990), 56–68.

17. Guarino Guarini, *Architettura civile* (Turin, 1737).

18. Ibid., trat. III, cap. xxi, os. 1–6.

19. H. A. Meek, *Guarino Guarini and His Architecture* (New Haven, 1988), 75.

20. Immanuel Kant, *Critique of Pure Reason*, trans. Norman Kemp Smith (London, 1929), 45.

21. Roger Scruton, *The Aesthetics of Architecture* (London, 1979), 43–52.

22. Luigi Moretti, "Strutture e seqenze di spazi," *Spazio: rassegna delle arti e dell'architettura*, 7 (December 1952–April 1953), 9–21.

23. Jacques Lacan, *The Four Fundamental Concepts of Pyscho-Analysis*, ed. J.-A. Miller, trans. Alan Sheridan (New York, 1978), 75.

Illustration Credits

All photographs were taken by the author, with the exception of those listed below. The author's family gratefully acknowledges the following individuals and institutions for permission to reproduce illustrations of works in their collections:

5 Bibliothèque de l'Institut de France, Paris; photographer Jean Loup Charmet

8 The Conway Library, Courtauld Institute of Art, London

11, 83, 88, 94, 113 The National Gallery, London

12, 112 The Metropolitan Museum of Art, New York

13 Kupferstichkabinett, Staatliche Museen zu Berlin; photographer Jörg P. Anders

16, 52 Musei Vaticani

20, 91 The Royal Collection, Windsor

22, 27 Collection, The Museum of Modern Art, New York

23, 27, 49, 148, 149, 156, 157, 158, 159, 167, 171, 184 © DACS 1995

24, 57 The British Library, London

25 Kunsthalle, Hamburg

26, 49 The Trustees of the Tate Gallery, London

26, 172 © ADAGP, Paris, and DACS, London, 1995.

31 The Busch-Reisinger Museum, Harvard University Art Museums, Gift of Walter Gropius

42, 60 Archiv der Akademie der Künste Berlin, Hans-Scharoun-Archiv

53, 181 The Board and Trustees of the Victoria and Albert Museum, London

58 The Provost and Fellows of Worcester College, Oxford

66, 87 Pinacoteca di Brera, Milan

67, 68, 72, 96, 97, 98 Galleria degli Uffizi, Florence

71, 90, 93 Archivi Alinari and Art Resource, New York

92, 95, 120 Musée du Louvre, Service Photographique de la Réunion des Musées Nationaux

111 Cooper-Hewitt Museum, Smithsonian Institution

125, 126, 127, 195 British Architectural Library, RIBA

141 Accademia Carrara di Belle Arti, Bergamo

149, 166, 168, 171, 173, 184 Fondation Le Corbusier

183 The Architectural Association Slide Library and Andrew Higgott

185, 186 The Stedelijk Museum, Amsterdam

189 The Fogg Art Museum, Harvard University Art Museums, Gift of Mrs. Irving M. Sobin

Index

Works ascribed to a specific artist are indexed under the artist's name. Buildings are primarily indexed by name, but are also entered under the name of the architect. Numbers in italics are figure numbers.

Aalto, Alvar, 70–74, 379n63
    Baker Dormitory (MIT), 71; *32, 33*
    Between Humanism and Materialism, 73
    "The Humanizing of Architecture," 70–73
    Jyväskylä University Primary School, 73–74
    on psychophysical function, 70–71, 74
    tuberculosis sanatorium (Paimio), 70
Acoustics, 96, 381nn114,119
Adami, Valerio
    Glas, 86
Agatharcus, 133
Aircraft, 305–307
Albers, Josef, 339
Alberti, Leone Battista, 17, 38, 42, 110–116, 141, 146, 247, 254
    De statua, 155
    map of Rome, 44–45
    on measurement, 155–156; *79*
    On Painting, 110, 133
    on perspective, 110–113, 120, 132–142, 147–149, 151–153, 173, 176–177, 254, 256, 384n33, 395n72; *51, 60*
    on proportion, 249–250, 265
    Santa Maria Novella (Florence), 248; *137*
    Sant'Andrea (Mantua), 116; *56*
    Ten Books of Architecture, 141, 375n80
Ambiguity, 9, 12, 30, 34, 41, 44, 339–344, 353, 376n94
Analytic geometry, 326–327, 330, 400nn9,19
Anamorphosis, 126, 290, 352
Anet, Château d' (Normandy)
    trompe at, 180–181, 183–189, 193–195, 200, 208, 237, 360, 388nn13,14,16; *101–106*
Anthropocentrism, 18, 21, 25, 38, 100, 374n38
Anthropometric dimensions, 274–275, 279, 282; *148*. See also Human figure, proportions of
Anthropomorphism, 102, 284–287, 291, 317
Apel, Willi, 244
Apollinaire, Guillaume, 65
    Les Peintres cubistes, 65
Apollonius, 41
Arcadia, 75–76
Architectural drawing, xxxi, xxxv, 60–62, 86, 94, 99–101, 107–110, 113–121, 151, 175–177, 195, 226, 238, 274, 302–305, 326–327, 333–334, 337, 341, 349, 359–360, 368, 382n20, 388n37, 392n119. See also Orthographic projection

axonometric (*see* Axonometric projection)
    elevations, 113–119, 254
    plans, 113–116, 118, 226–228
    sections, 118–120, 228
    sketches, 278, 282, 302, 337, 346–348, 368
    traits, 121, 179–181, 183–189, 193–201, 203–209, 214, 220–224, 231, 237, 387n1, 389nn39,42,49, 390n68; *102–104, 106, 115, 117, 118, 128* (see also Stereotomy, stonecutting; *Trompes*; Vaults)
Arch of Augustus (Fano), 117; *57*
Aristippus, 372n1
Aristoxenus, 266
Armillary spheres, 376n90
Arup, Ove, 380n104
Astronomy, 18–19, 27–30, 41–42, 51, 246, 268, 346, 374n59, 376n90; *19*
Asymmetry, 83
ATBAT consortium, 295, 301, 308–309
Atheneum (New Harmony, Indiana), 57, 92; *22*
Auden, W. H., 102
    Reading, 75
    Vespers, 75
Augustine (Saint), 27
Aureole, 25–26, 35. *See also* Light, as divine radiance
Auxiliary projection. *See* Orthographic projection, auxiliary
Axonometric projection, 60, 238, 256, 337–341, 344
    architecture of, 337–341, 401n8
    scale measure in, 337, 339
    visual ambiguity of, 339–344

Baker Dormitory (MIT), 71; *32, 33*
Baltard, J. L.
    slaughterhouse hall (La Villette, Paris); *39*
Banham, Reyner, 339
Baptisteries, 14
    Padua, 374n40
    Pisa, 14; *6*
Barbaro, Daniele, 147, 267
Barcelona Pavilion. *See* German Pavilion (International Exposition, Barcelona, 1928)
Bardi, Giovanni, 246
Barolsky, Paul, 163
Barthes, Roland, 85
Bartoli, Cosimo, 250
Bataille, Georges, 85, 87
Bath Abbey, 231
Battle scenes, depiction of, 167–175; *92, 94–96*
Bauhaus (Dessau), 57–59, 67; *23*
Baxandall, Michael, 143, 145, 256
Bek, Lise, 141–142

Bekaert, Geert, 181
Bentham, Jeremy
 panopticon, 102, 125, 130
Berdyaev, Nicholas, 102
Berenson, Bernard, 142
 The Italian Painters of the Renaissance, 143
Bergson, Henri, 317
Berkeley, George (Bishop), 351–352
Bernardino di Pietro da Carona, 117; *57*
Beza, Theodore, 36
Binchois, Gilles, 246
Blondel, François, 203, 209, 241, 265, 390n71
 Cours d'architecture; 136
Blouet, Abel, 225
Blunt, Anthony, 183, 388n13
Boccioni, Umberto, 377n5
Bodiansky, Vladimir, 305–306, 308
Boen, Johannes, 393n9
Boethius, Anicius Manlius Severinus, 103–104
Bohr, Niels, 101
Bois, Yve-Alain, 338
Bolyai, Janos, 62
Booker, Peter, 154
Borgia, Cesare, 45
Borromini, Francesco
 San Carlo alle Quattro Fontane (Rome), 121, 383n26
Boskovits, M., 147
Bosse, Abraham, 203
 La Pratique du trait, 201, 205–206; *117*
Botticelli, Sandro, 153
 Paradiso (illustration for Dante's *Divine Comedy*), 25; *13*
Botticini, Francesco
 Assumption of the Virgin, 21; *11*
Bradley, Thomas
 Elements of Geometrical Drawing; 180
Bradwardine, Thomas, 376n85
Bramante, Donato, 7, 110
 St. Peter's (Rome), proposals for, 34, 373n20, 382n12
 San Pietro in Montorio (Tempietto) (Rome), 9–11, 372nn11–12; *4*
 Santa Maria delle Grazie (Milan), 373n20; *7*
Braque, Georges, 57, 59–60, 377n10
 Mandola; 26
Brianchon, Charles-Julien
 principle of duality, 324
Bricolage, 76–79
Bristol Cathedral, 228
Brocone, Ruth, 163
Brooks, H. Allen, 307
Brown, Howard Mayer, 247
Brunelleschi, Filippo, 17, 30, 244, 254, 265, 376n93, 395n64
 Pazzi chapel (Santa Croce, Florence), 26–27
 on perspective, 113, 133, 175–176, 244, 251, 254, 256, 384n31
  demonstration panels, 136–138, 140, 176, 256, 384n31, 387n101
 San Lorenzo (Florence), 14, 26–27, 248, 253; *14*
 Santa Maria del Fiore (Florence cathedral), 14, 243–247, 373n21
 Santo Spirito (Florence), 253; *139*
Bryson, Norman, 132–133
Burckhardt, Jacob, 3–4
Burgess, Gelett, 59–60, 377n10
Burgin, Victor, 89
Burns, Howard, 355

Calvo, Marco Fabio
 map of ancient Rome, 47; *21*
Cambiaso, Luca
 figure study, 147; *72*
Camouflage, 57, 80, 83
Campo Santo (Pisa), 23
Candela, Felix, 312, 398n121
Canterbury Cathedral
 Lanfranc's Tower, 225
Capitalism, 80, 101–102
Capitoline palaces (Rome)
 Palazzo del Senatore, 281; *151*
Caramuel Lobkowitz, Juan, 265
Carpaccio, Vittore
 Birth of the Virgin; 141
Cartesian geometry, 125, 130, 326–327, 330, 339–341, 354
Cartography, 44–47, 135
Casa Batlló (Barcelona), 331; *182*
Casa Milà (Barcelona), 331
Cavalcaselle, G. B., 33
Centralized churches, 360, 377n95
 centers of, 6–11, 14–16, 19–22, 30, 35–36, 40, 102, 376n94; *3, 10*
 as characteristic architectural expression of the Renaissance, 3–47, 55, 103
 dome frescos of, 19–20, 22–23, 30, 35, 40; *10* (*see also* specific frescos indexed by artist)
 as representation of Renaissance cosmology, 16–22, 26–30, 42–44
 structure of, 14, 38, 373n20
 symbolism in, 16–19, 21, 27–29, 36–37, 40, 43 (*see also* Symbolism)
Centralized planning, 12, 18, 26, 96, 107. *See also* Baptisteries; Centralized churches
Cerceau, Jacques Androuet du. *See* Du Cerceau
Cervantes, Miguel de
 Don Quixote, 316
Cesariano, Cesare, 249
Chandigarh. *See* Le Corbusier, Chandigarh

Château d'Anet (Normandy). *See* Anet, Château d'
Chéreau, Jean, 389n42
Churches. *See also* specific churches indexed by name
 centralized (*see* Centralized churches)
 Latin cross plan, 42
 Lourtier, 293; *160*
Circles, xxv, 39–40, 48–53, 376nn83–84
 in cartography, 44–47
 motion in, 41–42, 376n90
 in Renaissance art and architecture, 14–47, 107 (*see also* Centralized churches)
 in Renaissance astronomy, 18–19, 23, 27–29, 41–42, 51; *19* (*see also* Astronomy; Cosmology)
Clark, John Willis, 225
Clark, Kenneth, 142
Classical orders, 181, 208–212, 214–220, 250, 270, 327
Claudel, Paul
 L'Annonce faite à Marie, 287
Clement VII (Pope), 111
Clouet, François, 76
Collage, 74–79, 92, 102
Color dimension, 339–341
Composition, xxv–xxvii, xxxi, xxxiii, 5, 17, 37–38, 40, 44, 102, 146, 267, 349, 360, 395n72; *137, 151. See also* Proportion
 centralized (*see* Centralized churches; Centralized planning; Perspective, central)
 fragmented (*see* Fragmentation)
 golden section (*see* Golden ratio)
 musical, 245 (*see also* specific compositions indexed by composer)
 structure of, 247–251, 267, 271, 281, 292, 394n39
Conoids. *See* Ruled surfaces
Conrad, Joseph
 The Secret Agent, xxv
Constantine (Emperor), 26
Constructivism, 67, 85, 92, 102, 327
Cooper, Douglas, 57
Coop Himmelblau, 93–94, 100
 Falkestrasse 6 (rooftop lawyer's office, Vienna), 93; *41*
Copernicus, Nicolaus
 Commentariolus, 42
 heliocentric astronomy of, 29–30, 41, 374n59
 De revolutionibus orbium coelestium, 376n90
Corbusier. *See* Le Corbusier
Cordemoy, Abbé de, 212
Corrado, Manuel, 339
Correggio (Antonio Allegri), 20
 Christ and the Apostles (San Giovanni Evangelista, Parma), 373n32

Cosmology, 344, 346
  Christian (*see* Cosmology, Renaissance; Theology, Christian)
  medieval, 23–24
  Renaissance, 16–25, 27–30, 40–43, 47, 51, 243, 374n38, 376n90; *19*
Cremer, Lothar, 381nn114,119
Critical theory, 104, 124, 126, 364, 379n78, 380n99
Crowe, J. A., 33
Cubism, 55, 57, 60, 63, 80, 91–92, 94, 101–102, 378n21, 380n101. *See also* specific artists indexed by name
  architecture of, 55, 57–59, 63–71, 378n21, 379n62
  multiple viewpoints of, 57–63
  and perception, 63–65, 378n21
  compared to technical drawing, 60–63, 94
  theory of, 57, 59, 62, 377n14
Cucci, Giorgio, 203
Cusanus, Nicolaus
  Of Learned Ignorance, 40

D'Alembert, Jean
  Encyclopédie, 209
Dalí, Salvador, 331–333
Damisch, Hubert, 133
Dante Alighieri
  Convivio, 19
  Divine Comedy, 19, 21, 24–25; *13*
David, Gérard
  Christ Nailed to the Cross; 113
Da Vinci. *See* Leonardo da Vinci
De Certeau, Michel, 78
Deconstruction, 83–86, 89–93, 102–104, 379n78
Dee, John
  Mathematical Praeface to the Elements of Euclid, 108–109, 382n7
Delacroix, Eugène
  Femmes d'Alger, 91
Delagardette (architect, 18th century), 209, 213
Delaunay, Robert, 377n5
Deleuze, Gilles, 89
Del Monte, Guidobaldo, 132
Delorme, Philibert, xxxiv–xxxv, 180, 183, 206–207, 249
  Château d'Anet (*see* Anet, Château d')
  on geometry, 197–200
  Hôtel Bullioud (Lyons), 182–183; *100*
  house for Patouillet (Paris), 181–182
  Premier tome de l'architecture, 180–181, 184, 187, 197–200, 205, 210, 221, 231; *101, 102, 105, 114, 119, 128*
  on ribbed vaults, 221–224, 226
  on stonecutting, 180–189, 195, 201, 209, 387n10, 388n16

De Machy, P. A.
  painting of Halle au Blé, 220
De' Medici, Cosimo, 169
Democracy
  architecture and urbanism of, 73, 75, 78–79
Derand, François, 203, 207, 213–214, 329, 390nn66,70
  L'Architecture des voûtes, 305, 390nn68–69; *118*
  Saint-Paul-Saint-Louis (Paris), 213–214; *121*
Derrida, Jacques, 84–89, 91–92, 359, 379nn70,77–79
  on light, 25
  Structure, Sign and Play, 53
  The Truth in Painting, 379n79
Desargues, Girard, 132, 201, 256, 324
  four-point involution (theorem of), 201, 256–258, 266; *142*
  on stonecutting, 201–203, 389n41; *115*
Descartes, René, 327, 354, 390n70, 400n9
  cogito, 130
  on perception, 125, 358
Descriptive geometry, 60, 94, 151, 208, 239, 303, 324, 326–328, 339; *177*. *See also* Orthographic projection; Technical drawing
  conic sections, 382n20
Desprez, Josquin, 247
  Missa Di dadi (Dice Mass), 393n28
De Stijl; *187*
Dews, Peter, 89
Diderot, Denis
  Encyclopédie, 209
Dionysius the Areopagite (pseudo-Dionysius)
  On the Celestial Hierarchies, 19
Doesburg, Theo van, 67, 348
  architectural projects, 67, 338, 341
  axonometric projection, 338–341, 344
  Color Construction in the Fourth Dimension of Space-Time, 339–341; *186*
  Counter-Compositions, 67, 338–341, 344, 349
  Counter-Composition V; 185
  on the fourth dimension, 339–344, 400n5
  hypercubes, 341–344; *187*
  maison particulière, 338
Dome frescos. *See* Frescos, dome
Donatello (Donato di Betto di Bardi), 266
Drouin, René
  La Première exposition des oeuvres de Pevsner; 172
Ducal Palace (Urbino), 387n102
Du Cerceau, Jacques Androuet, 131, 383n17
  Leçons de perspective positive, 131; *65*
Duchamp-Villon, Raymond
  Maison cubiste, 65; *28*

Dufay, Guillaume, 246–247, 262, 265, 393nn26,28
  Ave Regina Coelorum, 247
  L'homme armé, 247
  Se la face ay pale, 392n8
  Nuper rosarum flores, 243–247, 262, 265
Duffau, R.-F., 291
Dunstable, John, 246–247, 262, 265, 392n8, 393n28
  Preco preheminencie, 246
  Veni Sancte Spiritus, 246
Dupin, Charles, 400n19
Durand, J. N. L., 326–327, 400n8
  planning grids of, 326–327, 339; *178*
  Précis des leçons d'architecture, 325, 400n10; *178*
Duras, Marguerite, 87
Dürer, Albrecht, 147, 382n20, 386n66
  on human proportions, 254, 259; *140, 144*
  perspective machines of, 127–129, 140, 150–151, 154; *63, 64*
  on perspective proportion, 259–260; *143, 145*
  Underweysung der Messung, 127; *63, 64, 145*
  Vier Bücher von menschlicher Proportion, 259; *140, 143, 144*
Duyster (engineer), 298

Écluse de Kembs-Niffer
  customs post, 308
École Normale Supérieure, 324
École Polytechnique, 323–327, 339, 377n13, 399n1
Eesteren, Cornelis van, 338, 341
  maison particulière, 338
Egyptian architecture, 117
Einstein, Albert, 273, 344
  Geometry and Experience, 344
  relativity (special and general theories of), 52, 57, 341, 344–346
  on space, 344–346, 351–353; *190*
  on visualization, 344–345, 351–352; *190*
Einstein Tower (Potsdam), 346–349; *191*
Eisenman, Peter, 84, 86, 89–90, 100, 339
  Carnegie Mellon Research Institute project, 401nn7–8
Eliade, Mircea, 301
Engineering drawing. *See* Technical drawing
Erni, Hans
  Die neuen Ikarier; 153
Eucharist, 30, 33–37. *See also* Theology, Christian
Euclid, 108, 375n77, 394n54
  Elements, xxxii, 108, 200, 375n77, 394n54
  geometrical postulates of (*see* Euclidean geometry, axioms of)
  Optics, 130, 383n14
  on solid geometry, 324, 376n83

Euclidean geometry, xxvi–xxvii, xxxi–xxxii, 62, 124–125, 130, 133, 200–201, 255–256, 266, 324, 344–345, 351–353, 364–366, 375n77, 376n83, 389n39, 395n66, 401n5
  analytic (*see* Analytic geometry)
  axioms of, 62, 108, 133
  figures of, xxvi–xxvii, xxxii, 37–44, 195, 249, 287
    four-dimensional, 60; *24* (*see also* Hypercubes)
    plane (two-dimensional), 40, 42, 290–291 (*see also* Circles)
    solid (three-dimensional), 42, 44, 199, 291, 324, 328–330; *70, 177* (*see also* Ruled surfaces; Spheres)
  synthetic (*see* Synthetic geometry)
Eugenius IV (Pope), 243

Facteur Cheval, 76
Feininger, Lyonel, 379n62
Ferrari, Gaudenzio, 20
  Santa Maria dei Miracoli (Saronno), 374n40
Fibonacci series, 275, 290, 296, 395n7, 397n75. *See also* Modulor
Ficino, Marsilio, 21, 27–29, 38, 49, 198, 363, 375n79
Filarete (Antonio di Pietro Averlino), 396n50
Firminy chapel, 308, 312, 398–399n122, 401n8
Fischel, Oskar, 33, 375n61
Flaubert, Gustave
  Dictionary of Accepted Ideas, xxv
Florence cathedral. *See* Santa Maria del Fiore
Forum des Halles (Paris), 80–83; *36*
Foster, Peter, 237, 391n111
Foucault, Michel, 80, 85, 89, 91, 102, 357, 374n38
  The Order of Things, 100
  Panoptisme, 125
Fournier, Daniel
  Treatise on the Theory of Perspective; 50
Fourth dimension, 339–344, 348; *186*
  hypersolids, 60; *24* (*see also* Hypercubes)
Fragmentation
  in architectural composition, 55–57, 65–74, 83, 89, 91–102
  as characteristic expression of the twentieth century, 55–105
  in painting, 55, 57–63, 67, 91–92, 94, 102, 168
  urban, 74–80, 102
Frampton, Kenneth, 73
Frangenberg, Thomas, 131
Freeman, E. A., 229
Freemasonry, 200, 203, 205

French Pavilion (Exposition de l'Eau, Liège), 308–309
Frescos. *See also* specific frescos indexed by artist
  dome, 19–20, 22–23, 30, 35, 40, 373n32, 374n40; *10*
Freud, Sigmund, 89, 355
Freundlich, Erwin, 346
Freyssinet, Eugène, 308
Frézier, Amédée, 195–196, 208–210, 212, 220–221, 328, 387n1, 390n66
  conoidal *arrière-voussoirs; 179*
  corps régulièrement irrégulier, 329
  Dissertation sur les ordres d'architecture, 208–209, 389n59
  l'escalier vis de Saint-Gilles suspendu, 193; *109*
  La Théorie et la pratique de la coupe des pierres, 328, 388n29; *109, 179*
Friedrich, Caspar David
  The Wreck of the "Hope" (The Polar Sea); 25
Fry, Edward, 81
Functionalism, in architecture, 70–73, 75, 94, 100–101, 349
Futurism, 67, 327, 353
  Futurist House project, 66; *29*

Gabo, Naum, 310
Galilei, Galileo, 250
Galilei, Vincenzo, 246, 263
Gardiner, Stephen, 284
Gaudí, Antoni, 331–334
  Casa Batlló (Barcelona), 331; *182*
  Casa Milá (Barcelona), 331
  Güell chapel, 331–333
  Palau Güell, 331
  Parochial School of the Sagrada Familia (Barcelona), 334; *184*
  Sagrada Familia (Barcelona), 331–334; *183*
Gauricus, Pomponius, 394n49, 395n72
Gebser, Jean, 119–120, 359
Geometry, xxv–xxxi, xxxiii–xxxv, 44–47, 101, 127, 130, 197–199, 348, 354–355, 359–363, 366. *See also* specific geometries indexed by name
  as metaphor in art, 348–349, 354, 401n16
  as model of nature, 42–43, 47
German Pavilion (International Exposition, Barcelona, 1928), 360
Gerolamo da Vicenza
  Death and Assumption of the Virgin, 26
Geymüller, H. von, 34
Ghirardo, Diane, 86
Ghyka, Matila (Prince), 285
Gibson, J. J., 359
Giedion, Sigfried, 70, 94–95, 158, 348, 381n111
  on Aalto, 70, 73, 94–95

collage of the RCA Building, 58
  on cubism and architecture, 57–59, 67; *23*
  Space, Time and Architecture, 57–58, 67–70, 381n111; *23*
Gilbert, Creighton, 144
Gioseffi, Decio, 159–160, 162, 386n75
Giotto di Bondone, 163
Gleizes, Albert, 377n14
Gloucester Cathedral, 228
  choir, 226–229, 391n94; *129, 130*
  cloisters, 229–231, 391nn98,99; *131*
  Lady Chapel, 228
Golden ratio, 146, 274–275, 281, 285, 290–291, 394n51
Goltzius, Hendrick
  Job in Distress, 196, 388n31; *112*
  Man of Sorrows, 388n31
Gombrich, Ernst, 75, 168, 378n21
Goodman, Nelson, 126–127
  Languages of Art, 126
Griffin, David
  City of Composite Presence, 76–78; *35*
Gropius, Walter
  Bauhaus (Dessau), 57–59, 67; *23*
  Kapp-Putsch Monument (Memorial to the March Victims, Weimar), 67; *31*
Grosseteste, Robert, 23
Guarini, Guarino, xxix, xxxiv–xxxv, 207–208, 305–306, 360, 389n49
  Architettura civile, 305, 360, 372n5; *193, 196*
  on conoids, 305, 309, 329; *169*
  Euclides adauctus, 305
  Santa Maria della Divina Providenza (Lisbon); *195, 196*
  Santissima Sindone, chapel of (Turin cathedral), 360; *193, 194*
Guattari, Félix, 89
Güell chapel (Güell Colony), 331–333
Guido, Angel, 292
Guillerme, Jacques, 118

Habermas, Jürgen, 91
Hachette, Jean Nicolas Pierre, 330, 400nn17–19
Hacking, Ian, 104
Halfpenny, William, 225
Halle au Blé (Paris), 214–220, 390nn75–76; *125–127*
Halles, Les (Paris). *See* Forum des Halles
Hanning (co-worker of Le Corbusier)
  development of the Modulor, 279–281, 285; *150*
Hardy, G. H., xxix
Häring, Hugo, 94–95, 381nn110–111
Harmony, 241–245, 261–266, 268, 270, 315; *146*
  in human proportions, 254, 259; *140, 144*

in perspective measures, 244, 251–254, 258; *138, 143*
as universal order, 6, 242–243, 246, 265, 267, 394n63; *136*
Harries, Karsten, 102
Hartford Seminary (Connecticut), 83
Harvey, David, 80, 94
Harvey, John, 236
Hauser, Arnold, 376n91
Hegel, G. W. F., 209
Heidegger, Martin, 282
Henderson, Linda Dalrymple, 60, 348
Heninger, S. K., Jr.
  The Cosmographical Glass, 376n90
Henry VI
  tomb for, 238, 392n118
Hersey, George, 34
Hilton, Timothy, 67
Hinton, Charles Howard, 341, 400n5
Hipparchus, 41
Hire, Philippe de la, 203
Hitler, Adolf, 96
Hoesli, Bernhard, 291
Hogarth, William, 311
Holmes, Sir Charles, 33
Hôtel Bullioud (Lyons), 182–183
Hôtel de Ville (Arles), 213
Howlett, Robert
  stern of the *Great Eastern* (photograph); *181*
Human figure
  depiction of, 133–138, 142–147, 153–163, 167–168, 190, 248, 254, 259, 287, 290, 311, 317, 386n75
  measurement of (*see* Measurement)
  proportions of, 102, 254, 259, 274, 279, 284–287; *140, 144, 156* (*see also* Anthropometric dimensions)
Humanism, 6, 85, 100, 103. *See also* Anthropocentrism
Humanization of modernism, 55, 70–74, 78–80, 83, 94–95, 349
Hume, David, 357
Hyman, John, 358–359
Hypercubes, 341–344, 401n7; *187, 188*

Icons. *See* Signification; Symbolism
Imagination, 357–358, 363, 368–370
Institute of Architecture and Urban Studies (New York)
  "Idea as Model" exhibition, 90; *40*
Instruments
  drawing, 40, 118, 287
  measuring (*see* Measurement)
  musical, 263
Intuition, xxviii–xxix, 130, 135, 159, 162, 266–267, 270
Isometric projection, 339–341, 344. *See also* Axonometric projection

Ivins, William, xxxiii, 352

Jakobson, Roman, 101
Jameson, Fredric, 89
Janák, Pavel, 65
Janyns, Robert, 391n101
Jeanneret, Charles Édouard. *See* Le Corbusier
Jouffret, Esprit, 377n13
  polyhedroids, 60; *24*
  Traité élémentaire de géométrie à quatre dimensions; 24
Jousse, Mathurin, 203, 207, 209, 389n46
Julian (Emperor), 26
Julius II (Pope), 34
Jullian de la Fuente, Guillermo, 296, 316
Jyväskylä University Primary School, 73–74

Kahn, Louis, 74, 78
Kahnweiler, Daniel-Henry, 60
Kant, Immanuel, 363–366, 379n79
Karajan, Herbert von, 96
Kepler, Johannes, 42, 291
  Astronomia nova, 51
  celestial motion, 41–42, 51, 268
Kidson, Peter, 235
King's College chapel (Cambridge), 231
Koetter, Fred
  Collage City, 74–80, 102
Kolhoff, Hans
  The City of Composite Presence, 76–78; *35*
Koyré, Alexandre, 374nn38,59
  The Astronomical Revolution; 19
Krautheimer, Richard, 175–176
Kubovy, Michael, 127, 383n9
Kuhn, Thomas, 19, 374n38

Labatut, Jean, 277
Labrouste, Henri
  Temple of Hera (reconstruction, Paestum), 382n22
Lacan, Jacques, 89, 102, 125, 359, 369
Lachterman, David
  The Ethics of Geometry, 354
Lafaille, Bernard, 301
La Flèche, Jesuit academy at, 203, 214, 390n70; *107*
Lang, Susanne, 176
La Rue, J. B. de, 207–209, 387n1
  Traité de la coupe des pierres, 204–205; *116*
La Tourette. *See* Sainte-Marie de la Tourette
Laugier, Abbé, 212–214, 220, 390n66
  Essai, 213, 220, 390n64
Laurana, Luciano da
  Ducal Palace (Urbino), 387n102
Lavin, Marilyn, 144
Le Camus de Mézières, Nicolas, 214–218, 220
  Halle au Blé (Paris), 203, 214–218, 220; *125*

Le Corbusier (Charles Édouard Jeanneret), 67, 90, 276–277, 282–288, 291, 295–301, 307–309, 315–318, 320, 323, 399n1
  Aircraft, 305, 307; *170, 176*
  Apollo and Medusa; 157
  Après le cubisme, 67
  L'Art décoratif d'aujourd'hui, 284; *152*
  Chandigarh, 303, 398n91
    Palace of Justice (High Court), 308
    Museum of Knowledge, 308
    Palace of Assembly, 307–308, 312–314, 398n111
  on composition, 67, 248, 281, 291–292, 397nn61–63; *151*
  On Discovering Gaudí's Architecture, 331
  Écluse de Kembs-Niffer, 308
  Firminy chapel, 308, 312, 398–399n122, 401n8
  French Pavilion (Exposition de l'Eau, Liège), 308–309
  on Gaudí, 331–334; *184*
  on gender and form, 277, 284–288, 291; *155, 157*
  on handcraft, 282–284, 301, 307; *152*
  "House for Myself," 308
  Journey to the East, 282
  and the machine aesthetic, 284, 301, 305–307, 312, 328
  on mathematics, 275, 281, 285, 291
  Ministry of Health (Rio de Janeiro), 281
  Modulor I and II, 242, 273, 275, 278–279, 287, 291, 295–296, 314–315, 319, 394n51, 395n8; *148, 156* (*see also* Modulor)
  Notre-Dame du Haut (Ronchamp) (*see* Notre-Dame du Haut)
  Olivetti Computer Center (Milan), 308
  Philips Pavilion (Brussels International Exhibition, 1958) (see Philips Pavilion)
  Poème de l'angle droit, 288, 314; *158*
  ruled surfaces, 301, 305–306, 308–310, 312–315, 317, 329, 334, 398nn114–115, 399n125; *165, 168, 171*
  Sainte-Marie de la Tourette, monastery of, 296, 308; *162*
  Salvation Army Hostel (Paris), 401n8
  sculpture, 282, 301–302, 310–311, 396n31, 398n114
  secrets, 314–316, 399n134
  Unité d'Habitation (Marseilles), 295, 308, 398n108
  Vers une architecture; 151
  Villa Savoye (Poissy); *30*
  Villa Stein (Garches), 67, 292, 397n61, 399n148
  When the Cathedrals Were White, 315
Ledeur (Canon), 273

Ledoux, Claude-Nicolas, 103
Le Mans Abbey. *See* Notre-Dame de la Couture
Lemercier, Jacques, 213, 390n67
   Oratoire, 213–214
Lencker, Hans
   Perspectiva, 386n66
Leo X (Pope), 47, 107, 113
Leonardo da Vinci, 135–136, 142, 168, 251–253, 373n20
   Adoration of the Magi, 135; *67, 68*
   Battle of Anghiari, 168–169
   depiction of disorder, 168–169
   map of Imola, 44–45, 135; *20*
   Place for Preaching, 11, 372n13; *5*
   spheres; *18*
   A Town Overwhelmed by a Deluge; *91*
Leoni, Giacomo, 250
Lethaby, W. R., 236
Lévi-Strauss, Claude, 76, 79
   The Savage Mind, 76
Lhote, André, 62
Liberius (Pope), 355
Libeskind, Daniel, 100
Light, xxxiii, 108–109, 127, 346, 358, 383n12
   as divine radiance, 23–27, 30, 35
Lipton, Eunice, 59
Lissitzky, El (Lazar), 338, 348
   A. and Pangeometry, 394n51
   Proun Room, 342; *189*
Literary theory, xxxvi, 377n95
Lobachevsky, Nikolai, 62
Lobkowitz. *See* Caramuel Lobkowitz
Lomazzo, Giovanni Paolo, 20, 195, 255, 388n27, 394n47
   Choirs of Angels (Foppa Chapel, San Marco, Milan), 22, 30, 374n40; *9*
Longhi, Roberto, 143
Lonja (Barcelona), 213
L'Orme, Philibert de. *See* Delorme
Lotz, Wolfgang, 12–14, 107–108, 110
   The Rendering of the Interior in Architectural Drawing of the Renaissance, 107, 118
Lucretius, Titus, 63, 136
Luther, Martin, 34
Lyotard, Jean-François, 57
   Discours Figure, 91

Mach, Ernst, 352
Machine aesthetic, 284, 327–328, 349; *181*. *See also* Le Corbusier, and the machine aesthetic
Maillard, Elisa, 285–287, 291; *156*
Maisonnier, André, 278, 281, 285, 296
   on the construction of Ronchamp, 302–305, 309, 398n91; *168, 173*
Maisons-Lafitte, 213
Manetti, Antonio, 175–176, 384n31, 395n64

Mannerism, 195–196
Mansart, François
   Maisons-Lafitte, 213
Mansart, Jules Hardouin
   Hôtel de Ville (Arles), 213
Marchi, Vergilio
   Futurist House project, 66; *29*
Marinetti, Filippo, 67
Marolois, Samuel, 385n57
Martellange, Étienne, 213
   Jesuit chapel (La Flèche), 214
Martinell y Brunet, Cesar, 333–334
Martone, Thomas, 159
Marx, Karl, 381n2
Masolino
   The Foundation of Santa Maria Maggiore, 355; *192*
Masonry construction, 14, 18, 180, 203–205, 208–210, 214–231, 236–237, 239, 373n21. *See also* Stereotomy; Stonecutting; Structure
Mathematics. *See also* Geometry
   beauty in, xxix–xxx, 291 (*see also* Proportion, as measure of beauty)
   reasoning of, xxviii–xxx, 40
Matossian, Nouritza, 298
Matta-Clark, Gordon, 90, 380n95
   Conical Intersect, 90–91
   Idea as Model exhibition, 90–91; *40*
   Splitting: Four Corners, 89
Measurement, 135, 352. *See also* Surveying
   of the human body, 154–156, 162; *79*
   instruments of, 47, 226
   finitorum, 155–156; *79*
Meek, H. A., 360
Mei, Girolamo, 246, 266
Meier, Richard
   Atheneum (New Harmony, Indiana), 57, 92; *22*
   Hartford Seminary (Connecticut), 83
Menabuoi, Giusto de'
   cupola fresco (Padua baptistery), 374n40
Mendelsohn, Erich, 346–348
   Einstein Tower (Potsdam), 346–349; *191*
Mensuration, 37, 243–247, 296, 392n8, 393n28
   isorhythm, 244–247, 249, 261–262, 265, 392n8, 393nn9,26,28
Merleau-Ponty, Maurice, 125, 359
   The Phenomenology of Perception, 52
   The Visible and the Invisible, 125
Metzinger, Jean, 377n14
Meyer, Hansjörg, 291
Mézières military academy, 323–324
Michelangelo Buonarroti, 136, 195, 266
   Day and Night, 199
   Last Judgment, 249–250

Palazzo del Senatore (Capitoline Hill, Rome), 281; *151*
Michelozzo di Bartolommeo, 17
Mies van der Rohe, Ludwig, 67, 360
   Barcelona Pavilion, 360
Minkowski, Herman, 341
Mirandola, Pico della, 22, 375n78
Modulor, the, 242, 274–277, 284–285, 292, 296, 312–315, 394n51; *148*
   development of, 273–277, 279–281, 285, 287, 291, 395nn7–8, 396n23, 397n56; *150, 156*
   use of, 242, 273–279, 289–298, 312–314, 320; *159, 162*
Monge, Gaspard, 151, 323, 326–327, 339, 348, 385n57
   analytic geometry, 327, 330, 400nn9,18,19
   descriptive geometry, 324, 328–330, 339
   Géométrie descriptive, 324, 330, 400n17; *177*
   projective geometry, 324
   ruled surfaces, 329–330, 400nn17–19
Montclos, Jean-Marie Pérouse de, 220–221, 389n39
   L'Architecture à la française, 180–181, 387n4, 390n80
Moore, Richard, 291
Moretti, Luigi
   Spazio, 364; *195*
Mosso, Leonardo, 71
Mumford, Lewis, 73
Museum of Knowledge (Chandigarh), 308
Museum of Modern Art (New York)
   Deconstructivist Architecture exhibition, 89–91, 379n78
Musical notation, 198, 244–246, 296–298, 393n9
Musical scales, 241, 262–267, 270, 276, 296; *146*. *See also* Pitch

Nadreau, Jacques
   seminary chapel, La Flèche; *107*
Nash, John, 83
N-dimensional geometry, 341, 348, 364. *See also* Fourth dimension
Neolithic technology, 76
Nietzsche, Friedrich, 358–359
   The Genealogy of Morals, 52
Non-Euclidean geometry, 62, 344–345, 348, 352–353, 364
   Riemannian, 62, 89, 344–345, 352, 376n84; *190*
Norberg-Schulz, Christian, 73
Notre-Dame de la Couture, Abbaye de (Le Mans), 193–195, 214, 388n23; *110, 122*
Notre-Dame du Haut (Ronchamp), 273–274, 277, 281, 291, 315–316, 320, 346–348, 396n20; *147*

construction of, 282, 301–302, 307, 317, 399n123; *166, 168, 173*
duality in, 277, 295, 305, 316–320; *175*
form of, 274, 276–279, 282–284, 287, 290, 292–293, 301–317, 320, 323, 330–331, 346–348, 396n31, 397n87, 398nn96, 115; *149, 154, 155, 159, 161, 165–168, 171, 173, 174*
interior, 295; *161, 165, 174*
location, 293–295
models of, 302, 305, 312; *167*
modulor proportions of, 273–274, 276–277, 279, 284, 289–292, 312–314, 396n51, 399n125; *159*
roof of, 278, 284, 290, 302–306, 309, 314, 329, 398nn91,96,123; *154, 168, 171, 173, 174*

Ockeghem, Johannes, 262
Olek (co-worker of Le Corbusier), 301; *166*
Olivetti Computer Center (Milan), 308
Ortega y Gasset, José, 70, 102
Orthographic projection, 94, 108–110, 113–119, 121, 150–151, 154–156, 175–177, 179, 189, 196, 200, 254, 302–305, 324, 326–327, 360, 368, 382nn8–9,20, 383n17, 385n64; *50.* See also Architectural drawing; Descriptive geometry; Parallel projection; Technical drawing
auxiliary, 154, 156, 160–162, 386n66; *75*
and cubism, 62
depiction of the human figure, 154–156, 159–160, 386n75; *77, 80, 81, 83*
the perpendicular, 197–198, 200–201
scale measure in, 113, 256, 337, 359, 368
Ouvrard, René
Architecture harmonique, 241, 265, 269, 392n2, 394n63
Ozenfant, Amédée
Après le cubisme, 67

Pacioli, Luca, 177, 291
De divina proportione; 18, 70
Palace of Assembly (Chandigarh), 307–308, 312–314, 398n111
Palace of Justice (Chandigarh), 308
Palau Güell (Barcelona), 331
Palazzo Piccolomini (Pienza), 141
Palazzo Sacchetti (Rome)
Bathsheba Going to King David, 190–192, 195; *108*
Palladio, Andrea, 248
San Giorgio Maggiore (Venice); *8, 59*
San Petronio (Bologna), 117; *58*
Venetian churches, 117
villas of, 141, 247
Palmieri, Matteo
Città di Vita, 21

Panofsky, Erwin, 113, 126–129, 382n14, 394n51
Die Perspektive als symbolische Form, 126
Panopticon, 102, 125, 130
Pantheon (Rome), 12, 102
Paolo, Giovanni di
The Creation of the World and the Expulsion from Paradise, 23–25; *12*
Paracelsus (Philippus Aureolus Theophrastus Bombast von Hohenheim), 374n38
Parallel projection, 108–110, 113–118, 121, 177, 186, 189, 201, 355, 382nn8–9; *50.* See also Orthographic projection
Parc de La Villette (Paris), 84–87; *38, 39*
Parker, John Henry, 236
Pastor, Ludwig, 33
Pauly, Danièle, 277–278, 302, 396n16
Payne Knight, Richard, 83
Pélerin, Jean
De artificiali perspectiva, 140
on perspective, 140, 384n33; *69*
Pencreach, G.
Forum des Halles (Paris), 80–83; *36*
Perception, 123–130, 351–366, 369–370, 383n9, 401n5. See also Sensation; Space, perceptual; Vision
as pictures, 63–65, 351, 358–359
Perrault, Claude, 12, 268–271, 395n75
Perspective, 22–23, 34, 40, 99, 107, 110, 113–116, 119–120, 130–133, 144–147, 168, 198, 254–259, 266–267, 352, 360, 368–369, 373n32, 374n40, 382nn9,14, 394n49; *50, 63, 97–99*
as access to reality, 123–129, 136–138, 140, 255, 259, 382n14
as agency of power, 124–125, 129–130, 133–135, 140–142, 167
architecture of, 133–138, 141, 148–149, 151–153, 167, 176–177, 254
central, 102, 111, 131, 135–136, 146, 158, 167–169, 173, 176–177, 254, 259; *65*
Alberti's method, 110–113, 120, 132–136, 140–142, 147–149, 153, 173, 176–177, 254, 384n33; *51, 69*
Pélerin's method (tiers point method), 140, 384n33, 385n64; *69*
Piero's method, 148, 162–163; *73, 74*
as convention, 60, 102, 119, 123, 126–127, 140
criticism of, 60–62, 119, 123–130, 141–142
curvilinear, 127–129
depiction of the human figure, 133–138, 142–147, 153–163, 167–168, 254, 259; *72, 78, 82, 84, 85*
depiction of *mazzocchi,* 156, 173–174, 386n72; *98, 99*
and Euclidean geometry, 62, 124–125, 130, 133, 255–256

introduction of, 113, 382n14, 395n71; *55*
as model of vision, 60, 108, 119–120, 123–132, 138, 254–256, 358–359, 368–369; *64*
Piero's other method, 121, 131, 145, 147–159, 162–176, 385nn57,62,64, 386n72; *69, 76, 78*
proportional diminution, 148, 244, 251–254, 256–260, 265–267, 270, 385n60, 394n51, 395n65; *138, 142, 143, 145*
space of (see Space, perspectival)
stereographic projection (see Stereographic projection)
vanishing point, 23, 35, 111, 130–136, 147–148, 151, 156, 163, 167–173, 175–176, 382n9, 384n33, 385nn57,60
view point, 107–108, 110–111, 125–127, 130–132, 147, 383nn9,14
Perspective projection. *See* Perspective
Perugino, Pietro, 135, 384n26
Peruzzi, Baldassare, 7
Peruzzi, Sallustio
drawing of Sant'Eligio degli Orefici; *17*
Peterborough Cathedral
retrochoir, 231, 391n100; *132, 133*
Petit, Jean
Livre de Ronchamp, 316, 397n89, 398n92
Pevsner, Antoine, 309–310, 398n114
Construction dynamique; 172
Pevsner, Nikolaus, 18, 73, 277
Philharmonie (Berlin), 94–100, 119–121, 381nn112,114,118; *43, 44, 60*
auditorium, 96, 119–120, 381nn114,119; *45*
foyer, 95, 98, 121; *46–48*
Philips Pavilion (Brussels International Exhibition, 1958), 296–301, 306, 308, 312, 314, 398n111; *163*
Philosophy, 84, 103–105, 354–357, 359. *See also* philosophies indexed by name
aesthetics, 242, 379n79, 380n99
Photography, 123, 129, 386n67
Piazza della Signoria (Florence)
Brunelleschi's perspective depiction of, 138, 176 (*see also* Brunelleschi, on perspective, demonstration panels)
Picasso, Pablo, 57–60, 65, 91–92, 100–101, 379n62
L'Arlésienne, 57–58, 67; *23*
cardboard maquettes, 65; *27*
Houses on a Hill, 81
variations on Delacroix's *Femmes d'Alger,* 91
Weeping Woman; 49
Picturesque, the, 80–83, 103, 379n62
Pierce, C. S., 63, 373n26
Piero della Francesca, 142–147, 158–159, 163, 167, 176, 267, 387n102
The Baptism of Christ, 160; *83*

Brera altarpiece (Pinacoteca di Brera, Milan), 146, 163, 386n78; *87*
depiction of the human figure, 142–147, 153–163, 167, 384n44; *77, 78, 80–86*
De prospectiva pingendi, 131, 143, 146–159, 162, 267, 385nn60,61,65, 386nn68,78, 387n94; *73–78, 80, 81*
The Flagellation of Christ, 144–146, 162–163; *71, 86*
fresco cycle (choir of San Francesco, Arezzo), 386n74
Battle between Constantine and Maxentius, 167, 169, 386n85
Battle between Heraclius and Chosroes, 167, 169, 386n85
Dream of Constantine, 167, 386nn80,83; *90*
The Proving of the True Cross, 159–162; *82, 84, 85*
Libellus de quinque corporibus regularibus, 143, 148
Madonna del Parto, 167
on mathematics, 142–143, 148, 384n44
The Nativity, 163–167, 173, 386nn74, 79,82; *88, 89*
on orthographic projection, 150–155, 160, 385n64; *75, 77, 80, 81*
on perspective, 121, 131, 142–169, 175, 251, 385nn57,60, 386n67; *69, 73, 74, 76, 78, 84, 85, 89*
Trattato d'abaco, 148
Pinturicchio, Bernardino
The Life of Aeneas Silvius Piccolomini; *61*
Piranesi, Giovanni Battista, 98–99, 381n116
Carceri, 99
Pitch, musical, 242–245, 263–266, 268, 276, 296
Plato, 48–49, 199–200
Plotinus, 27, 393n29
Poincaré, Henri, xxxiii, 352–353
Poitiers, Diane de, 183
Polygons. *See* Euclidean geometry, figures of
Polyhedroids. *See* Euclidean geometry, figures of; Fourth dimension
Poncelet, Jean-Victor
principle of continuity, 324
Pope-Hennessy, John, 175
Posener, Julius, 98–99, 381n118
Posthumanism, 100
Postmodernism, 55, 83, 103. *See also* Deconstruction
Poststructuralism, 84–85, 103
Prague Creative Artists Collaborative, 65
Prix, Wolf, 93. *See also* Coop Himmelblau
Proclus, 108
Projection, xxxi–xxxv, 357–370, 382n8; *197*. *See also* specific types of projection indexed by name.

Projective geometry, xxxi–xxxiii, 147, 201, 324, 326, 351–352, 389n39, 395n66. *See also* Desargues, four-point involution
Proportion, 17, 256–261, 265–270, 292, 394n54. *See also* Fibonacci series; Modulor; Perspective, proportional diminution
as correspondence between architecture and music, 241–247, 251–253, 261–262, 265, 270, 275–276, 296–298, 394n63, 395n64; *136* (*see also* Modulor, use of)
geometric series, 290, 296
golden ratio (*see* Golden ratio)
harmony (*see* Harmony)
as measure of beauty, 241–242, 246–250, 254, 259, 265, 269, 271, 274–275, 281, 290–291, 394n63 (*see also* Composition)
mensural (*see* Mensuration)
superparticular ratios, 244
as universal order of the Renaissance, 242–243, 246–247, 249, 251, 253, 265, 267, 269, 274
Psychoanalysis, 89, 355
Psychological projection, 355
Psychology, 359
Ptolemy, astronomy of, 18–19, 27, 41; *19*
Puccio, Piero di
fresco in the Campo Santo (Pisa), 23
Puteaux cubist group, 60
Pythagoras, 315
musical consonances, 244–246, 248, 261–265, 269–270
musical scale, 262–266; *146*

Rabelais, François, 316
Gargantua and Pantagruel, 319
Ramsey, William
choir vault (Gloucester Cathedral); *129, 130*
Raphael (Raffaello Sanzio), 40–41, 47, 107, 110, 113, 118–119, 136, 142, 153, 377n99, 381nn2–3
Chigi chapel (Santa Maria del Popolo, Rome), 30, 375n61; *15*
Marriage of the Virgin, 132, 135; *66*
Sant'Eligio degli Orefici (Rome), 7–9, 17, 29–30, 35–36, 38, 44; *1–3*
Stanza della Segnatura (Vatican Palace, Rome), 23; *16, 52*
Disputà, 30–37; *16*
The School of Athens, 111, 175, 382n12; *52*
Villa Madama (Rome), 111–113, 141, 382n12; *53, 54*
Ratio. *See* Proportion
Rationality, 70, 271. *See also* Functionalism
RCA Building (New York), 58
Relativity, 52, 57, 341, 344, 346, 353

Representation, 357–359, 363, 368
Repton, Humphry, 83; *37*
Rheticus, Johannes
Narratio prima, 374n59
Rice, Peter, 92
Riemann, Georg Friedrich Bernhard
manifolds, 339–341
non-Euclidean geometry of, 62, 89, 344, 376n84
Robert, Hubert
L'Arc de Triomphe d'Orange, 211; *120*
Rodia, Sam, 76
Watt Towers (Los Angeles); *34*
Ronchamp. *See* Notre-Dame du Haut
Rondelet, Jean Baptiste
Traité de l'art de bâtir, 225
Rorty, Richard, 359
Rossellino, Bernardo
Palazzo Piccolomini (Pienza), 141
Rossi, Aldo
floating theater, 399n142
Rotan, Thurman, 58
Rousseau, Jean Jacques, 320, 399n149
Rowe, Colin, 67, 73, 79, 248
Collage City, 74–80, 102
Rubens, Peter Paul
Battle of Anghiari; *92*
Ruled surfaces, 298–315, 317, 328–330, 334, 398n108, 399n122, 400nn17–19; *163–165, 168, 172, 173, 180*
conoids, 298–301, 303–305, 309, 329, 398n96; *169, 179*
hyperbolic paraboloids, 298–301, 305, 312, 334, 398nn96,121
hyperboloids, 307, 312, 334, 398n111
Ruskin, John, 4, 236
Rykwert, Joseph, 200

Sade, Donatien-Alphonse-François de, 85
Sagrada Familia (Barcelona), 331–334; *183*
Parochial School of, 334; *184*
Sainte-Geneviève (Paris), 214, 220, 390n74; *123, 124*
Sainte-Marie de la Tourette, monastery of, 296, 308
pans de verre ondulatoires, 296; *162*
St. Mary's (Warwick), 229, 391n97
Saint-Paul-Saint-Louis (Paris), 213–214; *121*
St. Peter's (Rome)
Bramante's proposals for, 34, 373n20, 382n12
Saint-Pierre (Caen), 224
St. Saviour (London), 225
Saint-Sulpice (Paris), 195; *111*
Salviati, Francesco, 195
Bathsheba Going to King David (Palazzo Sacchetti, Rome), 190–192, 195, 388n20; *108*

San Carlo alle Quattro Fontane (Rome), 121, 383n26
San Domenico (Arezzo)
 Dragonelli Chapel; *55*
San Francesco (Arezzo)
 fresco cycle, 159–162, 167, 169, 386n74; *82, 84, 85, 90*
Sangallo, Antonio da (the Elder), 17
Sang des bêtes, 87
San Giorgio Maggiore (Venice); *8, 59*
San Giovanni (baptistery, Florence)
 Brunelleschi's perspective depiction of, 176, 384n31 (*see also* Brunelleschi, on perspective, demonstration panels)
San Lorenzo (Florence), 253
 sacristy of, 14, 26–27, 248; *14*
San Marco (Milan)
 Choirs of Angels (Foppa chapel), 22, 30, 374n40; *9*
San Michele (Fano), 117; *57*
San Petronio (Bologna), 117; *58*
San Pietro in Montorio (Tempietto) (Rome), 9–11, 372nn11–12; *4*
Santa Croce (Florence)
 Pazzi chapel, 26–27
Santa Maria del Fiore (Florence cathedral), 244–245
 dome of, 14, 243–246, 373n21
Santa Maria della Consolazione (Todi), 373n20
Santa Maria della Divina Providencia (Lisbon); *195, 196*
Santa Maria delle Grazie (Milan), 373n20; *7*
Santa Maria dei Miracoli (Saronno), 374n40
Santa Maria del Popolo (Rome)
 Chigi Chapel, 30, 375n61; *15*
Santa Maria Novella (Florence), 248; *137*
 Subsiding of the Flood (Chiostro Verde), 163, 169, 386n88; *93, 99*
Sant'Andrea (Mantua), 116; *56*
Sant'Eligio degli Orefici (Rome), 7, 17, 29–30, 36, 38, 44; *2*
 centers of, 7–9, 35–36; *3*
 dome of, 30, 38; *1*
 original project by Sallustio Peruzzi, 38; *17*
Santo Spirito (Florence), 253; *139*
Sartoris, Alberto, 339
 church at Lourtier, 293; *160*
 Encyclopédie de l'architecture nouvelle, 293, 397n64; *160*
Saussure, Ferdinand de, 101
Savina, Joseph, 282, 302, 396n31
Scévola, Guirand de, 80
Scharoun, Hans, 94–95, 100, 381n111
 Kirche als Fels, 94; *42*
 Philharmonie (Berlin) (*see* Philharmonie)
 Staatsbibliothek (Berlin), 94
Schön, Erhard, 147

Schumacher, Tom, 102
Schwitters, Kurt
 Merzbau, 65
Scruton, Roger, 248, 364–366
Scully, Vincent, 73
Sedlmayr, Hans, 102–103
Sensation, 351–353, 358, 401n5
Serlio, Sebastiano, xxix, 249, 372nn12–13
 Five Books of Architecture, xxvi–xxvii
Serralta, Justin, 281, 285
Serres, Michel, 89
Seznec, Jean, 375n61
Shearman, John, 375n61
Signification, 355–357, 373n26. See also Symbolism
Simonin (architect, 18th century)
 Traité élémentaire de la coupe des pierres, 209
Simultaneity, 57–59, 92
Sinding-Larsen, Staale, 372nn7,13,22
Siza, Alvaro
 Borges and Irmão Bank (Vila do Conde), 92
Slutzky, Robert, 67, 293
Socrates, 48
Solar paganism, 26
Soler i Faneca, Joan
 Lonja (Barcelona), 213
Soltan, Jerzy, 285, 395n7, 397n56
Sommer, Clemens, 34
Soufflot, J. G.
 Sainte-Geneviève (Paris), 214, 220, 390n74; *123, 124*
Space
 architectural, 107, 109–110, 117–118, 141, 363–364
 geometrical, xxix, 125, 339–345, 348, 351–353, 363–366, 401n5; *190*
 motor, 352, 364
 perceptual, 348, 351–353, 363–366, 401n5
 perspectival, 133–136, 141, 147–149, 158–159, 162, 167, 174–175, 368–369, 383n17 (*see also* Space, pictorial; Space, projective)
 pictorial, 55, 59, 63, 67, 81, 94, 107–110, 113–118, 132, 339, 342–344, 351–352, 363–364, 369, 382n14
 projective, xxxi, 357, 360, 364, 368–370
Speer, Albert, 96
Spheres, 39–42, 103, 198–199, 376nn83, 85,90; *18, 177*. See also Euclidean geometry, figures of; Stereographic projection
 in Renaissance art and architecture, 14–47 (*see also* Centralized churches)
Staatsbibliothek (Berlin), 94
Stein, Gertrude, 59, 67
Steinberg, Leo, 91, 100, 249–250, 383n26
Stereographic projection, 344–345, 352; *190*

Stereotomy, 179, 208, 327, 329, 387n1
 stonecutting, 60, 121, 179–181, 187–189, 201, 213–214, 305, 328, 331, 387n1; *179* (*see also* Architectural drawing, *traits*; *Trompes*; Vaults)
 architecture of, 179–180, 195–197, 205, 208, 213, 220, 225, 236–239, 389n49
 and Italianate classicism, 179–183, 189, 195, 209–210, 213–220, 238
 and Gothic architecture, 179–181, 209–210, 213–214, 220–238, 389n39, 392n119 (*see also* Stonecutting, Gothic)
 and Romanesque architecture, 220
Stevin, Simon, 132, 266
Stirling, James, 277, 301
Stonecutting, 200, 210. See also Architectural drawing, *traits*; Stereotomy, stonecutting
 dérobement, 195–196, 208
 Gothic, 203, 221–229
 jarrets, 214, 390n71
Structuralism, 101
Structural shells, 298–301, 309–310, 312–314, 328, 399n122
Structure, 8, 14, 18, 30, 38, 84, 92–94, 208, 373n20, 390n74
 and form, 180, 183, 190–195, 205, 208–212, 214–218, 224, 229, 236–238, 293, 298–301, 309, 312, 391nn113–114, 398n121
Sun, 29. See also Light; Solar paganism
 as metaphor for divinity, 26–30, 35, 50, 374n47
Surveying, 44–45, 108, 135. See also Measurement
Sydney Opera House, 93, 380n104
Symbolism, 17–18, 29, 34, 42–43, 47, 79–80, 85, 126, 290–291, 307, 317, 334, 349, 373n26, 374nn47,50
 in Renaissance art and architecture, 4–6, 16–19, 21–22, 24–29, 36–37, 40, 43, 146
Symmetry, 111, 118–119, 136, 141, 167, 177
Synthetic geometry, 330, 348; *180*

Taton, René, 285, 395n8
Technical drawing, 60–62, 92–93, 302, 324; *180*. See also Descriptive geometry; Orthographic projection
 development, 60, 187, 330; *177*
 instruments of, 40, 118
Tempietto. See San Pietro in Montorio
Temple of Hera (Paestum) (reconstruction), 382n22
Tewkesbury Abbey, 228–229, 391n96
Theology, Christian, 18–19, 21, 23–37, 40–41, 43, 198

Theresa of Avila (Saint)
The Interior Castle, 50
Thiersch, August, 281
Thomas Aquinas (Saint), 27
Tinctoris, Johannes
Proportions in Music, 245–246
Titus-Carmel, Gérard
Pocket Size Tlingit Coffin, 86
Tocqueville, Alexis de
Democracy in America, 78–79
Totalitarianism, architecture of, 57, 75
Traits. See Architectural drawing, *traits*
Transparency, 57–59, 67
Trompes, 180–189, 193–195, 200, 212, 237, 388nn14,16; *101–107*
Tschumi, Bernard, 84–87, 100
Parc de La Villette (Paris), 84–87; *38, 39*
Turin cathedral
chapel of Santissima Sindone, 360; *193, 194*

Uccello, Paolo, 175, 176
Battle of San Romano, 169–175, 387n90
The Counter-Attack of Micheletto da Cotignola; 95
Niccolò da Tolentino at the Head of the Florentines, 387n91; *94*
The Unhorsing of Bernardino della Ciarda; 96
Hunt in the Forest, 169
Monument to Sir John Hawkwood, 169
on perspective, 148, 169–175; *97–99*
Profanation of the Host, 169
Subsiding of the Flood (Santa Maria Novella, Florence), 163, 169, 386n88; *93, 99*
Unité d'Habitation (Marseilles), 295, 308
Urbanism, 74–80, 102, 141
Utopia, 75, 78
Utzon, Jørn
Sydney Opera House, 380n104

Van Doesburg, Theo. *See* Doesburg
Varèse, Edgard
Poème électronique, 298
Vasari, Giorgio, 20, 33–34, 136, 142, 145–146, 156, 167, 169–173, 175, 195
Florence cathedral, fresco, 22, 376n93
Le Vite, 375n68
Vasconi, C.
Forum des Halles (Paris), 80–83; *36*
Vatican Palace (Rome)
Stanza della Segnatura, 23, 30–37, 175; *16, 52*
Vaults, 14, 179, 181, 213, 373n20, 390n84
nu, 213–220, 237–238; *121–124, 127*
ribbed, 214, 221–228, 236–238, 391n94; *128–130*
fan, 229–238, 391nn96–100,113,114; *131–135*
net, 228–229

Venturi, Robert, 73
Vérin, Hélène, 118
Vertue, Robert
tomb for Henry VI, 238
Vertue, Robert and William
Henry VII Chapel (*see* Westminster Abbey)
Vico, Giambattista, 354
Vidler, Anthony, 83, 100, 102
Vignola, Giacomo da
Villa Giulia (Rome), 141
Villa Madama (Rome), 141, 382n12; *54*
loggia of, 111–113; *53*
Villa Reale (Poggio a Caiano), 141
Villa Savoye (Poissy); *30*
Villa Stein (Garches), 67, 292, 397n61, 399n148
Viollet-le-Duc, Eugène-Emmanuel
Dictionnaire raisonné, 225
Vision, xxxiii, 60, 108, 119–120, 123–132, 141, 250–251, 254, 256, 260, 270, 351–352, 357–363, 368–369, 383n14, 394n46, 401n5
Vitruvius Pollio, Marcus, 113, 200, 246, 250, 394n46
Vittone, Bernardo, 241
Von Moos, Stanislaus, 308
Vreedburgh (engineer), 298

Wailly, Charles de
pulpit for Saint-Sulpice (Paris), 195; *111*
Wang, Wilfried, 96
Warren, Charles, 243–245, 247, 262
Wastell, John, 231
retrochoir (Peterborough Cathedral); *132, 133*
Watts Towers (Los Angeles); *34*
Weber, Werner, 99, 120
Wells Cathedral, 226–229
Wenders, Wim
Wings of Desire, 94
West, George Herbert, 237
Westminster Abbey (London)
Henry VII Chapel, 231–238, 391nn101, 111,115, 392nn115–116; *134, 135*
White, John, 113, 382n14
Wigley, Mark, 89, 102–103
Willis, Robert
on Gothic vaults, 220, 224–231, 238, 390nn86,88, 391nn94,100, 392n115; *132*
on Henry VII Chapel (Westminster Abbey), 231, 236–237, 392n115; *135*
On the Construction of the Vaults of the Middle Ages, 220
Winchester Cathedral, 228
Wisniewski, Edgar
Staatsbibliothek (Berlin), 94
Wittkower, Rudolf, 14, 316

Architectural Principles in the Age of Humanism, 4–5, 103, 242
Brunelleschi and Proportion in Perspective, 244
on proportion, 251–254
on Renaissance centralized churches, 3–8, 17, 42, 103
on Renaissance symbolism, 4–6, 16–18, 37, 373n26
on Santa Maria Novella (Florence), 248; *137*
on universal harmony, 242–243
Wölfflin, Heinrich, 33, 36–37, 281, 372n6
Renaissance and Baroque, 3, 5
on Renaissance centralized churches, 3–9, 30, 102–103
on Santa Maria Novella (Florence), 248; *137*

Xenakis, Iannis, 295
architecture of, 296–298; *162, 163*
Metastasis, 296–298; *164*
musical compositions of, 295–298, 397n75; *164*
pans de verre ondulatoires, 296–298; *162*
Philips Pavilion (Brussels International Exhibition, 1958), 296–301, 306, 308, 312–314, 398n111; *163*
ruled surfaces, 298–301, 306, 309, 314, 398n111; *163, 164*
structural engineering, 295, 298, 397n79

Yeats, William Butler, 102

Zarlino, Gioseffo
musical scale; *146*
Zuccari, Federigo
Florence cathedral, fresco, 22